ADVANCES IN DATA SCIENCE AND COMPUTING TECHNOLOGY

Methodology and Applications

AAP Research Notes on Optimization & Decision-Making Theories

ADVANCES IN DATA SCIENCE AND COMPUTING TECHNOLOGY

Methodology and Applications

Edited by

Suman Ghosal, PhD

Amitava Choudhury, PhD

Vikram Kumar Saxena, MTech

Arindam Biswas, PhD

Prasenjit Chatterjee, PhD

First edition published 2023

Apple Academic Press Inc.
1265 Goldenrod Circle, NE,
Palm Bay, FL 32905 USA

760 Laurentian Drive, Unit 19,
Burlington, ON L7N 0A4, CANADA

CRC Press
6000 Broken Sound Parkway NW,
Suite 300, Boca Raton, FL 33487-2742 USA

4 Park Square, Milton Park,
Abingdon, Oxon, OX14 4RN UK

Library and Archives Canada Cataloguing in Publication

Title: Advances in data science and computing technology : methodology and applications / edited by Suman Ghosal, PhD, Amitava Choudhury, PhD, Vikram Kumar Saxena, MTech, Arindam Biswas, PhD, Prasenjit Chatterjee, PhD.

Names: Ghosal, Suman, editor. | Choudhury, Amitava, editor. | Saxena, Vikram Kumar, editor. | Biswas, Arindam, editor. | Chatterjee, Prasenjit, 1982- editor.

Description: First edition. | Series statement: AAP research notes on optimization and decision-making theories | Includes bibliographical references and index.

Identifiers: Canadiana (print) 20220220735 | Canadiana (ebook) 20220220808 | ISBN 9781774639979 (hardcover) | ISBN 9781774639986 (softcover) | ISBN 9781003277071 (ebook)

Subjects: LCSH: Computer systems. | LCSH: Electronic data processing.

Classification: LCC QA76 .A38 2023 | DDC 004—dc23

Library of Congress Cataloging-in-Publication Data

CIP data on file with US Library of Congress

ISBN: 978-1-77463-997-9 (hbk)
ISBN: 978-1-77463-998-6 (pbk)
ISBN: 978-1-00327-707-1 (ebk)

AAP RESEARCH NOTES ON OPTIMIZATION AND DECISION-MAKING THEORIES

Most real-world search and optimization problems naturally involve multiple criteria as objectives. Different solutions may produce trade-offs (conflicting scenarios) among different objectives. A solution that is better with respect to one objective may be a compromising one for other objectives. This compels one to choose a solution that is optimal with respect to only one objective. Due to such constraints, multi-objective optimization problems (MOPs) are difficult to solve since the objectives usually conflict with each other. It is usually hard to find an optimal solution that satisfies all objectives from the mathematical point of view. In addition, it is quite common that the criteria of real-world MOPs encompass uncertain information, which becomes quite a challenging task for a decision maker to select the criteria. Also, the complexities involved in designing mathematical models increase. Considering, planning, and appropriate decision-making require the use of analytical methods that examine trade-offs; consider multiple scientific, political, economic, ecological, and social dimensions; and reduce possible conflicts in an optimizing framework. Among all these, real-world multi-criteria decision-making (MCDM) problems related to engineering optimizations are categorically important and are quite often encountered with a wide range of applicability.

MCDM problems are basically a fundamental issue in various fields, including applied mathematics, computer science, engineering, management, and operations research. MCDM models provide a useful way for modeling various real-world problems and are extensively used in many different types of systems, including, but not limited to, communications, mechanics, electronics, manufacturing, business management, logistics, supply chain, energy, urban development, waste management, and so forth.

In the aforementioned cases, modeling of multiple criteria problems often becomes more complex if the associated parameters are uncertain and imprecise in nature. Impreciseness or uncertainty exists within the parameters due to imperfect knowledge of information, measurement uncertainty, sampling uncertainty, mathematical modeling uncertainty, etc. Theories like probability theory, fuzzy set theory, type-2 fuzzy set theory, rough set, grey theory, neutrosophic uncertainty theory available in the existing literature deal with such uncertainties. Nevertheless, the uncertain multi-criteria characteristics in such problems are not explored in depth, and a lot can be achieved in this direction. Hence, different mathematical models of real-life multi-criteria optimization problems can be developed on various uncertain frameworks with special emphasis on sustainability, manufacturing, communications, biomedical, electronics, materials, energy, agriculture, environmental engineering, strategic management, flood risk management, supply chain, waste management, transportations, economics, and industrial engineering problems, to name a few.

Coverage & Approach:

The primary endeavor of this series is to introduce and explore contemporary research developments in a variety of rapidly growing decision-making areas. The volumes will deal with the following topics:

- Crisp MCDM models
- Rough set theory in MCDM
- Fuzzy MCDM
- Neutrosophic MCDM models
- Grey set theory

- Mathematical programming in MCDM
- Big data in MCDM
- Soft computing techniques
- Modelling in engineering applications
- Modeling in economic issues
- Waste management
- Agricultural practice
- Material selection
- Renewable energy planning
- Industry 4.0
- Sustainability
- Supply chain management
- Environmental policies
- Manufacturing processes planning
- Transportation and logistics
- Strategic management
- Natural resource management
- Biomedical applications
- Future studies and technology foresight
- MCDM in governance and planning
- MCDM and social issues
- MCDM in flood risk management
- New trends in multi-criteria evaluation
- Multi-criteria analysis in circular economy
- Multi-criteria evaluation for urban and regional planning
- Integrated MCDM approaches for modeling relevant applications and real-life problems

Types of volumes:

This series reports on current trends and advances in optimization and decision-making theories in a wider range of domains for academic and research institutes along with industrial organizations. The series will cover the following types of volumes:

- Authored volumes
- Edited volumes
- Conference proceedings
- Short research (thesis-based) books
- Monographs

Features of the volumes will include recent trends, model extensions, developments, real-time examples, case studies, and applications. The volumes aim to serve as valuable resources for undergraduate, postgraduate and doctoral students, as well as for researchers and professionals working in a wider range of areas.

CURRENT & FORTHCOMING BOOKS IN THE SERIES

Multi-Criteria Decision-Making Techniques in Waste Management: A Case Study of India
Suchismita Satapathy, Debesh Mishra, and Prasanjit Chatterje

Applications of Artificial Intelligence in Business and Finance: Modern Trends
Editors: Vikas Garg, Shalini Aggarwal, Pooja Tiwari, and Prasenjit Chatterjee

Advances in Data Science and Computing Technology: Methodology and Applications
Editors: Suman Ghosal, Amitava Choudhury, Vikram Kr. Saxena, Arindam Biswas, and Prasenjit Chatterjee

Machine Learning Algorithms in Security Analytics: Applications, Principles, and Practices
Editors: Karan Singh, Latha Banda, and Manisha Manjul

Sustainable Development: A Geospatial and Statistical Perspective
Editors: Qazi Mazhar Ali, Rizwan Ahmad, Irfan Ali, Prasenjit Chatterjee, and Harjit Pal Singh

Decision Analysis and Optimization Models for Sustainable Development Goals 2030
Irfan Ali, Umar Muhammad Modibbo, and Bui Thanh Hung

Smart Sensors, Actuators, and Decision Support Systems for Precision Agriculture
Editors: Narendra Khatri, Ajay Kumar Vyas, Celestine Iwendi, and Prasenjit Chatterjee

Attacks on Artificial Intelligence: The New Facets of Cyber Ecospace
Editors: Kukatlapalli Pradeep Kumar, Vinay Jha Pillai, and Boppuru Rudra Prathap

ABOUT THE EDITORS

Suman Ghosal, PhD
Assistant Professor and Head of the Department of Electronics, Pailan College of Management and Technology, West Bengal, India, E-mail: ghosal.suman987@gmail.com

Suman Ghosal, PhD, is Assistant Professor and Head of the Department of Electrical and Electronics Engineering at Pailan College of Management and Technology, Kolkata, India. He is also Administrative In-Charge of Pailan Technical Campus, Kolkata. He has more than 12 years of academic experience in addition to industrial experience. His broad areas of interest include renewable energy, electronics devices, and soft computing. He has published many research papers in various national and international reputed journals and conferences. He obtained his BTech degree from Maulana Abul Kalam Azad University of Technology (formerly WBUT), Kolkata, and his MTech from Bengal Engineering and Science University, Shibpur. He is currently pursuing a PhD at the Indian Institute of Engineering Science and Technology, Shibpu, India.

Amitava Choudhury, PhD
Assistant Professor, School of Computer Science, University of Petroleum and Energy Studies, Dehradun, Uttarakhand, India

Amitava Choudhury, PhD, is Assistant Professor in the School of Computer Science, University of Petroleum and Energy Studies, Dehradun, India. He has more than ten years of experience in teaching and many years in research work. He serves as a reviewer for *IEEE Biomedical Transaction* and the journal *Medical & Biological Engineering & Computing*. He is member of IEEE, UP section, and the Institute of Engineers, India. His area of research interest is computational geometry in the field of micromechanical modelling, pattern recognition, character recognition, and machine learning. Currently, he is associated with the journal *Cluster Computing* as a guest editor. He earned his MTech from Jadavpur University and submitted his PhD thesis at the Indian Institute of Engineering Science and Technology, Shibpur, India.

Vikram Kumar Saxena, MTech
Assistant Professor, Pailan College of Management and Technology, West Bengal, India

Vikram Kumar Saxena is Assistant Professor at the Pailan College of Management and Technology Kolkata, India. He has been a teaching professional for more than eight years. His research area of interest includes electrical machines, power systems, and power electronics. He has published many research papers in national and international journals and conferences. He received his BTech from West Bengal University of Technology, Kolkata, and his MTech from the National Institute of Technology Patna, India.

Arindam Biswas, PhD
Associate Professor, School of Mines and Metallurgy, Kazi Nazrul University, West Bengal–713305, India, E-mail: mailarindambiswas@yahoo.co.in

Arindam Biswas, PhD, is Associate Professor in the School of Mines and Metallurgy at Kazi Nazrul University, Asansol, West Bengal, India. Dr. Biswas has 10 years of experience in teaching research and administration. He has published many technical papers in journals and conference proceedings as well as six books, an edited volume, and a book chapter. Dr. Biswas was awarded several research grants, such as from the Science and Engineering Research Board, Govt of India, and the Centre of Biomedical Engineering, Tokyo Medical and Dental University, in association with RIE, Shizouka University, Japan. Presently Dr. Biswas is serving as an Associate Editor of the journal Cluster Computing and as a guest editor of Nanoscience and Nanotechnology and Recent Patents on Materials Science. He was formerly a Visiting Professor at the Research Institute of Electronics, Shizouka University, Japan. He has organized and chaired several international conferences in India and abroad. Dr. Biswas acted as reviewer for journals and is a member of the Institute of Engineers (India) and regular fellow of Optical Society of India (India). His research interest is in carrier transport in low-dimensional systems and electronic devices, non-linear optical communications, and THz Semi-conductor sources. Dr. Biswas received an MTech degree in Radio Physics and Electronics from the University of Calcutta, India, and a PhD from NIT Durgapur, India. He was a postdoctoral tesearcher at Pusan National University, South Korea, with prestigious BK21PLUS Fellowship, Republic of Korea.

Prasenjit Chatterjee, PhD

Dean (Research and Consultancy) MCKV Institute of Engineering, West Bengal, India

Prasenjit Chatterjee, PhD, is the Dean (Research and Consultancy) at MCKV Institute of Engineering, India. He has published over 90 research papers in international journals and peer-reviewed conferences. He has received numerous awards, including Best Track Paper Award, Outstanding Reviewer Award, Best Paper Award, Outstanding Researcher Award, and University Gold Medal. He has been the guest editor of several special issues and has authored and edited several books on decision-making approaches, supply chains, and sustainability modeling. He is the lead editor of the book series AAP Research Notes on Optimization and Decision-Making Theories and Frontiers of Mechanical and Industrial Engineering with Apple Academic Press. Dr. Chatterjee is one of the developers of two "multiple-criteria decision-making methods called measurement of alternatives and ranking according to compromise solution" (MARCOS) and "ranking of alternatives through functional mapping of criterion sub-intervals into a single interval" (RAFSI).

CONTENTS

Contributors xvii

Abbreviations xxiii

Preface xxvii

Part I: Recent Application of Deep Learning and ANN Paradigm **1**

1. Deep Learning Supported Evaluation of Retinal Vessel Enhancement Techniques **3**
Mahua Nandy Pal, Minakshi Banerjee, and Shuvankar Roy

2. An Efficient and Optimized Deep Neural Network Architecture for Handwritten Digit Recognition **19**
Shuvankar Roy and Mahua Nandy Pal

3. Design and Development of a Driving Assistance and Safety System Using Deep Learning **35**
Shekhar Verma, Tushar Bajaj, Nidhi Sindhwani, and Abhijit Kumar

4. Prediction of the Positive Cases of COVID-19 for Individual States in India and Its Measure: A State-of-the-Art Analysis **47**
Amitava Choudhury, Kaushik Ghosh, and Suman Ghosal

5. Deep Residual Neural Network Based Multi-Class Image Classifier for Tiny Images **59**
Alok Negi, Prachi Chauhan, and Amitava Choudhury

6. Improved TDOA-Based Node Localization in WSN: An ANN Approach **69**
Arpan Chatterjee, Ashish Agarowala, and Parag Kumar Guha Thakurta

Part II: Role and Impact of Big Data in E-Commerce and the Retail Sector **81**

7. Impact of Big Data on E-Commerce Websites **83**
Divyang Bhartia and Mausumi Das Nath

8. A Review of the Revolutionizing Role of Big Data in Retail Industry **95**
Mausumi Das Nath and Madhu Agarwal Agnihotri

Part III: Algorithm for Load Balancing in Cloud Computing 107

9. Load-Balancing Implementation with an Algorithm of Matrix Based on the Entire Jobs 109
Payel Ray, Enakshmi Nandi, Ranjan Kumar Mondal, and Debabrata Sarddar

10. Load-Balancing Algorithm with Overall Tasks 119
Enakshmi Nandi, Payel Ray, Ranjan Kumar Mondal, and Debabrata Sarddar

11. Implementation of Load Balancing of Matrix Problem with the Row Sum Method 131
Payel Ray, Enakshmi Nandi, Ranjan Kumar Mondal, and Debabrata Sarddar

Part IV: Advances in Embedded System-Based Applications 143

12. Microcontroller-Based Wearable Location Tracker System with Immediate Support Facility.. 145
Sumanta Chatterjee, Dwaipayan Saha, Indrani Mukherjee, and Jesmin Roy

13. Design and Modifications of Eyedrop Bottle Holder Device: A Simplest Way to Install Ocular Drug Independently......................... 169
Sirajum Monira, Vishal Biswas, and Satyam Jha

14. Agro-Nutrition Alert (ANA).. 187
Rajneesh Kumar, Priya Anand, Debaparna Sengupta, and Monalisa Datta

Part V: Optimization Technique Using MATLAB Platform............................ 195

15. Selection of Optimal Mother Wavelet for Fault Analysis in Induction Motor Using Stator Current Waveform 197
Arunava Kabiraj Thakur, Palash Kumar Kundu, and Arabinda Das

16. High Efficiency High Gain SEPIC-Buck Boost Converter Based BLDC Motor Drive for Solar PV Array Fed Water Pumping.............. 219
K. M. Ashitha and D. Thomas

17. A Classy Fuzzy MPPT Controller for Standalone PV System 237
Rahul Kumar Singh, Rahul Kumar Prasad, Rahul Singh, Aayush Ashish, Debaparna Sengupta, Rishiraj Sarker, and Monalisa Datta

18. Active and Reactive Power Control in Variable Wind Turbine Using Simultaneous Working of STATCOM and Pitch Control 253
Arrik Khanna, Pushpanjali Singh Bisht, and Vikram Kumar Saxena

Part VI: Technique for Improving Information and Network Security 273

19. Information Security Using Key Management 275
Sanjib Halder, Bijoy Kumar Mandal, and Arindam Biswas

20. Strength-Based Novel Technique for Malicious Nodes Isolation........... 289
Prachi Chauhan and Alok Negi

21. Securing the Information Using Combined Method 305
Bijoy Kumar Mandal, Saptarshi Roychowdhury, Payel Majumder, and Anwesa Das

22. Implementation of DNA Cryptography in IoT Using Chinese Remainder Theorem, Arithmetic Encoding, and Asymmetric Key Cryptography 319
Ananya Satpati (Das), Soumya Paul, and Payel Majumder

23. Robust Technique to Overcome Thwarting Communication Under Narrowband Jamming in Extremely Low Frequency of CDMA-DSSS System on Coding- Based MATLAB Platform 333
Tanajit Manna, Rahit Basak, Mainak Das, and Alok Kole

Part VII: Miscellaneous Topics in Computing 347

24. The Utility of Regional Language Facility in Accounting Software for Retail Market of West Bengal.. 349
Madhu Agarwal Agnihotri and Arkajyoti Pandit

25. Comparative Analysis of Storey Drift of a G+9 RC-Structured Building Due to Seismic Loads in Various Seismic Regions 363
Budhaditya Dutta, Aditya Narayan Chakraborty, Ayushi Shah, Debarka Brahma, Amit Deb, and Tushar Kanti De

26. A Proposal for Disinfection of Contaminated Subject Area Through Light 383
Anubrata Mondal and Kamalika Ghosh

Index 397

CONTRIBUTORS

Ashish Agarowala
CSE Department, NIT Durgapur, West Bengal, India

Madhu Agarwal Agnihotri
Assistant Professor, Department of Commerce, St. Xavier's College (Autonomous), Kolkata, West Bengal, India, E-mails: madhu.cal@gmail.com; madhu.cal@sxccal.edu

Priya Anand
Department of Information Science and Engineering, AMC Engineering College, Bangalore–560083, Karnataka, India

Aayush Ashish
Techno International New Town, Kolkata–700156, West Bengal, India

K. M. Ashitha
Department of Electrical and Electronics Engineering, Amal Jyothi College of Engineering, Kanjirappally, Kerala–686518, India, E-mail: ashithakm@ee.ajc.i

Tushar Bajaj
Department of ECE, Amity School of Engineering and Technology, New Delhi GGSIPU, Delhi, India

Minakshi Banerjee
Computer Science and Engineering Department, RCC Institute of Information Technology, Kolkata–700015, West Bengal, India

Rahit Basak
Computer Science and Engineering Department, Pailan College of Management and Technology, Kolkata–700104, West Bengal, India

Divyang Bhartia
Student, St. Xavier's College (Autonomous), Kolkata, 30 Mother Teresa Sarani, Kolkata–700016, West Bengal, India

Pushpanjali Singh Bisht
Department of Electrical Engineering, DPGITM Gurugram, Haryana–122001, India

Arindam Biswas
Kazi Nazrul University, Asansol, West Bengal, India

Vishal Biswas
Department of Bachelor in Optometry, Pailan College of Management and Technology, Joka, Kolkata–700104, West Bengal, India

Debarka Brahma
Undergraduate Student, Department of Civil Engineering, Amity University Kolkata, Kolkata, West Bengal–700135, India

Aditya Narayan Chakraborty
Undergraduate Student, Department of Civil Engineering, Amity University Kolkata, Kolkata, West Bengal–700135, India

Arpan Chatterjee
CSE Department, NIT Durgapur, West Bengal, India

Sumanta Chatterjee
Department of Computer Science and Engineering, JIS College of Engineering, Kalyani, West Bengal, India, E-mail: Sumanta.chatterjee@jiscollege.ac.in

Prachi Chauhan
G.B. Pant University of Agriculture and Technology, Pantnagar, Uttarakhand, India, E-mail:cprachi664@gmail.com

Amitava Choudhury
School of Computer Science, University of Petroleum and Energy Studies, Dehradun, Uttarakhand, India, E-mail: a.choudhury2013@gmail.com

Anwesa Das
NSHM Knowledge Campus, Durgapur, West Bengal, India

Arabinda Das
Department of Electrical Engineering, Jadavpur University, Kolkata–700032, West Bengal, India

Mainak Das
Computer Science and Engineering Department, Pailan College of Management and Technology, Kolkata–700104, West Bengal, India

Monalisa Datta
Department of Electrical Engineering, Techno International New Town, Kolkata–700156, West Bengal, India

Tushar Kanti De
Associate Professor and Head of Department, Department of Civil Engineering, Budge Budge Institute of Technology, Budge Budge, West Bengal–700137, India

Amit Deb
Assistant Professor, Department of Civil Engineering, Amity University Kolkata, Kolkata, West Bengal–700135, India

Budhaditya Dutta
Undergraduate Student, Department of Civil Engineering, Amity University Kolkata, Kolkata, West Bengal–700135, India, E-mail: bdneil3103@gmail

Suman Ghosal
Department of Electrical and Electronics Engineering, Pailan College of Management and Technology, Kolkata, West Bengal, India

Kamalika Ghosh
Director, School of Illumination Science, Engineering and Design, Jadavpur University, Kolkata, West Bengal, India, E-mail: kamalikaghosh4@gmail.com

Kaushik Ghosh
School of Computer Science, University of Petroleum and Energy Studies, Dehradun, Uttarakhand, India

Sanjib Halder
The Bhawanipur Education Society College, Kolkata, West Bengal, India

Satyam Jha
Department of Bachelor in Optometry, Pailan College of Management and Technology, Joka, Kolkata–700104, West Bengal, India

Arrik Khanna
Department of Electrical Engineering, Chitkara University, Punjab–140401, India,
E-mail: arrik1433@gmail.com

Alok Kole
Electrical Engineering Department, RCC Institute of Information Technology, Kolkata–700015,
West Bengal, India

Abhijit Kumar
School of Computer Science, University of Petroleum and Energy Studies, Dehradun, Uttarakhand,
India, E-mail: abhijitkaran@hotmail.com

Rajneesh Kumar
Department of Electrical Engineering, Techno International New Town, Kolkata–700156, West Bengal,
India, E-mail: rajneeshkumar4044@gmail.com

Palash Kumar Kundu
Department of Electrical Engineering, Jadavpur University, Kolkata–700032, West Bengal, India

Payel Majumder
Department of Computer Science and Engineering, NSHM Knowledge Campus, Durgapur,
Arrah Shibtala Via Muchipara, Durgapur–713212, West Bengal, India

Bijoy Kumar Mandal
NSHM Knowledge Campus, Durgapur, West Bengal, India, E-mail: writetobijoy@gmail.com

Tanajit Manna
Electronics and Communication Engineering Department, Pailan College of Management and
Technology, Kolkata–700104, West Bengal, India, E-mail: tanajitmanna@gmail.com

Anubrata Mondal
Research Scholar, School of Illumination Science Engineering and Design,
(Jadavpur University, Kolkata) and Assistant Professor, Department of Electrical Engineering,
Global Institute of Management and Technology, West Bengal, India

Ranjan Kumar Mondal
Department of Computer Science and Engineering, University of Kalyani, West Bengal, India,
E-mail: ranjangcctt@gmail.com

Sirajum Monira
Department of Bachelor in Optometry, Pailan College of Management and Technology, Joka,
Kolkata–700104, West Bengal, India, E-mail: srisikhsa@gmail.com

Indrani Mukherjee
Department of Computer Science and Engineering, JIS College of Engineering, Kalyani, West Bengal,
India

Enakshmi Nandi
Department of Computer Science and Engineering, University of Kalyani, West Bengal, India,
E-mail: nandienakshmi@gmail.com

Mausumi Das Nath
Assistant Professor, Department of Commerce, St. Xavier's College (Autonomous), Kolkata,
30 Mother Teresa Sarani, Kolkata – 700016, West Bengal, India, E-mail: m.dasnath@sxccal.edu

Alok Negi
National Institute of Technology, Uttarakhand, India

Mahua Nandy Pal
Computer Science and Engineering Department, MCKV Institute of Engineering, Howrah–711204, West Bengal, India, E-mail: mahua.nandy@gmail.com

Arkajyoti Pandit
Assistant Professor, Department of Commerce, St. Xavier's College (Autonomous), Kolkata, West Bengal, India

Soumya Paul
Department of Computer Science and Engineering, NSHM Knowledge Campus, Durgapur, Arrah Shibtala Via Muchipara, Durgapur–713212, West Bengal, India

Rahul Kumar Prasad
Techno International New Town, Kolkata–700156, West Bengal, India

Payel Ray
Department of Computer Science and Engineering, University of Kalyani, West Bengal, India, E-mail: payelray009@gmail.com

Jesmin Roy
Department of Computer Science and Engineering, JIS College of Engineering, Kalyani, West Bengal, India

Shuvankar Roy
Computer Science and Engineering Department, MCKV Institute of Engineering, Howrah–711204, West Bengal, India, E-mail: shuvankarroy2@gmail.com

Saptarshi Roychowdhury
NSHM Knowledge Campus, Durgapur, West Bengal, India

Dwaipayan Saha
Department of Computer Science and Engineering, JIS College of Engineering, Kalyani, West Bengal, India

Debabrata Sarddar
Department of Computer Science and Engineering, University of Kalyani, West Bengal, India, E-mail: dsarddar1@gmail.com

Rishiraj Sarker
Jadavpur University, Kolkata–700032, West Bengal, India, E-mail: sarker.rishiraj88@gmail.com

Ananya Satpati (Das)
Department of Computer Science and Engineering, NSHM Knowledge Campus, Durgapur, Arrah Shibtala Via Muchipara, Durgapur–713212, West Bengal, India, E-mail: ananya.das@nshm.com

Vikram Kumar Saxena
Department of Electrical Electronics and Engineering, Pailan College of Management and Technology, West Bengal–700102, India

Debaparna Sengupta
Department of Electrical Engineering, Techno International New Town, Kolkata–700156, West Bengal, India

Ayushi Shah
Undergraduate Student, Department of Civil Engineering, Amity University Kolkata, Kolkata, West Bengal–700135, India

Nidhi Sindhwani
Department of ECE, Amity School of Engineering and Technology, New Delhi GGSIPU, Delhi, India

Rahul Kumar Singh
Techno International New Town, Kolkata–700156, West Bengal, India

Rahul Singh
Techno International New Town, Kolkata–700156, West Bengal, India

Arunava Kabiraj Thakur
Department of Electrical Engineering, Techno Main Salt Lake, Sector V, Bidhannagar, Kolkata–700091, West Bengal, India, E-mail: arunava.kabiraj007@gmail.com

Parag Kumar Guha Thakurta
CSE Department, NIT Durgapur, West Bengal, India

D. Thomas
Amal Jyothi College of Engineering, Kanjirappally, Kottayam, Kerala–686518, India

Shekhar Verma
Department of ECE, Amity School of Engineering and Technology, New Delhi GGSIPU, Delhi, India

ABBREVIATIONS

2D	two dimensional
ACC	adaptive cruise control
AGC	adaptive gamma correction
AHE	adaptive histogram equalization
ANA	agro nutrition alert
AWGN	additive white gaussian noise
BER	bit error rate
BS	base station
BT	body temperature
CDF	cumulative distribution function
CDMA	code division multiple access
CLAHE	contrast limited adaptive histogram equalization
CNN/ConvNet	convolutional neural network
COG	center of gravity
Coif	coiflets
CQA	classification quality assessment
CSA	control based pitch actuator system
CTC	connectionist temporal classification
db	Daubechies
DES	data encryption standard
DGR	directional greedy routing
DL	dead load
DNA	deoxyribonucleic acid
DNN	deep neural network
DSSS	direct sequence spread spectrum
DWT	discrete wavelet transform
EBGR	edge node based greedy routing
ESU	energy storage unit
EVCSs	electric vehicle charging stations
EVs	electric vehicles
FFT	fast Fourier transform
FLCs	fuzzy logic controllers
FN	false negative
FP	false positive

GPS	global positioning system
GSM	global system for mobile communications
GWEC	Global Wind Energy Council
GyTAR	greedy traffic-aware routing
HOG	histograms of oriented gradients
HR	heart rate
HRV	heart rate variability
HUD	heads up display
I/O	input/output
ICP	ideal customer profile
IDWT	inverse discrete wavelet transform
IoT	internet of things
IQA	image quality assessment
IR	infrared
IRB	institutional review board
KNN	k-nearest neighbor
LBMM	load balance Min-Min
LED	light emitting diode
LiDAR	light detection and ranging
LL	live load
LSTM	long short term memory
LVRT	low voltage ride through
MCT	minimum completion time
MM	min-min
MNIST	modified national institute of standards and technology
MOM	mean of maximu m
MPPT	maximum power point tracking
MSE	mean squared error
NB	negative big
NS	negative small
OLB	opportunistic load balancing
P&O	perturb and observe
PB	positive big
PS	positive small
PSD	power spectral density
PSNR	peak signal to noise ratio
ReLU	rectified linear unit
RMSE	root of mean square error
RNA	ribonucleic acid

RNN	recurrent neural network
ROI	region of interest
RSS	received signal strength
RSU	road side unit
SD	standard deviation
SGD	stochastic gradient descent
SGDM	stochastic momentum gradient descent
SISED	School of Illumination Science and Engineering
Sn	sensitivity
STATCOM	static compensator
STC	standard test conditions
STFT	short term Fourier transform
SVM	support vector machine
Sym	Symlets
TDOA	time difference of arrival
TN	true negative
TP	true positive
TSO	transmission system handling operators
UV	ultraviolet
V	voltage
V2V	vehicle to vehicle
VSI	voltage switching inverters
VUV	vacuum UV
WD	weighting distribution
WHO	World Health Organization
WSN	wireless sensor network
ZE	zero

PREFACE

This book aims to provide a sharing platform for researchers, academicians, and industry to throw in their expertise, acquaintance, and understanding in a wider domain of progressive and emerging technologies, including artificial intelligence, communication, cyber security, data analytics, Internet of Things (IoT), machine learning, power system, VLSI, embedded system, and much more. It aims to be a transdisciplinary and cross-cutting resource that present recent trends and advances and challenges.

We are delighted to deliver this book with the support and contributions from global academicians and researchers. This book is a manifestation of various interesting and important aspects of data science and computing technologies and their applications and methodologies in a wider range of applications including deep learning, DNA cryptography, classy fuzzy MPPT controller, driving assistance, and safety systems.

This book consists of 26 chapters, covering discussion on innovative algorithms and their applications for solving cutting-edge computational and data science problems with an interdisciplinary research perspective. All 26 chapters are divided in following seven parts containing interdisciplinary applications and methodologies of data science and computing technologies:

- Part I: Recent Application of Deep Learning and ANN Paradigm
- Part II: Role and Impact of Big Data in E-Commerce and Retail Sector
- Part III: Algorithm for Load Balancing in Cloud Computing
- Part IV: Advances in Embedded System Based Application
- Part V: Optimization Technique Using MATLAB Platform
- Part VI: Technique for Improving Information and Network Security
- Part VII: Miscellaneous Topics in Computing

In summary, in this 21st century, there is a genuine need for technology upgradation, and this book provides interdisciplinary research and a technological overview that the editors hope will lead to open a new dimension of understanding, which will be useful for uncovering new in this growing area.

The editors of this book are thankful to all authors, reviewers, and Apple Academic Press for their valuable input and contributions to make this book possible.

—Editors

PART I

Recent Application of Deep Learning and ANN Paradigm

CHAPTER 1

DEEP LEARNING SUPPORTED EVALUATION OF RETINAL VESSEL ENHANCEMENT TECHNIQUES

MAHUA NANDY PAL,[1] MINAKSHI BANERJEE,[2] and SHUVANKAR ROY[1]

[1]*Computer Science and Engineering Department, MCKV Institute of Engineering, Howrah–711204, West Bengal, India, E-mail: mahua.nandy@gmail.com (M. N. Pal)*

[2]*Computer Science and Engineering Department, RCC Institute of Information Technology, Kolkata–700015, West Bengal, India*

ABSTRACT

Hypertension, diabetes, glaucoma, etc., are different diseases, the symptoms of which prevails in retinal fundus images at preliminary stage. Retinal vessel segmentation is a significant prior step for different automated disease analysis system of retinal fundus images. Further, image enhancement has a fundamental role in retinal vessel segmentation. The better is the enhancement the better segmentation result is expected. In this work, three enhancement techniques are considered which are very frequently used before retinal vessel segmentation in contemporary literatures. The competency of each of the enhancement techniques is evaluated using two different categories of image quality assessment (IQA) metrics, peak signal to noise ratio (PSNR) and absolute mean brightness error (AMBE). In this way, morphological operation performs the best when compared with other two enhancement techniques. This observation has been further evaluated from a reverse view

Advances in Data Science and Computing Technology: Methodology and Applications. Suman Ghosal, Amitava Choudhury, Vikram Kumar Saxena, Arindam Biswas, & Prasenjit Chatterjee (Eds.)

point of effective deep convolutional neural network (CNN) vessel classification result on three differently enhanced image dataset which supports the previous observation in terms of classification efficiency metrics as well. Better vessel classification metrics are obtained from the same classification model when input datasets are better enhanced to represent vessel.

1.1 INTRODUCTION

Retinal fundus image is characterized both by anatomical and pathological artifacts present in its surface. The key anatomical artifacts are blood vessels, optic disc, and fovea. Pathological artifacts are usually symptoms of different optical and cardio-vascular diseases. These are manifested at an early stage in retinal image surface. Automated retinal image analysis system helps medical professionals in fast screening of huge real life dataset. Challenges present in automated retinal diagnostic system are presence of noise, low contrast, uneven illumination, minute sizes of medically significant local artifacts such as micro aneurysms, hemorrhages, etc. In process of automated retinal analysis system development, retinal vessel segmentation plays a significant role, Moreover, retinal image enhancement is the pre requisite of retinal vessel segmentation.

1.2 LITERATURE SURVEY

Adaptive histogram equalization (AHE) is an image enhancement technique. But AHE reports two major problems namely slow speed and over enhancement of noise. These problems are addressed by Pizer et al. (1987). Zuiderveld in 1994 presented contrast limited adaptive histogram equalization (CLAHE), to acquire required results overcoming the disadvantages Nandy and Banerjee (2017) made use of CLAHE for enhancement of images before vessel segmentation. Huang, Cheng, and Chiu (2013) dealt with AGC based image enhancement. They presented image transformation through gamma correction and probability distribution of luminance pixels. Rahman et al. (2016) discussed application of AGC for different types of image enhancement. An adaptive gamma correction (AGC) based on image contrast is adopted in Nandy and Banerjee (2019). Application of morphological operation has been utilized in medical image enhancement in 2015 (Hassanpour et al., 2015). Hassan et al. (2015) used different morphological operators and clustering for vessel segmentation. An entropy based image quality metric has been worked out in 2012 (Nandy and Banerjee, 2012). Sheba and Gladston (2016) evaluated different image enhancement techniques in the domain of mammogram images.

Almotiri et al. (2018) presented a survey on different retinal vessel segmentation algorithms. Roychowdhury et al. (2015) iteratively thresholded the residual retinal image to obtain higher vessel segmentation accuracy. Fu et al. (2016); Liskowski et al. (2016); and Mo and Zhang (2017) exploited deep learning architecture to identify vessels from retinal fundus images. Fu et al. (2016) used convolutional neural network (CNN) which follows vessel probability map. In this chapter, fully connected conditional random fields are combined with probability map to segment vessels. Liskowski et al. (2016) applied several preprocessing steps and training was conducted on a large number of samples. Generally large volume of training data is required for effective learning of deep architecture. These processes are extremely computationally exhaustive and the training is also very time consuming with requirements of expensive resources.

Ronneberger et al. (2015) proposed U-net CNN model which is efficient in segmenting bio medical images.

The contribution is to find the effective enhancement technique quantitatively following two diverse viewpoints. We have experimented with some of the enhancement techniques, frequently used for retinal vessel segmentation such as, CLAHE, AGC and morphological enhancement techniques. On the basis of IQA metric values, we have established with the notion that morphological enhancements are the best among them in an application oriented way. To arrive at a conclusive result from a different dimension, we experimented with the efficiency of deep vessel classification on three differently enhanced image datasets also. In this phase, the better the classification evaluation metric values the more informative the input images are. The results obtained from deep classification experiments are decently aligned with the observations obtained from image quality assessment (IQA) too.

The work representation is structured as follows. Section 1.3 presents enhancement methods normally used before vessel segmentation of retinal fundus. IQA evaluation metrics are there in Section 1.4 and deep classification based evaluation metrics. Section 1.5 presents database description. Section 1.6 explores deep neural network (DNN) architecture, training details and implementation requirements used to evaluate our observation. Section 1.7 describes experimental results of IQA based evaluation and deep CNN classification based evaluation respectively. Section 1.8 demonstrates the experimental results found in Section 1.7. Section 1.9 is the conclusion of the work.

1.3 IMAGE ENHANCEMENT TECHNIQUES

1.3.1 CONTRAST LIMITED ADAPTIVE HISTOGRAM-BASED ENHANCEMENT

General image transformation function is:

$$s = T(r); 0 \leq r \leq l_{max} \tag{1}$$

The l_{max} is maximum intensity value.

Pixel intensity levels of an image may be thought of as random variables. Random variables are fundamentally described by probability density function. Thus, the transformation function is:

$$s = T(r) = \int_0^r p_r(w)\,dw \tag{2}$$

where; w is a dummy integration variable. Eqn. (2) represents the cumulative distribution function (CDF) (digital image processing, Gonzalez). Following discrete representation, the probabilities of occurrence of intensity level r_k is:

$$P_r(r_k) = \frac{n_k}{MN} \tag{3}$$

where; n_k is the number of pixels having intensity level r_k. So, the discrete variation of the function is:

$$s_k = T(r_k) = \frac{(l_{max})}{MN\sum_{j=0}^{k} n_j} \tag{4}$$

where; $k = 0, 1, \ldots, l_{max}$

Eqn. (4) represents histogram equalized image. Pizer et al. (1987) elaborated AHE application. Zuiderveld (1994) modified it to CLAHE to obtain better enhancement result. CLAHE follows local region based approach. All the local regions contrast enhancement takes place independently and are combined with bilinear interpolation. Thus, boundary effects between the region can be removed.

1.3.2 ADAPTIVE GAMMA CORRECTION (AGC)-BASED ENHANCEMENT

AGC with weighting distribution (WD) (Huang, Cheng, and Chiu, 2013) is depicted as follows. Gamma correction formula is:

$$T(l_{max}) = (l_{max})\left(\frac{l}{l_{max}}\right)^{\gamma} \tag{5}$$

where; l_{max} is the maximum intensity. The gamma value may be greater than 1 or may be less than 1 (Huang et al., 2013). AGC with weight distribution is as follows (Huang et al., 2013):

$$T(l) = l_{max}\left(l / l_{\max}\right)^{1-cdf(l)} \tag{6}$$

The underlined concept is that the process tries to increase lower intensity level and at the same time tries to resist decrease of higher intensity level. The WD function has the following formula:

$$pdf_w(l) = pdf_{max}\left(\frac{pdf(l) - pdf_{min}}{pdf_{max} - pdf_{min}}\right)^a \tag{7}$$

where; a is the adjusted parameter. Thus, cdf becomes:

$$cdf_w(l) = \sum_{l=0}^{l_{max}} \frac{pdf_w(l)}{\sum pdf_w} \tag{8}$$

where; Σpdf_w is defined as:

$$\sum pdf_w = \sum_{l}^{l_{max}} pdf_w(l) \tag{9}$$

At the end, the gamma (γ) value is obtained in the following way:

$$\gamma = 1 - cdf_w(l) \tag{10}$$

1.3.3 MORPHOLOGICAL OPERATION-BASED ENHANCEMENT

Morphological operations make use of structuring elements while enhancing retinal artifacts. Disk shaped masks are generally used among researchers in biomedical image data enhancement. Tophat and bottom-hat transformations are represented (Digital Image Processing, Gonzalez) in Eqns. (11) and (12), respectively:

$$Tophat(I) = \left(I - \left(I\left(open\right)b\right)\right) \tag{11}$$

$$Bottomhat(I) = \left(\left(I\left(close\right)b\right) - I\right) \tag{12}$$

where; b is structuring element and morphological operation ($I(open)b$)is erosion of I by b, followed by dilation with b and ($I(close)b$ is dilation of I by b, followed by erosion with b. According to Hassanpour et al. (2015), subsequent applications of top-hat and bottom-hat transformation lead to contrast enhancement between light and dark arenas.

$$En = I + Tophat(I) - Bottomhat(I) \quad (13)$$

When Eqn. (13) is tried with different structural elements of suitable size enhancement result varies. The output enhanced retinal image is (En).

1.4 EVALUATION METRICS

We used two types of evaluation metrics for the assessment of retinal image enhancement techniques. They are IQA metrics and classification quality assessment (CQA) metrics.

1.4.1 *IQA METRICS*

IQA metrics are categorized in two classes: statistical error based metrics and visual information based metrics. Ideally, a human observer only can provide most consistent IQA metric to assess an image enhancement performance. It is qualitative evaluation. Visual information may not be properly represented by statistical error based metrics always instead human observer perception can be quite aptly captured by visual information based metrics. PSNR and AMBE are the mostly used statistical error based IQA metric and visual information based IQA metric, respectively.

1.4.1.1 PEAK SIGNAL TO NOISE RATIO (PSNR)

PSNR is defined as:

$$PSNR = 10log\frac{(I_{max})^2}{MSE} \quad (14)$$

MSE is mean squared error (MSE) between the original image(I) and the enhanced image(En) and is defined as:

$$MSE = \frac{1}{MN}\sum_{m,n}\left(En(m,n) - I(m,n)\right)^2 \quad (15)$$

Higher PSNR value specifies more information content.

1.4.1.2 ABSOLUTE MEAN BRIGHTNESS ERROR (AMBE)

AMBE is the absolute difference between enhanced image expectation (E(En)) and original image expectation (E(I)).

$$AMBE = \left|\left(E(En) - E(I)\right)\right| \tag{16}$$

Lower AMBE value specifies better image enhancement.

1.4.2 DEEP CLASSIFICATION METRICS

The significances of true positive (TP), false positive (FP), true negative (TN), and false negative (FN) with respect to vessel classification are as follows, TP is the proportion of actual vessel pixels in the whole population whereas, TN is the proportion of actual non-vessel pixels in the whole population. Similarly, FP is the proportion of falsely identified vessel pixels in the whole population whereas FN is the proportion of falsely recognized non-vessel pixels in the whole population. Accuracy and sensitivity are two important evaluation metrics in case of medical image processing.

1.4.2.1 SENSITIVITY

Sensitivity (Sn) is TP rate. The closer the value of sensitivity to 1, the better is the identification power of vessel pixel of the method.

$$Sn = TP / (TP + FN) \tag{17}$$

1.4.2.2 ACCURACY

Accuracy (Acc) provides proportion of both TP and true negative.

$$Acc = (TP + TN) / (TP + TN + FP + FN) \tag{18}$$

1.5 DATASET DESCRIPTION

DRIVE (Digital retinal image for vessel extraction, Image Sciences Institute) retinal image vessel segmentation dataset is used for method evaluation as it is the most frequently used dataset among researchers. Canon

CR5 non-mydriatic 3CCD camera with 45° field of view has been used to capture retinal images. Each image was preserved with 8 bits per color plane and at 565 x 584 pixels' resolution. In this dataset ground truths are also available.

1.5.1 PATHOLOGICAL SAMPLES

Around 40 photographs have been randomly considered, 33 are healthy and 7 show symptoms of mild early diabetic retinopathy. Among all those 7 images 3 belongs to training set whereas 4 belongs to test set. A brief description of the abnormalities is as available online (https://drive.grand-challenge.org/):

- **25 Training:** Pigment epithelium changes, probably buttery maculopathy with pigmented scar in fovea, or choroidopathy, no diabetic retinopathy or other vascular abnormalities.
- **26 Training:** Background diabetic retinopathy, pigmentary epithelial atrophy, atrophy around optic disk.
- **32 Training:** Background diabetic retinopathy.
- **03 Test:** Background diabetic retinopathy.
- **08 Test:** Pigment epithelium changes, pigmented scar in fovea, or choroidopathy, no diabetic retinopathy or other vascular abnormalities.
- **14 Test:** Background diabetic retinopathy.
- **17 Test:** Background diabetic retinopathy.

1.6 DEEP NEURAL NETWORK (DNN)

A deep CNN model has been implemented and then tested with differently pre-processed image datasets. CNN architecture has been formed using U-net type connections. U-net is a CNN, efficient in the field of medical image segmentation. The characteristic of U-net is that it utilizes the same feature maps for contraction of the image to a feature vector and for expansion of the vector to a reconstructed image. Thus, structural integrity of image data is retained. U-net model is trained on DRIVE trainset for vessel identification. Application of U-net model on appropriately enhanced image dataset leads to pixel level classification of vessel and non-vessel. Classification efficiency measurement of the same model with differently enhanced image dataset indicates better image enhancement method.

1.6.1 ARCHITECTURE

U-net architecture Ronneberger et al. (2015) is used here for CNN based input IQA. Cross-entropy loss function and stochastic gradient descent (SGD) optimization algorithm is used in implementation. Rectified linear unit (ReLU) is used as activation function. A drop out of 0.2 has been employed between two consecutive convolutional layers. The summary of the train model is represented in Figure 1.1.

Layer (type)	Output Shape	Param #	Connected to
input 1 (InputLayer)	(None, 1, 48, 48)	0	
conv2d 1 (Conv2D)	(None, 32, 48, 48)	320	input 1[0][0]
dropout 1 (Dropout)	(None, 32, 48, 48)	0	conv2d 1[0][0]
conv2d 2 (Conv2D)	(None, 32, 48, 48)	9248	dropout 1[0][0]
max pooling2d 1 (MaxPooling2D)	(None, 16, 24, 48)	0	conv2d 2[0][0]
conv2d 3 (Conv2D)	(None, 64, 24, 48)	9280	max pooling2d 1[0][0]
dropout 2 (Dropout)	(None, 64, 24, 48)	0	conv2d 3[0][0]
conv2d 4 (Conv2D)	(None, 64, 24, 48)	36928	dropout 2[0][0]
max pooling2d 2 (MaxPooling2D)	(None, 32, 12, 48)	0	conv2d 4[0][0]
conv2d 5 (Conv2D)	(None, 128, 12, 48)	36992	max pooling2d 2[0][0]
dropout 3 (Dropout)	(None, 128, 12, 48)	0	conv2d 5[0][0]
conv2d 6 (Conv2D)	(None, 128, 12, 48)	147584	dropout 3[0][0]
up sampling2d 1 (UpSampling2D)	(None, 256, 24, 48)	0	conv2d 6[0][0]
concatenate 1 (Concatenate)	(None, 320, 24, 48)	0	conv2d 4[0][0] up sampling2d 1[0][0]
conv2d 7 (Conv2D)	(None, 64, 24, 48)	184384	concatenate 1[0][0]
dropout 4 (Dropout)	(None, 64, 24, 48)	0	conv2d 7[0][0]
conv2d 8 (Conv2D)	(None, 64, 24, 48)	36928	dropout 4[0][0]
up sampling2d 2 (UpSampling2D)	(None, 128, 48, 48)	0	conv2d 8[0][0]
concatenate 2 (Concatenate)	(None, 160, 48, 48)	0	conv2d 2[0][0] up sampling2d 2[0][0]
conv2d 9 (Conv2D)	(None, 32, 48, 48)	46112	concatenate 2[0][0]
dropout 5 (Dropout)	(None, 32, 48, 48)	0	conv2d 9[0][0]
conv2d 10 (Conv2D)	(None, 32, 48, 48)	9248	dropout 5[0][0]
conv2d 11 (Conv2D)	(None, 2, 48, 48)	66	conv2d 10[0][0]
reshape 1 (Reshape)	(None, 2, 2304)	0	conv2d 11[0][0]
permute 1 (Permute)	(None, 2304, 2)	0	reshape 1[0][0]
activation 1 (Activation)	(None, 2304, 2)	0	permute 1[0][0]

Total params: 517,090
Trainable params: 517,090
Non-trainable params: 0

FIGURE 1.1 Summary of the deep CNN model.

1.6.2 TRAINING

DRIVE images undergo three types of image enhancement techniques to obtain three different datasets as discussed in Section 1.3. Training is performed on image patches of differently preprocessed image datasets. Data augmentation has not been performed. During deep classification based evaluation, 48 x 48 random overlapping patches are generated from image dataset for proper training with less number of image samples. 90% of patch dataset has been used for training whereas 10% of patch dataset has been used for validation. Mini batch size is 32 patches.

1.6.3 IMPLEMENTATION REQUIREMENTS

All the experiments have been implemented in Python 3.6.8. High level APIs of Google's deep learning library TensorFlow have been used to implement and evaluate the work. Implementations have been executed in Google online Colab cloud environment where Tesla K80 GPU, 12 GB VRAM, 4* Intel(R) Xeon(R) CPU @ 2.20 GHz, 13.51 GB RAM, 358.27 GB HDD are available as deep network execution resources.

1.7 EXPERIMENTAL RESULTS

1.7.1 IMAGE QUALITY ASSESSMENT (IQA) METRIC BASED EVALUATION

During enhancement, green channel of retinal image is only considered (Nandy and Banerjee, 2012) as the intensity in green channel facilitates retinal structures to be observable. Retinal images are almost saturated in the red channel and have very low contrast in blue channel. Green channel images are contrast enhanced by applying CLAHE. Another effective image enhancement tool is Gamma correction. Green channel images undergo adaptive linear gamma correction where all the pixels are linearly weighted by some predefined factor (Huang et al., 2013). One more exhaustively used retinal image enhancement tool is application of morphological operators (Hassanpour et al., 2015). All four images and their histograms are given in Figure 1.2. Average PSNR and AMBE for DRIVE images are shown both graphically and in tabular form in Figure 1.3 and Table 1.1, respectively.

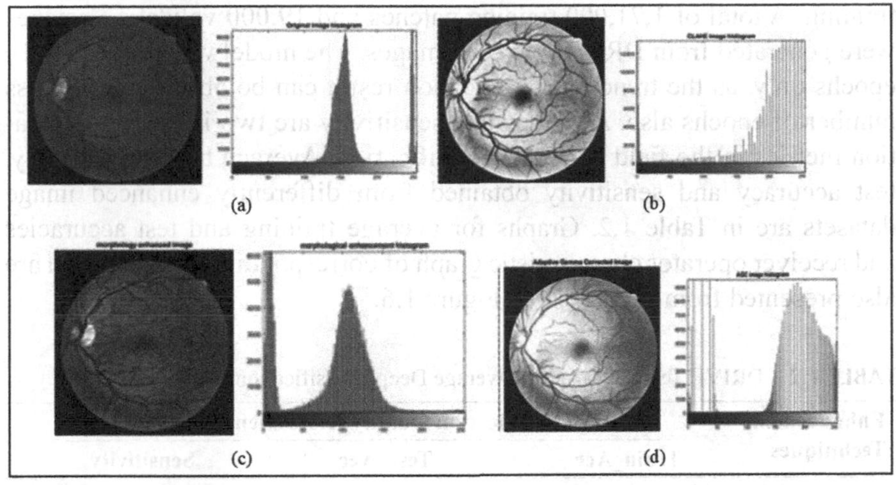

FIGURE 1.2 (a) Retinal green channel and histogram; (b) CLAHE and stretched histogram; (c) morphology and stretched histogram; (d) AGC and stretched histogram.

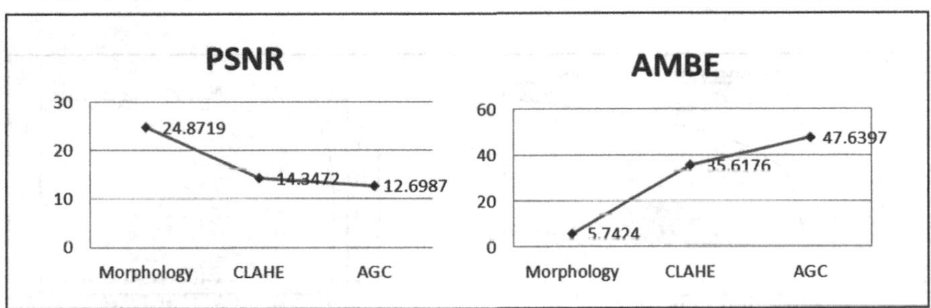

FIGURE 1.3 Average PSNR and AMBE graph for DRIVE images.

TABLE 1.1 DRIVE Test Set Images – Average IQA Metrics Value

Enhancement Techniques	Avg Image Quality Assessment Metrics Value	
	PSNR	AMBE
Morphology	24.8719	5.7424
CLAHE	14.3472	35.6176
AGC	12.6987	47.6397

1.7.2 DEEP CLASSIFICATION METRIC BASED EVALUATION

48 x 48 random and overlapping image patches are generated from image dataset. U-net type deep CNN architecture has been fed with patches for

training. A total of 1,71,000 training patches and 19,000 validation patches were generated from DRIVE train set images. The model was trained for 10 epochs only, as the trend of classification result can be observed from less number of epochs also. Accuracy and sensitivity are two important evaluation metrics in the field of vessel classification. Average training accuracy, test accuracy and sensitivity obtained from differently enhanced image datasets are in Table 1.2. Graphs for average training and test accuracies and receiver operator characteristic graph of corresponding classification are also presented from Figure 1.4 to Figure 1.6.

TABLE 1.2 DRIVE Test Set Images – Average Deep Classification Metrics

Enhancement Techniques	Avg Classification Quality Assessment Metrics Value		
	Train_Acc	**Test_Acc**	**Sensitivity**
Morphology	0.9576	0.9524	0.7573
CLAHE	0.9568	0.9525	0.7564
AGC	0.9510	0.9484	0.7349

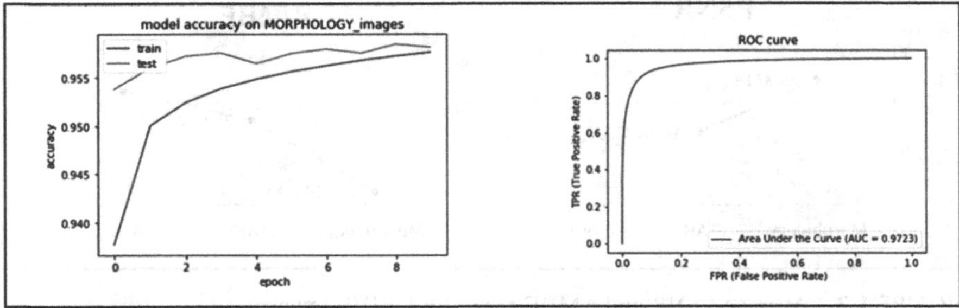

FIGURE 1.4 ROC and accuracy curves of morphologically enhanced images.

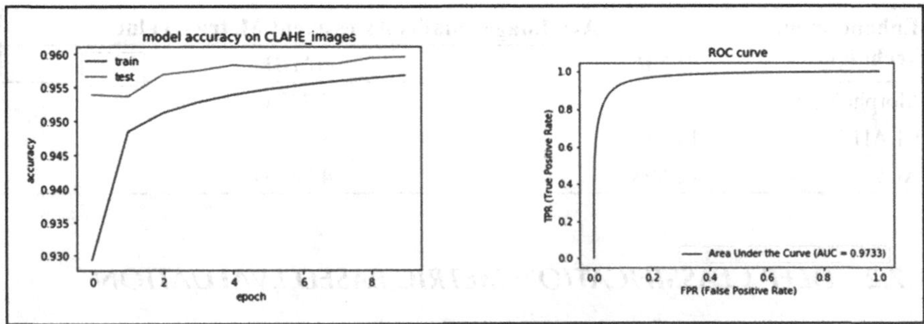

FIGURE 1.5 ROC and accuracy curves of CLAHE images.

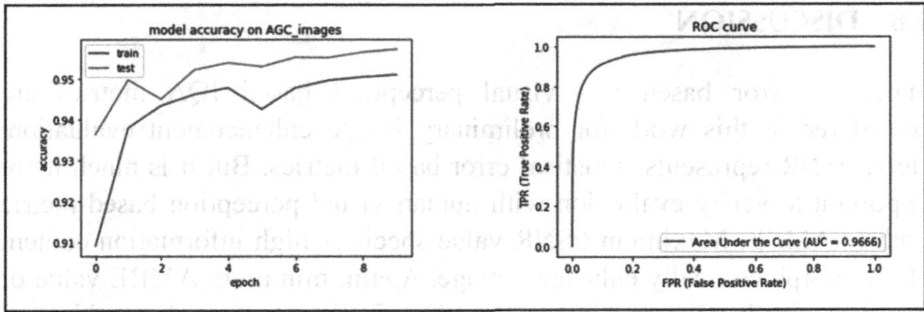

FIGURE 1.6 ROC and accuracy curves of AGC enhanced images.

1.7.2 IMAGE QUALITY ASSESSMENT (IQA) OF PATHOLOGY SAMPLES

Four images among 20 drive test images have mild early DR signs as pathological symptoms. Table 1.3 shows individual performance measurements obtained by applying deep CNN classification on those pathology samples separately. This is a reverse view point to assess the applicability of different enhancement techniques on abnormal images also. In most of the cases morphologically enhanced abnormal images are characterized with higher metric values.

TABLE 1.3 Deep CNN Classification Metrics on DRIVE Test Dataset Abnormal Images

Symptoms	Enhancement Techniques	Deep Classification Assessment Metrics	
		Accuracy	Sensitivity
Hard exudate	Morphology	0.9420	0.6921
	CLAHE	0.9420	0.6928
	AGC	0.9420	0.7670
Hard exudate	Morphology	0.9464	0.6580
	CLAHE	0.9458	0.6489
	AGC	0.9365	0.633
Hemorrhage	Morphology	0.9574	0.8164
	CLAHE	0.9571	0.8106
	AGC	0.9535	0.7987
hard exudate	Morphology	0.9493	0.6864
	CLAHE	0.9515	0.7153
	AGC	0.9487	0.7082

1.8 DISCUSSION

Statistical error based and visual perception based IQA metrics are considered in this work for preliminary image enhancement evaluation. Here, PSNR represents statistical error based metrics. But it is much more important to verify evaluation with human visual perception based metric such as AMBE. Maximum PSNR value specifies high information content of the morphologically enhanced image. Again, minimum AMBE value of morphologically enhanced image also identifies it as better enhanced image. Table 1.1 data represents the fact properly. Average values are computed on DRIVE test dataset images.

This observation is further estimated from the view point of deep CNN classification as vessel enhancement is a very important prior sub part of retinal vessel segmentation. So, better enhanced images tend to produce better classification efficiency metrics. Maximum value of average segmentation accuracy indicates the superior quality of morphologically enhanced image data with application to retinal fundus images. In the field of vessel segmentation use of sensitivity as performance measurement index is highly significant. Maximum average sensitivity has also been obtained from morphologically enhanced test images. Table 1.2 represents average values of train accuracy, test accuracy and sensitivity on DRIVE images. Different ailment symptoms are evident in retina at a very early phase. Table 1.3 provides individual test accuracy and sensitivity of pathology samples of DRIVE test set separately. In individual evaluation of pathology samples also, most of the classification efficiency values are in favor of morphological operation.

Though morphological enhancement performed very well in both IQA based evaluation as well as deep classification based evaluation, performing image enhancement by combining enhancement methods in some specific way may produce better result as well.

1.9 CONCLUSION

Three single image enhancement techniques are experimented in this work for their efficiency in image enhancement with application to retinal vessel segmentation. These single enhancement techniques are applications of CLAHE, AGC, and morphological enhancements. From the view point of IQA metric values it is decided quantitatively that morphological enhancement performs well for this purpose. This claim has been further established by deep CNN vessel classification based evaluation also as better classification metrics are expected to obtain from better vessel enhanced input image datasets.

In vessel segmentation literature, these prevalently used enhancement techniques are used in different arbitrary combinations also. It is intended to enrich this work by analyzing combinations of these techniques in different possible orders and evaluating which one is supposed to yield even better results. Deep learning based result verification may be extended to cross dataset evaluation, too. In this way, the best combination of enhancement techniques relevant to the pertinent field can be decided. This may be considered as the future scope of this work.

KEYWORDS

- **adaptive gamma correction (AGC)**
- **contrast limited adaptive histogram equalization (CLAHE)**
- **convolutional neural network (CNN)**
- **image quality assessment (IQA)**
- **morphology**

REFERENCES

Almotiri, J., Elleithy, K., & Elleithy, A., (2018). Retinal vessels segmentation techniques and algorithms: A survey. *Applied Sciences, 8*.

Cai, J., Gu, S., & Zhang, L., (2018). Learning a deep single image contrast enhancer from multi-exposer images. *IEEE Transactions on Image Processing, 27*, 2049–2062.

Fu, H., Xu, Y., Lin, S., Wong, D., & Liu, J., (2016). Deep vessel: Retinal vessel segmentation via deep learning and conditional random field. *Proceedings of Medical Image Computing and Computer Assisted Intervention MICCAI*, 132–139.

Gonzalez, R. C., & Woods, R. E. (2015). *Digital Image Processing* (3rd edn.). Pearson.

Hassan, G., El-Bendary, N., Hassanien, A., Fahmy, A., Shoeb, A., & Snasel, V., (2015). Retinal blood vessel segmentation approach based on mathematical morphology. *Procedia Computer Science, 65*, 612–622.

Hassanpour, H., Samadiani, N., & Mahdi, S. S. M., (2015). Using morphological transforms to enhance the contrast of medical images. *The Egyptian Journal of Radiology and Nuclear Medicine, 46*, 481–489.

Huang, S., Cheng, F., & Chiu, Y., (2013). Efficient contrast enhancement using adaptive gamma correction with weighting distribution. *IEEE Transaction on Image Processing, 22*.

Liskowski, P., & Krawiec, K., (2016). Segmenting retinal blood vessels with deep neural networks. *IEEE Transactions on Medical Imaging, 35*, 2369–2380.

Mo, J., & Zhang, L., (2017). Multi-level deep supervised networks for retinal vessel segmentation. *International Journal of Computer Assisted Radiology and Surgery, 12*, 2181–2193.

Nandy, M., & Banerjee, M., (2012). Retinal vessel segmentation using Gabor filter and artificial neural network. *Proceedings of IEEE International Conference on Emerging Applications of Information Technology – EAIT*, 157–160.

Nandy, M., & Banerjee, M., (2019). Automatic diagnosis of micro aneurysm manifested retinal images by deep learning method. *Proceedings of ICETST*.

Nandy, P. M., & Banerjee, M., (2017). A comparative analysis of application of niblack and sauvola binarization to retinal vessel segmentation. *Proceedings of CINE*. IEEE Xplore.

Pizer, S., Amburn, E., Austin, J., Cromartie, R., Geselowitz, A., Greer, T., Ter, H. R. B., et al., (1987). Adaptive histogram equalization and its variations. *Computer Vision, Graphics and Image Processing, 39*, 355–368.

Rafael C. Gonzalez & Richard E. Woods, (2008). *Digital Retinal Images for Vessel Extraction (DRIVE) Dataset*. Image Sciences Institute, http://www.isi.uu.nl/Research/Databases/DRIVE/ (accessed on 8 December 2021).

Rahman, S., Rahman, M., Abdullah-Al-Wadud, M., Al-Quaderi, G., & Shoyaib, M., (2016). An adaptive gamma correction for image enhancement. *EURASIP Journal on Image and Video Processing, 35*, 1–13.

Ronneberger, O., Fischer, P., & Brox, T., (2015). *U-Net: Convolutional Networks for Biomedical Image Segmentation*. arXiv:1505.04597v1 [cs.CV].

Roychowdhury, S., Koozekanani, D., & Parhi, K. K., (2015). Iterative vessel segmentation of fundus images. *IEEE Transactions on Biomedical Engineering, 62*, 1738–1749.

Sheba, K., & Gladston, R., (2016). Objective quality assessment of image enhancement methods in digital mammography-a comparative study. *Signal and Image Processing: An International Journal, 7*.

Singh, S., & Bovis, K., (2005). An evaluation of contrast enhancement techniques for mammographic breast masses; Chapter VIII. 5. *Graphics Gems, IV., 9*, 109–119.

Staal, J., Abrano, A., Niemeijer, M., Viergever, M., & Ginneken, B., (2004). Ridge based vessel segmentation in color images of the retina. *IEEE Transactions on Medical Imaging, 23*, 501–509.

Zuiderveld, K., (1994). Contrast limited adaptive histogram equalization. Chapter VIII.5, *Graphics Gems, IV*, 474–485.

CHAPTER 2

AN EFFICIENT AND OPTIMIZED DEEP NEURAL NETWORK ARCHITECTURE FOR HANDWRITTEN DIGIT RECOGNITION

SHUVANKAR ROY and MAHUA NANDY PAL

Computer Science and Engineering Department, MCKV Institute of Engineering, Howrah–711204, West Bengal, India, E-mail: shuvankarroy2@gmail.com (S. Roy)

ABSTRACT

This work presents a new, application oriented and hyper parameter optimized architecture of deep neural network (DNN) for hand written digit recognition. Though a substantial amount of work has been done in this field, yet it is an open area of research as there is scope to determine the suitable deep architecture for a particular problem domain and to improve the efficiency of deep classifier through optimization. Evaluating the efficiency of deep classification process and further improvement of its performance in the relevant field leads to the development of a competent network. In this chapter, A DNN architecture, has been developed which is capable of recognizing hand written digits in a very efficient way. Classification accuracy is used as evaluation metric. The proposed DNN architecture is capable of achieving almost 96.5% accuracy before hyper parameter optimization and shows a further improved accuracy value of 98.12% after hyper parameter optimization. These results are indicative of its relevance in the particular domain.

Advances in Data Science and Computing Technology: Methodology and Applications. Suman Ghosal, Amitava Choudhury, Vikram Kumar Saxena, Arindam Biswas, & Prasenjit Chatterjee (Eds.)

2.1 INTRODUCTION

Before establishment of proper technologies, we had been heavily relied on hand written materials and data, which is error prone and even difficult to store and access. Due to the wave of digitization, the previously necessary hand written documents are needed to be digitized. Handwritten digit recognition is a subpart of the same problem. Digitized handwritten documents can be made usable by adopting a complete highly accurate handwritten text recognition tool. Hand written character and digit recognition make the whole process of storage, organization, and access of hand written data comparatively simpler. Adoption of a highly accurate handwritten text recognition tool makes user authentication problem easier.

As the system has immense practical importance, deep learning has been applied to it to improve its efficiency. Currently deep learning application has become state-of-the-art for various classification problem for its outstanding performance in the relevant field. But not all network architectures are suitable for various problem domains. So, deciding proper network architecture with optimized hyper parameters is very important with respect to a particular application domain. Though training of deep neural network (DNN) is computationally expensive and very time consuming, optimization of hyper parameters is expected to yield better network performance.

2.2 LITERATURE SURVEY

Now-a-days deep learning exploration has become the recent trend of research in various fields. With advent of high-end architecture suitable to implement deep learning environment, researchers try to exploit the benefit of this to a greater extent. Among different handwritten digit or character classification methods, Hochreiter and Schmidhuber (1997), and Voigtlaender, Doetsch, and Ney (2016) describe applications of long short-term memory (LSTM). Graves, Fernandez, Gomez, and Schmidhuber (2006) and Graves et al. (2009) trained recurrent neural network (RNN) with connectionist temporal classification (CTC). Graves and Schmidhuber (2009) also used a multidimensional RNN. Pham, Bluche, Kermorvant, and Louradour (2014) adjusted neural network parameters to improve the result. Bluche and Messina (2017) deals with multi lingual hand writing recognition. Puigcerver (2017) verifies the applicability of different recurrent layers whereas Castro, Bezerra, and Valena (2018) establishes the effectiveness of boosting multidimensional

LSTM for hand written recognition system. DNNs outperform on many machine learning problems, but they are very sensitive to hyperparameters setting (Domhan, Springenberg, and Hutter, 2015).

Digit recognition system has been evaluated using different automated learning algorithms namely, Multilayer Perceptron, support vector machine (SVM), Naïve-Bayes, Bayes Net, Random Forest, J48 and Random Tree using WEKA in 2018 (Shamim, Miah, Sarker, Rana, and Jobair, 2018). The chapter (Zohra and Rao, 2019) presented performance analysis of SVM, K-nearest neighbor (KNN) and convolutional neural network (CNN).

Deep learning is a powerful technique in current generation. The chapter (Mishra, Malathi, and Senthilkumar, 2018) dealt with handwritten digit recognition on the well-known image database using Convolution neural network. Deep learning increases accuracy. In the process, the model becomes highly efficient to predict distorted data also. Additionally, a comparative analysis of the model has been performed on multicore CPU and General GPU and computation times are recorded. In this chapter (Vijayalaxmi, Rudraswamimath, and Bhavanishankar, 2019), the Handwritten Digit Recognition using Deep learning methods has been implemented. The maximum used machine learning algorithms were compared by training and testing on same dataset. The classifiers used are KNN, SVM, RFC, and CNN. Utilizing these deep learning techniques, a high amount of accuracy can be obtained. The goal of the chapter (Siddique, Sakib, and Siddique, 2019) is to analyze the effective number of hidden layers and epochs and corresponding accuracies. They performed experiments on Modified National Institute of Standards and Technology (MNIST) dataset to evaluate their work. Lastly, Baldominos, Saez, and Isasi, (2019) has provided an exhaustive review of the state of the art for both the MNIST and EMNIST databases.

In industries and academic fields, the demand for the type written and hand written character or digit recognition is always there. Character recognition is a combination of detecting, segmenting, and identifying characters from an image. The ultimate goal of hand written character recognition is to simulate the human reading capabilities so that machine can read, understand, edit, and work with text similar to human. Handwritten character recognition is more difficult than type written one. Type written texts are equally spaced and have a specified size of each character, which makes character segmentation a lot easier. Whereas handwritten texts may not be accurately spaced, may contain skewness and non-horizontal line, malformed font style and even non-human understandable characters, which makes the process of character extraction a tougher job. A big challenge in multi-lingual hand written text recognition is non-availability of large corpus of hand written data. The results of hand written digit or character recognition using deep learning produces better results compared to the same using traditional machine learning in most of the time.

The main focus of this chapter is to provide a framework for deep learning architecture in the field of hand written digit recognition. The current literature indicates that the challenges present in the field of deep learning are proper application-oriented network architecture, parameter optimization, minimization of time complexity and computational complexity, requirement of huge sample data for network learning, etc. So, determining proper application specific deep architecture and its hyper parameter optimization for hand written digit recognition is very significant in this respect.

Following sections represent theoretical background, implementation, and results, discussion, and conclusion of the chapter, respectively.

2.3 THEORETICAL BACKGROUND

2.3.1 CLASSIFIER OPTIMIZATION

Classifier optimization problem deeply relies on formulating four factors. They are identification of parameter search space, minimization or maximization of objective function, the basic model on which search space values will be evaluated and stored outcomes of each evaluation. In case of DNN optimization, different network hyper parameters that are intended to optimize in this chapter are learning rate, learning algorithm used, number of hidden layers, batch size and dropout. These terms are explained in the following subsections. The search space for each individual parameter is defined. The objective functions used here are validation accuracy and test accuracy.

2.3.2 DEEP NEURAL NETWORK (DNN)

A DNN is an artificial neural network having more than one hidden layer in between input and output layer. The neurons of the network are inter connected with weighted connections. The advantage of using deep network is that the network itself is capable of extracting features of the input patterns. Thus, programmer does not require to establish efficient handpicked features to represent input patterns. The DNN computes and adjusts the correct mathematical weights to convert the input into the output. The impulse or signal passes through the neural network layers finding the probability of each output.

2.3.3 NETWORK LAYERS

Neural network has three types of layers made up of a number of neurons. They are the input layer, hidden layer, and output layer.

2.3.3.1 INPUT LAYER

Input layer accepts initial information from outside the network. The consecutive nodes of the network process these inputs to draw some inference.

2.3.3.2 HIDDEN LAYER

These are the layers in between input layer and output layer of a neural network. The nodes of theses layers are responsible for processing the inputs and all intermediate computations.

2.3.3.3 OUTPUT LAYER

Output layer is responsible for producing the output of the neural network by computing individual class probability.

2.3.4 NETWORK HYPER PARAMETERS

Network hyper parameters are some values which are required to be decided before starting training process of the network such as learning algorithm, leaning rate, number of hidden layers, batch size, dropout value, etc. Deciding the best evaluated value of hyper parameters is known as hyper parameter tuning. Hyper parameter tuning may be done either manually or by applying grid search over the search space.

Following are some of the network hyper parameters which have been tried to be optimized or tuned in the following subsections for network performance improvement.

2.3.4.1 LEARNING RATE

In machine learning and statistics, the learning rate is a tuning parameter that defines the step size to be taken at each iteration of learning algorithm while moving towards a minimum of loss function. The lower the value of learning rate, the slower we travel towards minimum and the lower is the chance of missing local minima, but low learning rate also means the model learning phase will be time consuming and will take long time to converge if the model is stuck in plateau region.

2.3.4.2 OPTIMIZER

Optimizer or optimization algorithm can be used interchangeably. Optimization algorithm determines the speed and efficiency of a machine learning algorithm. Different optimization algorithms are stochastic gradient descent (SGD), mini-batch gradient descent, gradient descent with momentum, Adam optimizer, etc.

SGD method iteratively calculates stochastic approximation of gradient descent optimization. SGD replaces the actual gradient value, calculated from the entire data set by an estimated value, calculated from a randomly selected subset of data.

Adam optimization algorithm can be considered as a combination of RMSprop and SGD with momentum. It utilizes squared gradients to scale up the learning rate like RMSprop and uses of momentum by using moving average value of the gradient instead of gradient itself like SGD with momentum.

Adagrad is a learning algorithm with parameter-specific learning rates. The algorithm depends on frequency of updating of a parameter during training. Learning rate value is inversely proportional to the update frequency that is the greater number of a parameter receives update, the smaller is the value of learning rate.

2.3.4.3 HIDDEN LAYERS

Features of input layers are propagated through hidden layers. Number of hidden layers should also be determined for obtaining proper output.

2.3.4.4 BATCH SIZE

It indicates number of training samples presented in one iteration. The entire dataset is divided into number of small batches. Ideally the batch size may be within the range from 1 to the whole training sample set.

2.3.4.5 DROPOUT

Dropout is a regularization method and is used to prevent overfitting during neural network training. By drop out some neurons are randomly removed from each layer and not used in forward or back propagation. The significance of the dropout hyperparameter is the probability of training a given node in a layer, where 1.0 means no dropout, and 0.0 means no outputs from the layer.

2.4 IMPLEMENTATION AND RESULTS

2.4.1 DATASET DESCRIPTION

MNIST (modified national institute of standards and technology database) is a large collection of handwritten digits that is commonly used for training and testing in the fields of various image processing systems as well as machine learning systems. The images in the MNIST database are from two NIST's databases: Special Database 1 and Special Database 3. Special Database 1 and Special Database 3 digits were written by high school students and employees of the United States Census Bureau, respectively. The dataset contains 60,000 training sample images and 10,000 test sample images of 28 pixels in height and 28 pixels in width, for a total of 784 pixels in total. Each pixel is associated with an intensity value between 0 and 255 inclusive, indicating lightness or darkness of that pixel. The original black and white (binary) images were size normalized to 20×20 pixel while preserving the aspect ratio. The normalization algorithm uses anti-aliasing technique and thus the resulting images contain gray levels. The images were centered in a 28×28 image by computing the center of mass of the pixels, and translating the image so as to position this point at the center of the 28×28 field. Figure 2.1 represents some examples of handwritten digits of MNIST dataset.

FIGURE 2.1 Example of handwritten digits in MNIST database.

Each row of the csv file represents one labeled example. As this is a 10-class classification problem, column 0 represents the label that a human observer has assigned for it. Columns 1 through 784 represent the feature values for each pixel of 28×28=784-pixel values. The pixel values are in the range 0–255. 0 represents white, 255 represents black. Most of the pixel values are 0. Numbers of training samples and test samples are 48,000 and 12,000, respectively. The ratio of train set and validation set is 80:20.

2.4.2 EVALUATION METRICS

Validation accuracy and test accuracy have been used as evaluation metrics in this process. Validation set is a part of training set, used to validate the model during its generation. Validation phase avoids overfitting of the model. Test set is the unknown set of data, which the already built model is supposed to classify.

True positive (TP) is the proportion of truly positive samples in the whole population whereas, true negative (TN) is the proportion of truly negative samples in the whole population. Similarly, false positive (FP) is the proportion of falsely positive samples in the whole population whereas false negative (FN) is the proportion of falsely negative samples in the whole population.

Accuracy can be defined as the proportion of true results, both TP and TN, in a population. It can be defined in terms of TP, TN, FP and FN as:

$$\text{Accuracy} = (\text{TN} + \text{TP})/(\text{TN} + \text{TP} + \text{FN} + \text{FP}) \quad (1)$$

2.4.3 IMPLEMENTATION REQUIREMENTS

All the experiments have been implemented in Python 3.6.8 using TensorFlowv1 at backend. Implementations have been executed in Google online Colab cloud environment where Tesla K80 GPU, 12 GB VRAM, 4* Intel(R) Xeon(R) CPU @ 2.20 GHz, 13.51 GB RAM, 358.27 GB HDD are available as deep network execution resources.

High level APIs of Google's deep learning library TensorFlow have been used to implement and evaluate the work.

The input layer contains 784 weights corresponding to the input images shape of 28×28 pixel. The first hidden layer will contain 784×n weights where n represents the number of nodes in the layer. First hidden layer captures low level features and consecutive layers capture more sophisticated features which can be visualized for perception also.

2.4.4 NETWORK ARCHITECTURE

The network architecture followed to evaluate the work consists of one input, one output layer and 4 hidden layers. Rectified linear unit (ReLU) has been used as nonlinear activation function in hidden layers. One dropout layer has been inserted between 4th hidden layer and output layer with dropout value of 0.2. Different layers, their output shape and number of parameters in each layer have been depicted in Figure 2.2.

2.4.5 HYPER PARAMETER OPTIMIZATION

2.4.5.1 LEARNING RATE EVALUATION

Performance of the model has been evaluated using three different learning rates and the results obtained are represented in Table 2.1. Best validation accuracy and test accuracy are secured using learning rate = 0.05. Table 2.1 represents learning rate evaluation.

2.4.5.2 OPTIMIZATION ALGORITHM EVALUATION

Performances of using three optimization algorithms during training model generation have been evaluated keeping best performing learning rate = 0.05

as constant. These optimization algorithms are Adagrad, Adam, and Gradient Descent. Table 2.2 represents optimization algorithm evaluation.

```
Layer (type)                 Output Shape              Param #
=================================================================
input_1 (InputLayer)         (None, 784)               0
_________________________________________________________________
Hidden_Layer_1 (Dense)       (None, 300)               235500
_________________________________________________________________
Hidden_Layer_2 (Dense)       (None, 100)               30100
_________________________________________________________________
Hidden_Layer_3 (Dense)       (None, 100)               10100
_________________________________________________________________
Hidden_Layer_4 (Dense)       (None, 200)               20200
_________________________________________________________________
Dropout_Layer_1 (Dropout)    (None, 200)               0
_________________________________________________________________
Output_Layer (Dense)         (None, 10)                2010
=================================================================
Total params: 297,910
Trainable params: 297,910
Non-trainable params: 0
```

FIGURE 2.2 Summary of the model.

TABLE 2.1 Learning Rate Evaluation

Learning Rate	Validation Accuracy	Test Accuracy
0.5	0.61	0.61
0.1	0.95	0.95
0.05	**0.95**	**0.96**
0.01	0.93	0.93

TABLE 2.2 Optimization Algorithm Evaluation

Optimizer	Validation Accuracy	Test Accuracy
Adagrad optimizer	0.95	0.96
Adam optimizer	0.86	0.87
Gradient descent optimizer	0.94	0.94

2.4.5.3 NUMBER OF HIDDEN LAYERS

Network performance has been evaluated by taking 1, 2, 3, 4, and 5 number of hidden layers while previously discussed parameters are kept fixed at their best performing values. Table 2.3 represents number of hidden layer evaluation.

TABLE 2.3 Optimal Number of Hidden Layer Evaluation

Number of Hidden Layers	Validation Accuracy	Test Accuracy
1	0.9792	0.9788
2	0.9828	0.9815
3	0.9823	0.9797
4	0.9835	0.9812
5	0.9818	0.9791

2.4.5.4 BATCH SIZE EVALUATION

Batch size of 30, 60, 100, and 150 have been experimented for network performance evaluation. Table 2.4 shows all the metric values. Best performance is provided if batch size considered is 60.

TABLE 2.4 Batch Size Evaluation

Batch Size	Validation Accuracy	Test Accuracy
30	0.9834	0.9811
60	0.9837	0.9821
100	0.9835	0.9812
150	0.9827	0.9793

2.4.5.5 DROPOUT EVALUATION

Performances of the model has been evaluated by introducing a dropout layer after hidden layers keeping best performing learning rate = 0.05, number of hidden layers = 4, batch size = 60 and Adagrad as best performing optimization algorithm. Best Accuracy has been achieved with dropout value of 0.2 beyond which performance of the network decreases drastically.

Test accuracy value of 0.98 using dropout is more significant than the same accuracy obtained without dropout as drop out decreases time and computational complexity of deep learning training phase to a great extent.

2.4.6 GRAPHICAL REPRESENTATION OF HYPER PARAMETER EVALUATION

LogLoss curves are plotted to represent the performances of the classifier model. Figures 2.3(a–c) represent the same for three different optimization algorithms. They are Adagrad, Adam, and SGD respectively. During evaluation of optimization algorithm performances learning rate is set to the value of 0.05. Figure 2.3(d) represents the application of regularization method. The LogLoss curves for dropout introduction with a value of 0.2 is also very interesting as the same accuracy have been obtained after introduction of dropout too. These plots have been obtained while keeping learning rate, number of hidden layer and batch size fixed to the optimized value and taking Adagrad as optimization algorithm.

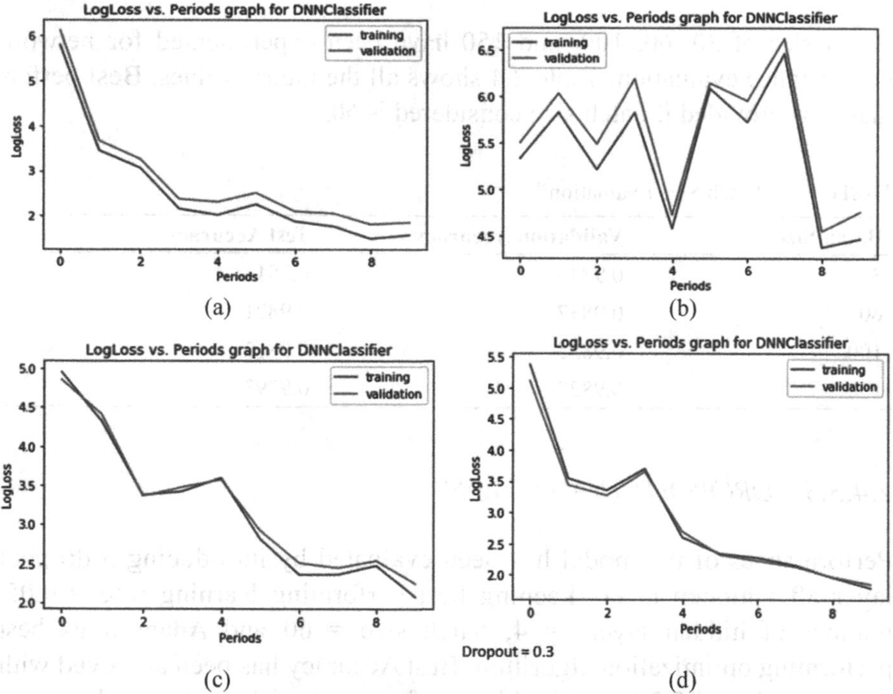

FIGURE 2.3 LogLoss curves (a) Adagrad; (b) Adam; (c) SGD; (d) Adagrad with lr = 0.05 and dropout = 0.3.

Finally, the curves for training/validation accuracy and training/validation loss with optimized classifier have been shown separately in Figures 2.4(a) and (b), respectively.

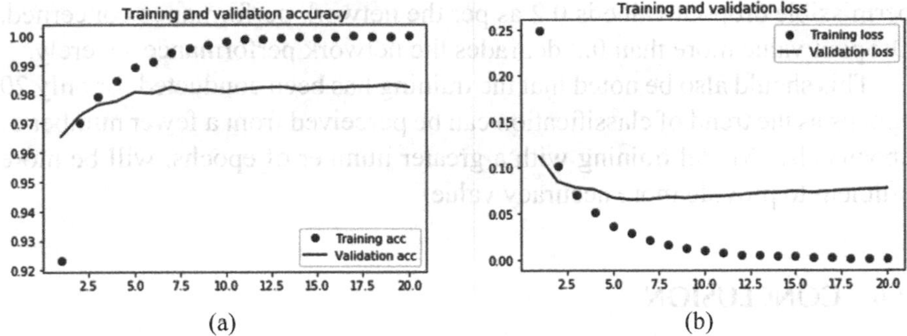

(a) (b)

FIGURE 2.4 (a) Training and validation accuracy of optimized network; (b) training and validation loss of optimized network.

2.5 DISCUSSION

Among different DNN hyper parameters, learning rate (lr), learning algorithm, number of hidden layers, dropout value and batch size have been addressed to frame the optimization problem. Optimizations of these hyper parameters are evaluated by maximization of objective function. The objective functions used here are test and validation accuracies. The stored outcome of best evaluation indicates optimized hyper parameter values.

The search space of the optimization problem has been defined as follows. Four lr values (0.5, 0.1, 0.05, and 0.01) have been evaluated. Among them best accuracy was obtained with lr = 0.05. Different learning algorithms such as SGD, Adam, and Adagrad have been evaluated with lr = 0.05. Adagrad algorithm performed best to the specific application of hand written digit recognition.

This classifier has been further optimized for number of hidden layers. Five variations (1, 2, 3, 4, and 5 hidden layers) have been considered for evaluation. Following the experimental values obtained for objective functions, it can be decided that the network architectures with two or four hidden layers perform the best. Moreover, the classifier has been experimented with variations of batch sizes (30, 60, 100, and 150). The best performing batch size is 60.

With these four network hyper parameter optimizations, the model is further optimized in terms of time and computational complexity by applying regularization method, i.e., introducing dropout layers in between last hidden layer and output layer as dropout layer randomly drops out some definite percentage of hidden neurons from the network. In this process, the same accuracy has been obtained with 0.2 dropout value also. Maximum permissible dropout value is 0.2 as per the network performance concerned. Dropout value more than 0.2 degrades the network performance severely.

This should also be noted that the training has been conducted for only 20 epochs as the trend of classification can be perceived from a fewer number of epochs also. Model training with a greater number of epochs, will be more efficient to provide more accuracy value.

2.6 CONCLUSION

In this chapter, hyper parameter optimization problem of DNN environment for hand written digit recognition has been addressed. It can be concluded from this work that the Adagrad optimizer with learning rate of 0.05 may be effective selections for the relevant field. Additionally, four hidden layers and the batch size of 60 are efficient for the problem domain. Though accuracy has not improved with introduction of 0.2 dropout value, yet inclusion of dropout reduces computational and time complexity of DNN.

The work may be enriched by extending it for multi lingual digit recognition. It may be a good future direction to automate and speed up network hyper parameter optimization so that the process can cover up a large search space in restricted time frame.

KEYWORDS

- **batch size**
- **deep neural network**
- **dropout**
- **handwritten digit recognition**
- **hidden layer**
- **learning rate**
- **optimization algorithm**

REFERENCES

Baldominos, A., Saez, Y., & Isasi, P., (2019). A survey of handwritten character recognition with MNIST and EMNIST. *Applied Sciences, 9*(15), 3169. doi: 10.3390/app9153169.

Bluche, T., & Messina, R., (2017). Gated convolutional recurrent neural networks for multilingual handwriting recognition. In: *2017 14th IAPR International Conference on Document Analysis and Recognition (ICDAR).* doi: 10.1109/icdar.2017.111.

Castro, D., Bezerra, B. L. D., & Valenca, M., (2018). Boosting the deep multidimensional long-short-term memory network for handwritten recognition systems. In: *2018 16th International Conference on Frontiers in Handwriting Recognition (ICFHR).* doi: 10.1109/icfhr-2018.2018.00031.

Domhan, T., Springenberg, J. T., & Hutter, F., (2015). Speeding up automatic hyperparameter optimization of deep neural networks by extrapolation of learning curves. *Twenty-Fourth International Joint Conference on Artificial Intelligence (IJCAI).*

Graves, A., & Schmidhuber, J., (2009). Offline handwriting recognition with multidimensional recurrent neural networks. *Advances in Neural Information Processing Systems,* 545–552.

Graves, A., Fernández, S., Gomez, F., & Schmidhuber, J., (2006). Connectionist temporal classification. *Proceedings of the 23rd International Conference on Machine Learning – ICML 06.* doi: 10.1145/1143844.1143891.

Graves, A., Liwicki, M., Fernandez, S., Bertolami, R., Bunke, H., & Schmidhuber, J., (2009). A novel connectionist system for unconstrained handwriting recognition. *IEEE Transactions on Pattern Analysis and Machine Intelligence, 31*(5), 855–868. doi: 10.1109/tpami.2008.137.

Hochreiter, S., & Schmidhuber, J., (1997). Long short-term memory. *Neural Computation, 9*(8), 1735–1780. doi: 10.1162/neco.1997.9.8.1735.

Mishra, S., Malathi, D., & Senthilkumar, K., (2018). Digit recognition using deep learning. *International Journal of Pure and Applied Mathematics, 118*(22), 295–302. ISSN: 1314-3395.

Pham, V., Bluche, T., Kermorvant, C., & Louradour, J., (2014). Dropout improves recurrent neural networks for handwriting recognition. In: *2014 14th International Conference on Frontiers in Handwriting Recognition.* doi: 10.1109/icfhr.2014.55.

Puigcerver, J., (2017). Are multidimensional recurrent layers really necessary for handwritten text recognition? In: *2017 14th IAPR International Conference on Document Analysis and Recognition (ICDAR).* doi: 10.1109/icdar.2017.20.

Shamim, S. M., Miah, M. B. A., Sarker, A., Rana, M., & Jobair, A. A., (2018). Handwritten digit recognition using machine learning algorithms. *Indonesian Journal of Science and Technology, 3*(1), 29. doi: 10.17509/ijost.v3i1.10795.

Siddique, F., Sakib, S., & Siddique, M. A. B., (2019). *Handwritten Digit Recognition Using Convolutional Neural Network in Python with Tensor Flow and Observe the Variation of Accuracies for Various Hidden Layers.* doi: 10.20944/preprints201903.0039.v1.

Vijayalaxmi, R. R., & Bhavanishankar, K., (2019). Handwritten digit recognition using CNN. *International Journal of Innovative Science and Research Technology, 4*(6). ISSN: 2456-2165.

Voigtlaender, P., Doetsch, P., & Ney, H., (2016). Handwriting recognition with large multidimensional long short-term memory recurrent neural networks. In: *2016 15th International Conference on Frontiers in Handwriting Recognition (ICFHR).* doi: 10.1109/icfhr.2016.0052.

Yann LeCun (1998). *The MNIST Database.* http://yann.lecun.com/exdb/mnist/ (accessed on 08 December 2021).

Zohra, M., & Rao, D. R., (2019). A comprehensive data analysis on handwritten digit recognition using machine learning approach. *International Journal of Innovative Technology and Exploring Engineering (IJITEE), 8*(6). ISSN: 2278–3075.

CHAPTER 3

DESIGN AND DEVELOPMENT OF A DRIVING ASSISTANCE AND SAFETY SYSTEM USING DEEP LEARNING

SHEKHAR VERMA,[1] TUSHAR BAJAJ,[1] NIDHI SINDHWANI,[1] and ABHIJIT KUMAR[2]

[1]*Department of ECE, Amity School of Engineering and Technology, New Delhi GGSIPU, Delhi, India*

[2]*School of Computer Science, University of Petroleum and Energy Studies, Dehradun, Uttarakhand, India,*
E-mail: abhijitkaran@hotmail.com

ABSTRACT

Road conditions are one of the major cause of accidents around the world. Distress on road and combination of human ignorance/error may be really fatal. This chapter proposes a system that can help to prevent road accidents caused by distress on roads and human error. The proposed system minimizes the occurrence of road accidents through real-time monitoring of road conditions and the driver's activities. A robust system has been proposed based on image processing, computer vision and deep learning. This system employs two cameras, one fixed in front of the vehicle for capturing the input video frames which is being processed in real time using image processing and computer vision algorithms, this detects the potholes or any sort of distress on the roads which will be projected on the driver's windshield using a heads up display (HUD). Another camera will check the driver's attention, whether or not driver is operating radio or checking phone, talking over

Advances in Data Science and Computing Technology: Methodology and Applications. Suman Ghosal, Amitava Choudhury, Vikram Kumar Saxena, Arindam Biswas, & Prasenjit Chatterjee (Eds.)

mobile, etc. These two systems works along with a decision maker algorithm which decides the severity of the situation and generate an alert signal to the driver, if the road conditions are poor and if, the driver is distracted. A deep learning based architecture applied to enhance the efficiency for monitoring driver's attention, which used StateFarm dataset.

3.1 INTRODUCTION

The number of vehicles operating in the world is increasing day by day. There are millions of reported road accidents in the world in which thousands of people lose their life every day. In a country like India, it was reported in 2019, that over 151,000 people died in road accidents (India Today, https://www.indiatoday.in/india/story/over-9300-deaths-25000-injured-in-3-years-due-to-potholes-1294147-2018-07-24; Guardian, https://www.theguardian.com/world/2018/jul/25/more-deadly-than-terrorism-potholes- responsible-for-killing-10-people-a-day-in-india). In times like this, the safety of passengers is of utmost importance. The accidents might be caused by not following the traffic rules, bad situations of the roads, weather, or human errors. In another report, it was stated that over 30% of the accidents caused in India were due to the poor road conditions, i.e., Potholes, Disintegration, etc. Potholes are circular shape openings or depressions that can be caused by water in the underlying soil surface and weathering of the road surface. They appear when the top layer of the road, has been eroded by busy traffic thus exposing the concrete base or structure underneath. Potholes are hazardous for drivers and vehicles when moving at high speeds because it is very hard for the driver to see the potholes on the road surface. Moreover, they possess a threat to the car tires as they can rupture from the sudden impact leading to accidents. Techniques, such as Machine Learning, Artificial Intelligence and computer vision can be used in the vehicles to avoid and prevent these situations. Vehicles should come with some sort of integrated infrastructure in a way that this chapter suggests in order to reduce the chance so accidents due to human error and road conditions. This system uses an approach of determining the state of roads as well as the state of driver simultaneously and producing alerts in real time which could even help in providing data of the roads for improvement along with significant help in driving and ensuring safety. The images captured by the road camera and the dashboard camera will be processed using an onboard computer (for example, a SOC like raspberry pi) which will also act as an interface for the Audio-visual alert and HUD system. The pothole detection is entirely based on Image processing and computer rather than machine learning to reduce the latency and make the system as real time as possible. The driver's attention monitoring is done using a deep learning approach implemented using Tensor flow (an open source machine learning library), which will monitor the gesture of the driver and will decide whether he/she is paying attention to driving or not. The algorithms and techniques are implemented using python 3.7 as the choice of programming language and Open CV (an open source computer vision library) and Tensor flow libraries.

The outline of this chapter consists of six sections. Section 3.1 illustrates the introduction and motivation of the proposed work. Section 3.2 of this chapter discusses the glimpse of state of art exists in the proposed area. Section 3.3 discusses the proposed system, and workflow of the model. Section 3.4 emphasis on the results and related discussions. Finally, Section 3.5 concluded the proposed work.

3.2 EXISTING APPROACHES

To contextualize this work, several previous studies are outlined on the detection of potholes and distress on the road. There are various approaches to detect and identify the potholes on the road. Kang, Byeong Ho Choi, and Su Il developed a pothole tracking system and method through 2D LiDAR and Camera (Kang and Choi, 2017). They joined various heterogeneous sensor systems to improve the accuracy of the overall system. Information regarding angle and distance can be obtained by LiDAR (light detection and ranging). The algorithm of pothole detection consists of removing noise, pre-processing, clustering, line segment extraction and gradient of pothole data function. This approach is very useful but not very cost effective due to added cost of multiple heterogeneous sensors. Another way is to solely use image processing to identify and locate the potholes in the image. Images are captured by a camera mounted on the dashboard of the moving vehicle and then pre-processed on the basis of difference of Gaussian-Filtering and image segmentation methods based on clustering as proposed by Vigneshwar and Kumar (2016). Based on various performances, it was observed that the segmentation based on the K-Means clustering is desirable due to its fastest computing time whereas edge segmentation based on edge detection is desirable for its specificity. Computer Vision based techniques such as seeing the appearance and shape of the potholes uses histograms of oriented gradients

(HOG) features being of pothole or not. Normalized graph cut segmentation scheme is used to locate the pothole in the detected pothole images (Azhar et al., 2016). To monitor the driver's attention some techniques, divide the entire view which the driver sees while driving (Windshield, speedometer, ORVMs, etc.), into various sectors (seven sectors mostly) and eye/gaze tracking to see where the driving is paying the most attention (Mizuno et al., 2017). Other ways include using video processing and computer vision to retrieve the information from changing environments or the monitoring of the physical state of human operators. It uses the head orientation of the driver to get the state of visual attention of the driver and the smart phone's camera is used to collect the video data for this process (Mihai et al., 2015). Choudhury et al. (2013, n.d.) exploits object recognition towards recognition of number plate of the vehicle running over the road.

3.3 PROPOSED SYSTEM MODEL

The system is divided into two independent parts:

- pothole detection; and
- driver's attention tracking.

The two independent systems use different input sources (Cameras) but a common processing environment to process the inputs and make the decisions. The quality and performance of the camera use will directly affect the overall functioning of the system. A decent camera (preferably night vision enabled) is required. All the input images are resized to 1280 × 720 pixels to make the processing faster without compromising on the efficiency and accuracy of the system. The pothole detector and the driver's attention monitor will run in parallel using multiprocessing techniques on the same onboard processing environment (a SOC) and will be controlled by a decision making algorithm that will decide when to alert the driver. The driver can be alerted using multiple Ways, but it is observed that Visual clues are the easiest to detect by the human brain, for example flashing lights, or blinking LEDs, etc.

This system model described in Figure 3.1 shows the two independent systems working in parallel. The pre-processing for both the systems is quite different as they work in different manner as shown in the next section of this chapter. As the driver's attention monitor is based on Deep learning, it performs significantly better on GPUs rather than CPU.

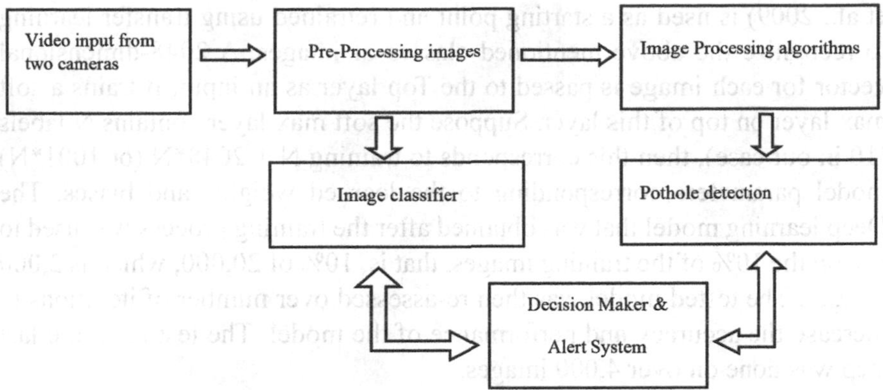

FIGURE 3.1 Model of proposed system.

3.3.1 DRIVER'S ATTENTION MONITOR

The model that this chapter uses for the detection of the attention of the driver is based on the image classification implemented using tensor flow. In this, a deep learning model is trained to classify images among various classes and then recognize them based on the training received. The state farm dataset (Abouelnaga, Eraqi, and Moustafa, 2017) which contains snapshots from a video captured by a camera mounted in the car is used in this project. The training set has ~22.4 K samples with equal distribution among the classes and 79.7 K unlabeled test samples.

There are 10 classifications in the state farm dataset:

- Safe driving;
- Texting using right hand;
- Talking on the phone using right hand;
- Texting using left hand;
- Talking on the phone using left hand;
- Operating radio/multimedia system;
- Drinking;
- Reaching behind;
- Hair and makeup;
- Talking to passenger(s).

The driver's attention monitor was trained using tensor flow transfer learning (Chris, n.d.). Inception v3 (It is a widely-used image recognition model) (Szegedy et al., 2016) which is trained on Image net dataset (Deng et al., 2009) is used as a starting point and retrained using transfer learning to recognize the above mentioned classes of images. A 2048-dimensional vector for each image is passed to the Top layer as an input. It trains a soft max layer on top of this layer. Suppose the soft max layer contains N labels (10 in our case), then this corresponds to training N + 2048*N (or 1001*N) model parameters corresponding to the learned weights and biases. The Deep learning model that was obtained after the training process was used to test on the 10% of the training images, that is, 10% of 20,000, which is 2,000 images. The tested model was then re-assessed over number of iterations to increase the accuracy and performance of the model. The testing at the last step was done on over 4,000 images.

3.3.2 POTHOLE DETECTOR

The flowchart described in Figure 3.2 shows the technique and algorithm used. It uses a mix of computer vision and image processing techniques to find the potholes/distress in the input feed. Firstly, the given input frames are converted from BGR to Gray scale to improve the performance simply because BGR images have three channels and gray scale images have one channel. The Gray Scale image is passed through a 3×3 Gaussian filter to remove noise before passing it to canny edge algorithm to detect the edges in the image which is the base of our classification (Canny, 1986). The detected edges serves 2 purposes, one is to find the lanes in the road to extract the region of interest (ROI) from the road which helps to remove unnecessary parts from the images which greatly improves the performance and reduces latency. The lanes are detected using Hough Lane detection, which converts the edges into lanes which are filtered to get the ROI. This system uses a 10 second timer to reset the position of lanes so that it can counter changing road conditions. The ROI is superimposed on the edges to find the contours only on the surface of the road. These detected contours are filtered based on their shape and size to avoid false detections. The filtered contours are then projected back.

3.4 RESULTS AND DISCUSSION

The system has been developed using AMD A8-APU running at 2.2 Ghz with 8 GB of RAM for testing both the systems. Testing on the video feeds gave us a speed of 30 Frames per second making it more real-time. In our experimentation, we have found that our method with computer vision was more real-time. Computer vision methods are easy to implement and manufacture as well provide with better real-time results. The number of frames one system can process in a second with desired accuracy was chosen to measure the performance of both the systems. The results of the two models are further explained in subsections.

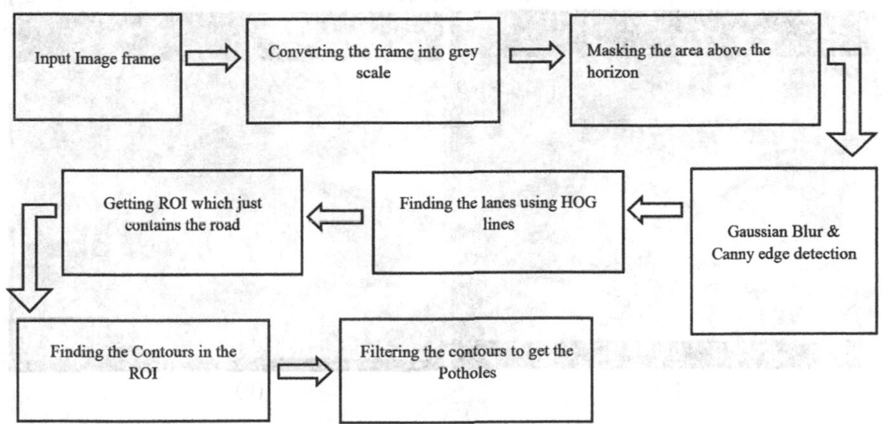

FIGURE 3.2 Pothole detector algorithm.

3.4.1 DRIVER'S ATTENTION MONITOR

The driver's attention monitor was tested on us in a live video stream in an automobile. It predicted correct results in each of the 10 classifications on which it was based. The efficiency of the model was around 68%. Figure 3.3 shows a few samples of result obtained from the driver's attention model. Figure 3.3(a) correctly depicts the user talking on phone with his right hand. Figure 3.3(b) depicts the user reaching behind. Figure 3.3(c) depicts the user is drinking something. Figure 3.3(d) shows that the user is careful.

3.4.2 POTHOLE DETECTOR

The pothole detector, when worked on a sample video feed, works efficiently Processing 30 Frames per second on an average. Some of the result samples are given below Figure 3.4 describes the process of pothole detection and the output. Figure 3.4(a) is the input frame of the road image. Figure 3.4(b) shows the initial ROI, that is, just the road, obtained from the frame of Figure 3.4(a). Figure 3.4(c) depicts the output of the canny edge detector that is used to detect the lane in the road as shown in Figure 3.4(d). Figure 3.4(e) further reduces the final ROI to the driving lane at the same time, detects the edges for the pothole contours. Figure 3.4(f) shows the bounding box around the pothole contour that are being detected. This image is also being projected to the driver.

FIGURE 3.3 Output of driver's attention model: (a) depicts the user talking on phone using his right hand; (b) depicts the user reaching behind; (c) depicts the user is drinking something; and (d) shows that the user is careful.

3.5 CONCLUSIONS

The approach and procedure suggested in this chapter has not been tested practically in real life scenario, hence this just acts as a prototype system.

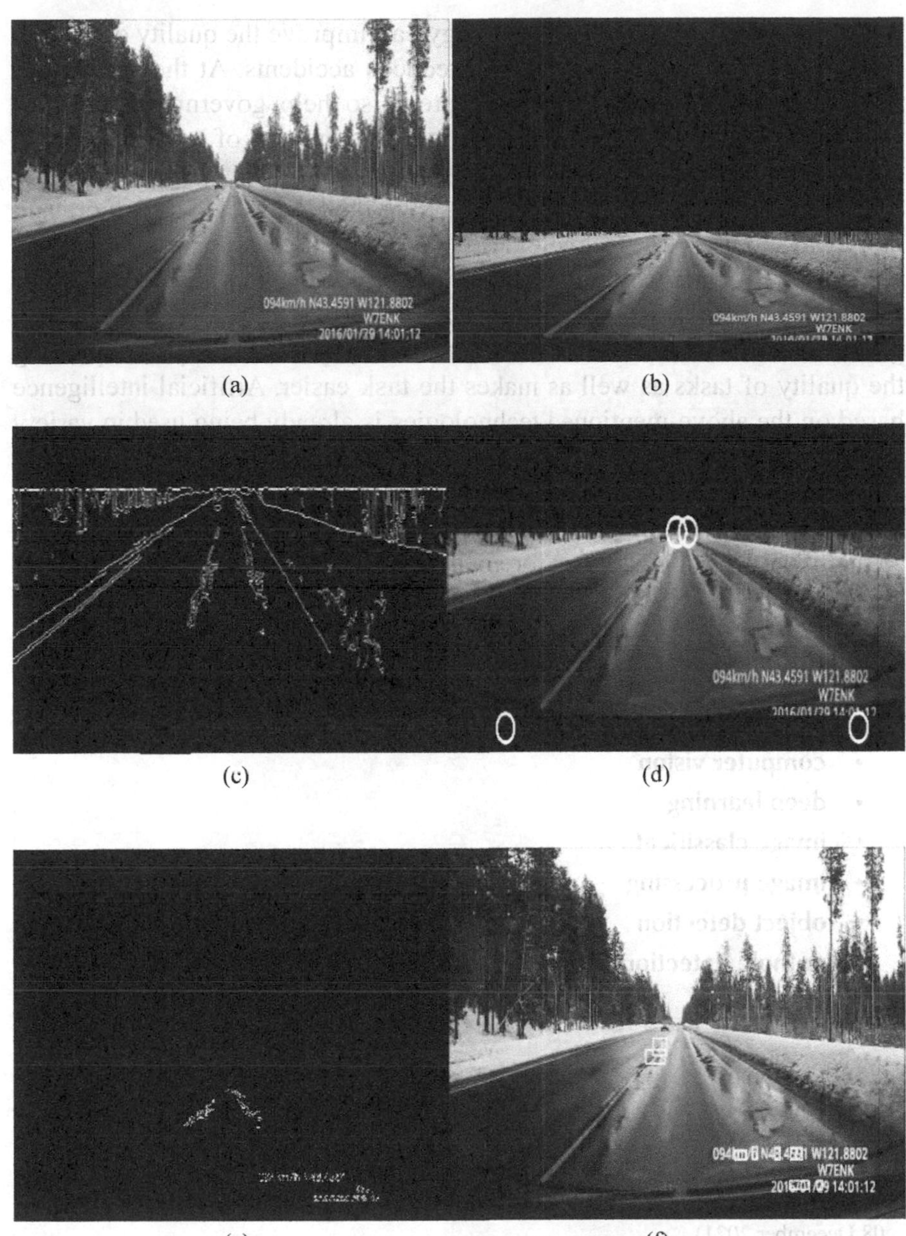

FIGURE 3.4 Output of pothole detector at various stages: (a) is the input frame of the road image; (b) shows the initial region of interest (ROI); (c) depicts the output of the canny edge detector that is used to detect the lane in the road as shown in (d); (e) reduces the final region of interest (ROI) to the driving lane at the same time, detects the edges for the pothole contours; and (f) shows the bounding box around the pothole contour that will are being detected.

If implemented properly, this technology can improve the quality of driving making it easier and save lives as it reduces accidents. At the same time, the techniques suggested in this chapter also help government agencies around the world to improve and monitor the condition of roads. Computer vision and machine learning are the future and consistent improvements are being made in these fields with each passing day. We should seek more and more ways to use these technologies in our daily lives to make an impact on the society. Techniques such as computer vision, image processing, deep/machine learning are the basic fundamental blocks for the artificial intelligence on which the future will be shaped as it helps the society in improving the quality of tasks as well as makes the task easier. Artificial intelligence based on the above-mentioned technologies is already being used in variety of fields such as medical, mining, robotics, space studies, industries, etc., and its uses will increase as the years pass by to various other fields. Hence, this chapter suggests that this model should be used in automobile industries so that it can provide various aforementioned to the users.

KEYWORDS

- **computer vision**
- **deep learning**
- **image classification**
- **image processing**
- **object detection**
- **pothole detection**

REFERENCES

Abouelnaga, Y., Eraqi, H. M., & Moustafa, M. N., (2017). *Real-Time Distracted Driver Posture Classification.* No. Nips, [Online]. Available: http://arxiv.org/abs/1706.09498 (accessed on 08 December 2021).

Azhar, K., Murtaza, F., Yousaf, M. H., & Habib, H. A., (2016). Computer vision based detection and localization of potholes in asphalt pavement images. In: *2016 IEEE Canadian Conference on Electrical and Computer Engineering (CCECE)* (pp. 1–5). doi: 10.1109/CCECE.2016.7726722.

Canny, J., (1986). A computational approach to edge detection. *IEEE Trans. Pattern Anal. Mach. Intell., PAMI-8*(6), 679–698. doi: 10.1109/TPAMI.1986.4767851.

Choudhury, A., & Negi, A. (2016). A new zone based algorithm for detection of license plate from Indian vehicle In: *2016 Fourth International Conference on Parallel, Distributed and Grid Computing (PDGC)* (pp. 370–374). IEEE.

Chris, D. (2018). *Tensor flow Image Classification.* https://github.com/Microcontrollers And More/TensorFlow_Tut_2_Classification_Walk-through (accessed on 08 December 2021).

Deng, J., Dong, W., Socher, R., Li, L. J., Li, K., & Fei-Fei, L., (2009). ImageNet: A large-scale hierarchical image database. In: *CVPR09.*

Guardian, T. (2018). *More Deadly Than Terrorism: Potholes Responsible for Killing 10 People a Day in India.* https://www.theguardian.com/world/2018/jul/25/more-deadly-than-terrorism-potholes-responsible-for-killing-10-people-a-day-in-india (accessed on 08 December 2021).

India Today. (2018). *Over 9300 Deaths, 25,000 Injured in 3 Years Due to Potholes.* https://www.indiatoday.in/india/story/over-9300-deaths-25000-injured-in-3-years-due-to-potholes-1294147-2018-07-24 (accessed on 08 December 2021).

Kang, B. H., & Choi, S. I., (2017). Pothole detection system using 2D LiDAR and camera. In: *International Conference on Ubiquitous and Future Networks* (pp. 744–746). doi: 10.1109/ICUFN.2017.7993890.

Mihai, D., Dumitru, A., Postelnicu, C., & Mogan, G., (2015). Video-based evaluation of driver's visual attention using smartphones. In: *2015 6th International Conference on Information, Intelligence, Systems and Applications (IISA)* (pp. 1–5). doi: 10.1109/IISA.2015.7387983.

Mizuno, N., Yoshizawa, A., Hayashi, A., & Ishikawa, T., (2017). Detecting driver's visual attention area by using vehicle-mounted device. In: *2017 IEEE 16th International Conference on Cognitive Informatics Cognitive Computing (ICCI*CC)* (pp. 346–352). doi: 10.1109/ICCI- CC.2017.8109772.

Roy, S., Choudhury, A., & Mukherjee, J., (2013). An approach towards detection of Indian number plate from vehicle. *International Journal of Innovative Technology and Exploring Engineering (IJITEE), 2*(4), 241–244.

Sharma, S. K., & Sharma, R. C., (2019). Pothole detection and warning system for Indian roads. In: *Advances in Interdisciplinary Engineering, Lecture Notes in Mechanical Engineering* (pp. 511–519).

Szegedy, C., Vanhoucke, V., Ioffe, S., Shlens, J., & Wojna, Z., (2016). Rethinking the inception architecture for computer vision. *Proc. IEEE Comput. Soc. Conf. Comput. Vis. Pattern Recognit.*, 2818–2826. doi: 10.1109/CVPR.2016.308.

Vigneshwar, K., & Kumar, B. H., (2016). Detection and counting of pothole using image processing techniques. In: *2016 IEEE International Conference on Computational Intelligence and Computing Research (ICCIC)* (pp. 1–4). doi: 10.1109/ICCIC.2016.7919622.

CHAPTER 4

PREDICTION OF THE POSITIVE CASES OF COVID-19 FOR INDIVIDUAL STATES IN INDIA AND ITS MEASURE: A STATE-OF-THE-ART ANALYSIS

AMITAVA CHOUDHURY,[1] KAUSHIK GHOSH,[1] and SUMAN GHOSAL[2]

[1]*School of Computer Science, University of Petroleum and Energy Studies, Dehradun, Uttarakhand, India,*
E-mail: a.choudhury2013@gmail.com (A. Choudhury)

[2]*Department of Electrical and Electronics Engineering, Pailan College of Management and Technology, Kolkata, West Bengal, India*

ABSTRACT

India too is been affected by the global pandemic of COVID-19. It is therefore, necessary to predict the number of cases well in advance such that the country remains prepared well in advance to face the situation. Many predictions have come up for the entire country, however, for a country like India, it is more suitable to consider the situation of individual states separately. In this chapter, we have thus predicted the number of expected cases of COVID-19 for India along with a deeper analysis of data for majorly affected states. The results have shown that a state wise analysis is going to give better results for prediction with a lower margin of error, as compared to a countrywide analysis. Moreover, we have also identified three best states that have performed better than others in terms of controlling the rate of transmission among people.

Advances in Data Science and Computing Technology: Methodology and Applications. Suman Ghosal, Amitava Choudhury, Vikram Kumar Saxena, Arindam Biswas, & Prasenjit Chatterjee (Eds.)

4.1 INTRODUCTION

The pandemic caused by COVID-19 virus has both shook and united the world like never before. Never in the recent past, has any epidemic concurrently affected governments across the globe with the same sense of urgency. With the global casualty figures crossing over 200,000 (worldometers.info/coronavirus, 2020)—different countries and agencies are revisiting their warfare strategies against this virus. Originated in China (Huang et al., 2020), the pandemic has caused havoc in other countries like Italy and USA as well (Graselli et al., 2020). The entire Europe and North America for that matter are yet to control the pandemic. India too is fighting her war against this pandemic. Figures 4.1(a) and (b) have shown the number of active cases of COVID-19, worldwide, and India respectively.

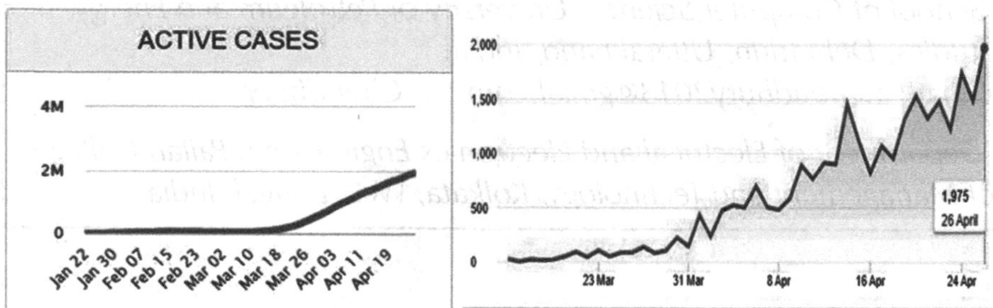

FIGURE 4.1 (a) Active cases worldwide (Worldometers, 2020); (b) active cases with in India (Wikipedia, 2020).

The first case of Novel Corona virus (COVID-19) in India was reported on 30th January, 2020 (PIB, 2020). WHO declared it as a global pandemic (Wang et al., 2020) and as a result the Indian government declared an initial lockdown of 21 days (Advisory Trade Facilitation, GoI, 2020) on 24th of March, 2020, in order to prevent community transmission. On 14th April, India entered into the second phase of lockdown (till 3rd May, 2020), adhering to the advice of different agencies. Different medical and administrative bodies in the country (and outside too) proposed social distancing as one of the means to deal with the pandemic (Singh and Adhikari, 2020).

As the end of second phase of lockdown in India is approaching, the country has an approximate 30,000 (mygov.in, 2020) reported cases of COVID-19. Although this is a large number and the numbers are expected to increase further, yet it must be said that in a country like India, with a population more than 1.35 billion (Wikipedia, 2020) and a population density of 464/Km^2 (worldometers.info, 2020), the country has not yet entered into the stage of community transmissions. Even if India is lucky enough to avoid community transmissions, the total infected population is sure to be a mammoth one due to the reasonably high population density of the country. As a result, a prediction mechanism with an acceptable degree of accuracy is required to estimate beforehand the number of people who might be infected by the virus. This is required in order to keep both the government and the health workers prepared for the upcoming months, such that they do not get overwhelmed with the number of cases all of a sudden.

Tomar and Gupta (2020) have done predictions of the number of affected patients countrywide, for a period of 30 days. However, a monolithic prediction of this type is not appropriate for a country like India where the number of infected persons vary by a huge margin for different states. To mention, Maharashtra has the highest number of affected persons and Sikkim has none.

For example, till 26th April, 2020, the total number of reported cases in Maharashtra was more than 8,000, whereas, for at least 12 states/union territories the number of confirmed cases were yet to reach the three-figure mark (covid19india.org, 2020). One of the reason for this is a huge variation in population density among the states of the country. In our results, we have shown why a countrywide prediction will give a greater margin of error as compared to predictions done for the individual states.

In this chapter, we have predicted the number of confirmed/active cases, for a period of next 30 days, for the entire country. For the purpose, we have used algorithm comprising of long short-term memory (Hochreiter and Schmidhuber, 1997) and have compared the results for the major states and union territories of India. On analyzing the results, we have found that different states are performing differently while combating the current situation.

Our main contributions in this chapter are as follows:

- Predicting the total number of COVID-19 cases in India for coming 30 days;
- Showing that a countrywide prediction is bound to give greater margin of error when compared to state wise prediction; and
- Identifying the best states in terms of preventive actions taken against the outbreak of the pandemic.

We have trained our system with data ranging from 30th January, 2020 to 20th April, 2020. Thereafter, we have validated our results for the period of 21st April, 2020 to 25th April, 2020.

The prediction made in this chapter is for the period of one month, i.e., 26th April, 2020 to 26th May, 2020. The rest of the chapter is organized in the following way: Section 4.2 contains methodology of our work. Section 4.3 contains the results along with the data sets and finally we have concluded in Section 4.4.

4.2 METHODOLOGY

The exact reason for the occurrence of COVID-19 is yet to be known and hence it is quite difficult to categories the COVID-19 symptoms. As per the guidelines shared by World Health Organization (WHO), COVID-19 symptoms are very similar to normal flu or influenzas. Hence, to create a mathematical model for predicting the number of affected persons by the virus is really difficult. In scenarios like this, recurrent neural network (RNN) algorithm is very useful. Generally, this algorithm is used to predict time series analysis based on large historical information. However, the main disadvantage of such a model is the time consumed in training a large dataset. This is due to the gradient exploding problem (Jiang and Schotten, 2020; Connor et al., 1994). As a solution, long short term memory (LSTM) is often very useful and this model is having the capabilities to learn order dependence in sequential prediction problems (Tomar et al., 2020; Hochreiter and Schmidhuber, 1997).

4.2.1 LONG SHORT TERM MEMORY (LSTM)

As discussed, LSTM has a similar process flow mechanism as RNN. However, the difference is in the operations within the LSTM internal structure. The key component of LSTM consists of cell state. It is the transport highway and is used to transfer relative information to all other blocks. The other components are various gates, such as forget gate, input gate and output gate.

Gates contains sigmoid activations as shown in Figure 4.2. A sigmoid activation is similar to the tanh activation. Instead of squishing values between –1 and 1, it squishes values between 0 and 1. That is helpful to update or forget data because any number multiplied by 0 is 0; causing values to disappear or to be "forgotten."

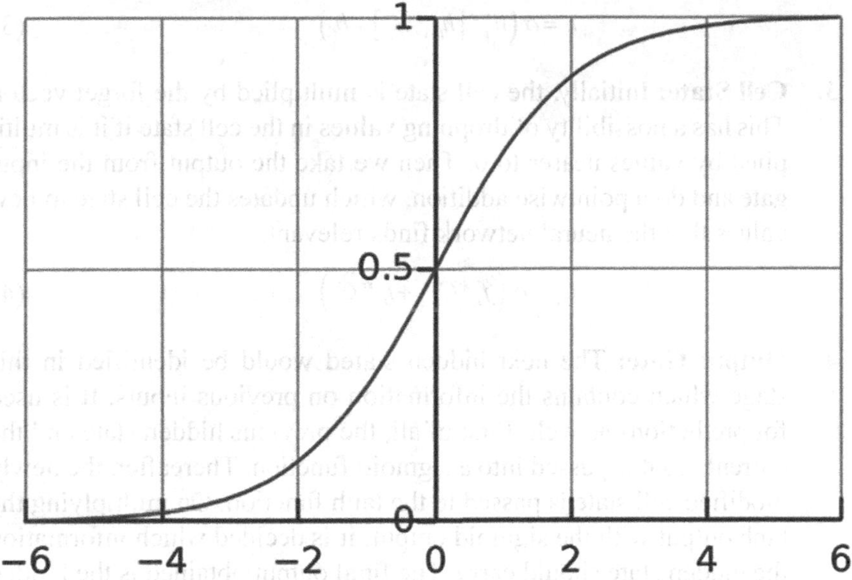

FIGURE 4.2 Sigmoid squishes values to be between 0 and 1.

A sigmoid function can be calculated as:

$$\varphi(z) = \frac{1}{1+e^{-z}} \tag{1}$$

1. **Forget Gate:** The functionality of this gate to decide whether an information is to be thrown away or kept. Information from the previous hidden state and information from the current input is passed through the sigmoid function. Predicted values come out between 0 and 1. Closer the value to 0, it is to forget, and closer it is to 1, it is to be kept. It is formulated in Eqn. (2):

$$f_t = \sigma\left(W_f * [h_{t-1}, X_t] + b_f\right) \tag{2}$$

2. **Input Gate:** To update the cell state, input gate is used. First, the previous hidden state and current input are pass into a sigmoid function as. That decides which values will be updated by transforming the values between 0 and 1 which indicates not important and important measure respectively. A hidden state and current input also pass into the tanh function to squish values between −1 and 1 to regulate the network. It is formulated in Eqn. (3).

$$i_t = \sigma\left(W_i * [h_{t-1}, X_t] + b_i\right) \quad (3)$$

3. **Cell State:** Initially, the cell state is multiplied by the forget vector. This has a possibility of dropping values in the cell state if it is multiplied by values nearer to 0. Then we take the output from the input gate and do a pointwise addition, which updates the cell state to new values that the neural network finds relevant.

$$C_t = \sigma\left(f_i * C_{t-1} + i_t * \tilde{C}_t\right) \quad (4)$$

4. **Output Gate:** The next hidden stated would be identified in this stage which contains the information on previous inputs. It is used for predictions as well. First of all, the previous hidden state and the current input is passed into a sigmoid function. Thereafter, the newly modified cell state is passed to the tanh function. On multiplying the tanh output with the sigmoid output, it is decided which information the hidden state should carry. The final output obtained is the hidden state. The new cell state and the new hidden state is to be carried over to the next time step.

$$O_t = \sigma\left(W_o * [h_{t-1}, X_t] + b_o\right) \quad (5)$$

4.3 RESULTS AND DISCUSSION

In this chapter, we used COVID-19 pandemic dataset collected from official website of Govt. of India (COVID-19.in, 2020). To implement the work, we have used python programming with few useful libraries such as pandas for data preprocessing, scikit learn machine-learning analysis, matplotlib, and seaborn for plotting data. The data has been collected from 30th January, 2020 to 20th April, 2020 and validated for the period 21st April, 2020 to 25th April, 2020. In this study, we tried to predict the probability of getting infected by COVID-19 virus. Along with this, we tried to get a better understanding of the effect of lockdown and other precautionary measures on the number of reported cases. Firstly, we made the predictions for entire India and then, in order to get an in-depth understanding, we did a state wise prediction and analysis. We have also discussed the effect of the lockdown and other measures in 12 major states of India. Figure 4.3 shows the comparison between predicted value and actual value of confirmed cased for the period 21st April 2020 to 25th April 2020, for entire India.

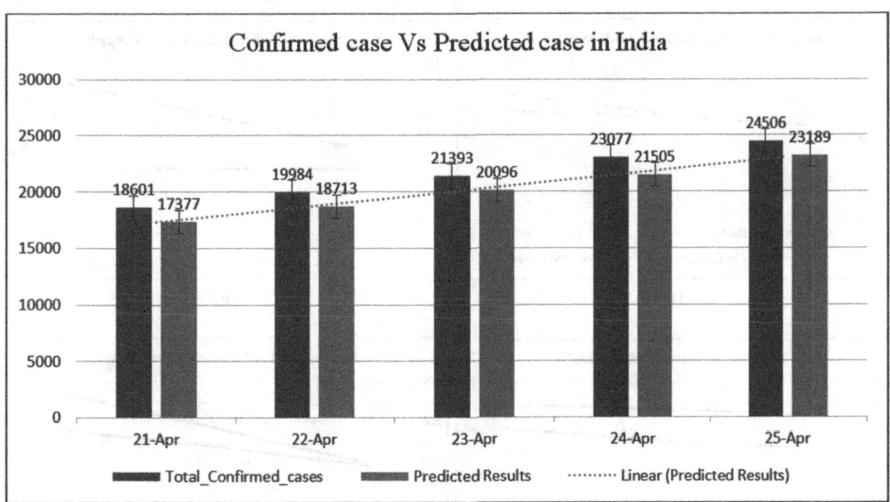

FIGURE 4.3 COVID-19 confirm cases actual vs. predicted from 21st April 2020 to 25th April 2020.

A similar study has also been conducted by Tomar et al. (2020). However, India is a country with a population more than 1.35 billion, with wide variation in population density among different states. Hence, it is very difficult to equate the condition of different states and predict the number of affected persons. Hence, a state wise prediction and analysis is also essential. Figure 4.4 has shown the state wise prediction results along with validation.

From Figure 4.4, it is clearly visible that the prediction results are varied from state to state. In Figure 4.4(a, h, l), the prediction results are very close to the actual report given by the government. It can therefore, be inferred that these three states (Andhra Pradesh, Gujarat, and West Bengal) have taken adequate medical measures and people of these states have followed the lockdown in a rigorous manner. This ensures that the state government has been able to reduce the rate of transmission by taking proper measures. Whereas in Figure 4.4(c, d, e, f) the actual number of cases have superseded the predicted results. Hence, it may be concluded that these states, viz. Haryana, Maharashtra, Tamil Nadu, and Uttar Pradesh still need to put more efforts in controlling the pandemic. May be an extension of the lockdown period for these four states is recommendable in braking the transmission chain of the infection. A tabular validation of the prediction results is given for entire India and West Bengal respectively in Tables 4.1 and 4.2, along with the percentage of error.

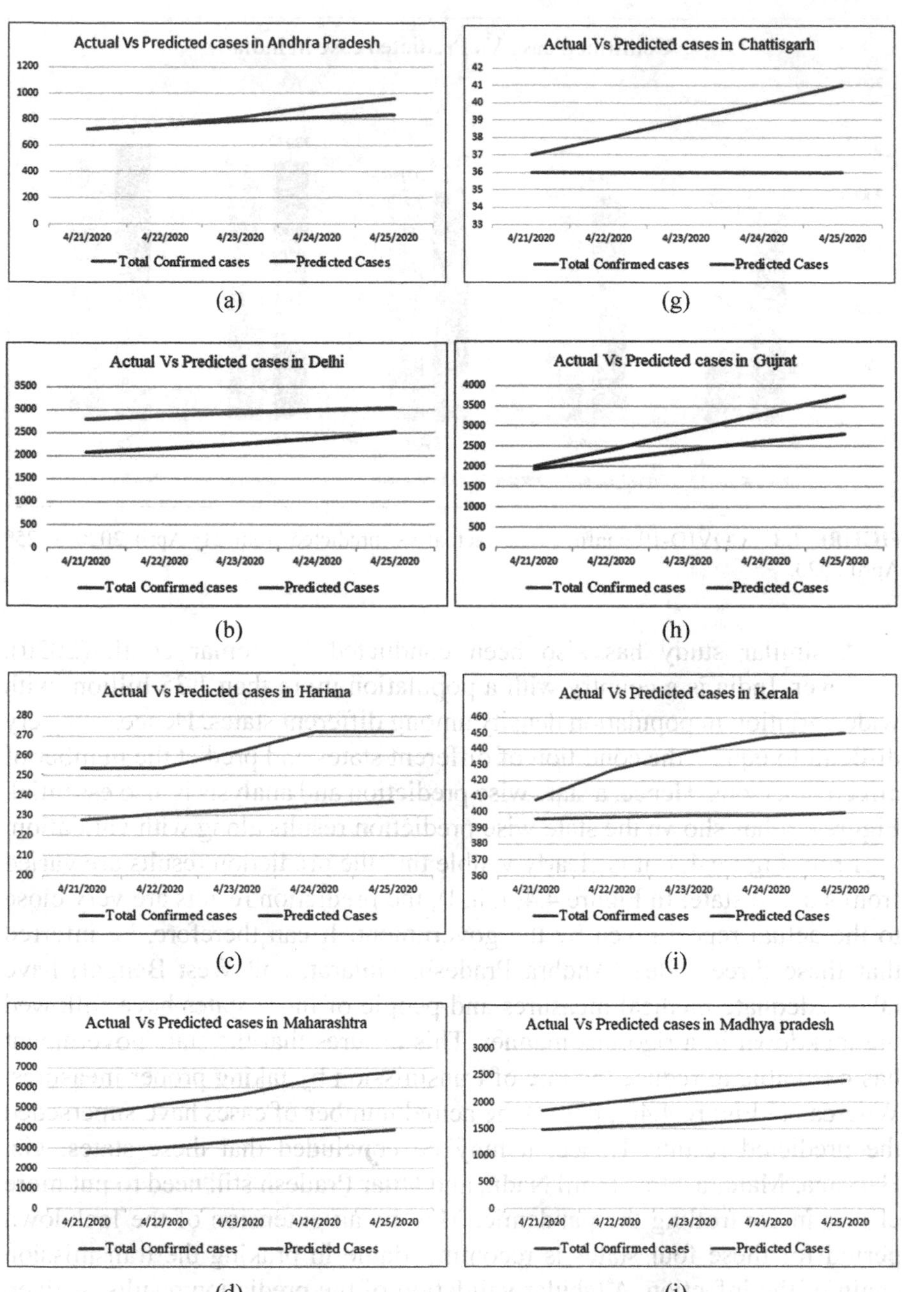

FIGURE 4.4(a–l) *(Continued)*

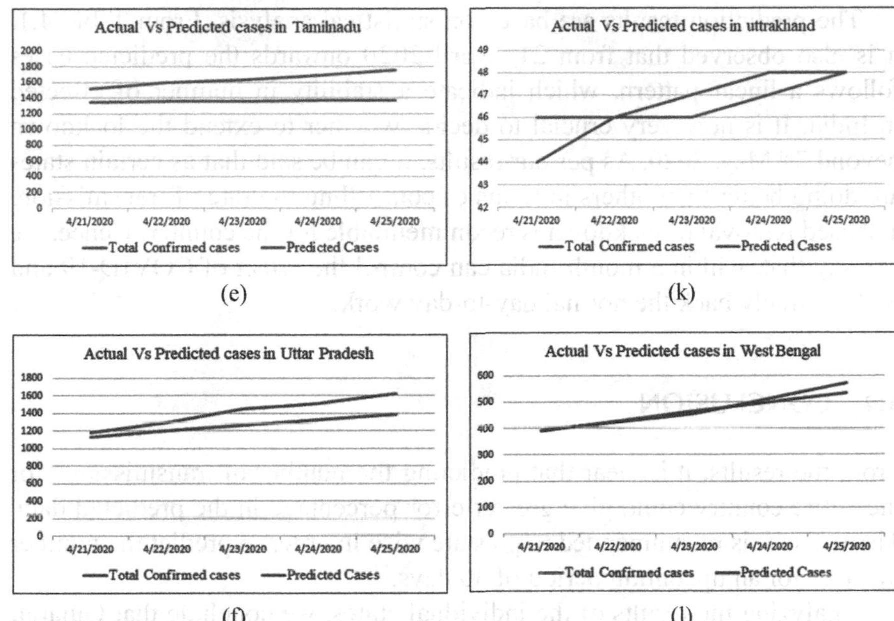

FIGURE 4.4 (a–l) State wise COVID-19 cases prediction and validation between April 20th 2020 to April 25th 2020. Particularly for Gujarat we can see that the number of actual cases is less than that of the predicted cases.

TABLE 4.1 Comparison of Reported and Predicted Cases for India

Date	Actual Data Cited in Government Website	Predicted Data	Error Percentage
21st April 2020	18,601	17,377	6.58
22nd April 2020	19,984	18,713	6.36
23rd April 2020	21,393	20,096	6.06
24th April 2020	23,077	21,505	6.81
25th April 2020	24,506	23,589	5.37

TABLE 4.2 Comparison of Reported and Predicted Best Case (West Bengal)

Date	Actual Data Cited in Government Website	Predicted Data	Error Percentage
21st April 2020	392	391	0.255102
22nd April 2020	423	429	–1.41844
23rd April 2020	456	466	–2.19298
24th April 2020	514	501	2.529183
25th April 2020	571	535	6.304729

The prediction results are based on statistical analysis. From Table 4.1, it is also observed that from 21st April 2020 onwards the predicted cases follows a linear pattern, which indicate a stability in number of effected in India. It is now very crucial to decide whether to extend the lockdown beyond 3rd May, 2020. As per our results, it can be said that as certain states are doing better than others in terms of controlling the rate of transmission, a phased removal of lockdown is recommendable for the country. Hence, we can say that, within a month India can control the effect of COVID-19 and will definitely back the normal day-to-day work.

4.4 CONCLUSION

From the results, it is clear that predicting the number of transmissions for the entire country could give greater error percentage in the predicted data. Therefore, it is recommended to go state wise in order to predict the number of cases for an upcoming period of 30 days.

Analyzing the results of the individual states, we conclude that Gujarat, West Bengal, and Andhra Pradesh have been able to keep the rate of transmission lower as compared to any other state.

On the other hand, in states like Haryana, Maharashtra, Tamil Nadu and Uttar Pradesh the rate of transmission of the infection has increased than expected. May be an extension of the lockdown period could reduce the rate of transmission in these states.

KEYWORDS

- **COVID-19**
- **digital India**
- **long short-time memory**
- **machine learning**
- **pandemic**
- **prediction**

REFERENCES

Advisory, Trade Facilitation, (1945). Government of India. 2017–2020.

Connor, J. T., Martin, R. D., & Atlas, L. E., (1994). Recurrent neural networks and robust time series prediction. *IEEE Trans. Neural Netw., 5*(2), 240–254.

Grasselli, G., Pesenti, A., & Cecconi, M., (2020). Critical care utilization for the COVID-19 outbreak in Lombardy, Italy: Early experience and forecast during an emergency response. *JAMA*.

Hochreiter, S., & Schmidhuber, J., (1997). Long short-term memory. *Neural Comput., 9*(8), 1735–1780.

Huang, C., Yeming, W., Xingwang, L., Lili, R., Jianping, Z., Yi, H., Li, Z., et al., (2020). Clinical features of patients infected with 2019 novel coronavirus in Wuhan, China. *The Lancet, 395*(10223), 497–506.

Jiang, W., & Schotten, H. D., (2020). Deep learning for fading channel prediction. *IEEE Open J. Commun. Soc., 1,* 320–332.

MyGov. (India), (2020). https://www.mygov.in/covid-19/ (accessed on 08 December 2021).

PIB, (2020). https://pib.gov.in/pressreleaseiframepage.aspx?prid=1601095 (accessed on 08 December 2021).

Singh, A. K., Singh, A., Shaikh, A., Singh, R., & Misra, A., (2020). Chloroquine and hydroxychloroquine in the treatment of COVID-19 with or without diabetes: A systematic search and a narrative review with a special reference to India and other developing countries. *Diabetes & Metabolic Syndrome: Clinical Research & Reviews*.

Singh, R., & Adhikari, R., (2020). *Age-Structured Impact of Social Distancing on the COVID-19 Epidemic in India*. arXiv preprint arXiv:2003.12055.

Tomar, A., & Gupta, N., (2020). Prediction for the spread of COVID-19 in India and effectiveness of preventive measures. *Science of the Total Environment*, 138762.

Wang, L. S., Wang, Y. R., Ye, D. W., & Liu, Q. Q., (2020). A review of the 2019 Novel Coronavirus (COVID-19) based on current evidence. *International Journal of Antimicrobial Agents*, 105948.

Wikipedia, (2020). https://en.wikipedia.org/wiki/ (accessed on 08 December 2021).

Wikipedia, (2020). https://en.wikipedia.org/wiki/Demographics_of_India (accessed on 08 December 2021).

Worldometers, (2020). https://www.worldometers.info/coronavirus/ (accessed on 08 December 2021).

Worldometers, (2020). https://www.worldometers.info/world-population/india-population/ (accessed on 08 December 2021).

CHAPTER 5

DEEP RESIDUAL NEURAL NETWORK BASED MULTI-CLASS IMAGE CLASSIFIER FOR TINY IMAGES

ALOK NEGI,[1] PRACHI CHAUHAN,[2] and AMITAVA CHOUDHURY[3]

[1]*National Institute of Technology, Uttarakhand, India*

[2]*G.B. Pant University of Agriculture and Technology, Pantnagar, Uttarakhand, India*

[3]*University of Petroleum and Energy Studies, Dehradun, India, E-mail: a.choudhury2013@gmail.com*

ABSTRACT

With the rapid growth of digital object recognition in over the year, automatic image classification and detection has become the most difficult challenge in computer vision areas. Contrary to human experiences, automated perception, and interpretation of tiny images by machine is difficult. Numerous researches have been conducted throughout the existing identification model to address problem, but still the performance has been restricted. Nevertheless, those approaches lack appropriate classification method. In this proposed work, deep learning algorithms is used to obtain the predicted outcomes in the field, such as computer visions and features a deep learning algorithm often used to classify the images automatically. In order to classify tiny images, we use the image-net data set as a standard. The tiny images need more computational power to classify images. By training the images using deep network we get the better accuracy response with in experimental phase it demonstrates that our approach provides the high precision in image

Advances in Data Science and Computing Technology: Methodology and Applications. Suman Ghosal, Amitava Choudhury, Vikram Kumar Saxena, Arindam Biswas, & Prasenjit Chatterjee (Eds.)

classification. Accuracy and loss curve based qualitative analysis performed for proposed work.

5.1 INTRODUCTION

Image recognition and identification are most important issues in computer vision. These are the origin of several other complex vision problems, along with optimization, detection, and action interpretation. Image classification and identification (Barbedo, 2018a, b) are widely used throughout the several areas, including facial recognition and intelligent video processing, image classification, traffic scenario for vehicles in traffic counting, regressive motion sensors and so on. The purpose of this proposed research is to create various neural networks that classify different tiny images (Bastidas 2017) in advanced convolution. In deep learning, we considered the neural networks which recognize the image on the basis of its characteristics. Deep learning depends on large quantities of high-quality analysis to model future patterns and trends of behavior. The data sets must be descriptive of the classes we expect to forecast otherwise the method will generalize the distribution including its weighted classes as well as the bias will ruins the model. The most common methods of deep learning models are Convolution neural network (Schmidhuber, 2015), recurrent neural network (RNN), auto encoders, etc.

Convolutional neural networks (CNNs) that are nowadays considered narrowly the dominant method just for object classification. The deep CNN are trained by tuning the parameters of the network (Alom et al., 2018) to enhance linking during the training process and performs conscientiously in tiny image classification. In computer vision concept, CNNs were known to the strong visual models which produce feature hierarchies that allow for accurate and remarkable segmentation and also execute predictions comparatively faster than other more algorithms, and simultaneously retaining competitive performance.

In bygone models of image classification, there is a use of actual raw pixels to classify or identify the images. Images can be categorized by color histogram and edge detection. This approach has been exclusively successful but before more complex variants are encountered. That is where the classical image recognition keeps failing, because some other features are not taken into consideration by the model and that is why consider convolutional neural network (ConvNet). CNN is a type of model of the neural network that helps to extract feature or higher image representations (Srivastava et al., 2014). CNN takes the raw pixel data from the image, trains the model, and then automatically extracts features such as margins. Rashad et al. (2011) highlighted patterns from the image for better classification.

5.2 LITERATURE REVIEW

Chandrarathne et al. (2020) conducted a systematic analysis evaluating the efficiency of classifying CNN for limited data. This research is performed on two datasets that are publicly accessible and tests the efficiency of two key approaches: scratched training on transfer learning. Classification prediction accuracy of scratch training are calculated by adjusting the number of convolutional layers for various architectures. Image classification efficiency is calculated based on the number of fine-tuned layers tested. Based on this analysis, we can infer that the better way to identify the tiny data is by fine tuning. It gives high precision because the training and test data are quite close. Furthermore, researchers reported that by learning the target layers with a far lower learning rate, accuracies of small datasets can indeed be increased. For many real world applications, those findings would be very beneficial.

Liu et al. (2020) proposed a deep learning method dependent on stack-sparse coding, that incorporates the concept of sparse expression into the machine learning network layout and the extensive use of sparse interpretation of better multidimensional linear data decomposed capability and deep structural benefits of multilayer nonlinear transformation. This addresses the problem of approximating complex operations and builds a deep learning approach with the potential to adapt. Alternatively, a sparse representation classifier is proposed to optimize kernel functions to address the issue of poor classification performance in learning models. The whole sparse classifier for representation will enhance the precision of the classification of images.

Xin et al. (2019) explained that deep ConvNet has been used to classify images that are variable in scaling, localization, and other types of distortion. The learning network can obtain parallel processing functionality by W, and apparently its parameters and high computational are lower than that of the conventional neural network. Based on the findings of the error back-propagation algorithm, authors suggested an advanced deep neural network (DNN) training standard for a minimal classification loss at maximum intervals. The cross entropy and M3CE are simultaneously evaluated and merged to achieve better results. The analytical results demonstrated that M3 CE can improve cross-entropy, which is an important addition to the criteria of cross-entropy. M3CE-CEc performed well in both databases.

Antun et al. (2019) concluded that the new approach of learning reconstruction algorithm for reconstruction of images in the inverse problem produces unstable approaches usually through deep learning. In addition, the test shows growing anomalies of uncertainty, obstacles, and novel ideas towards science. Specifically: minor disruptions result in a variety of different objects. Various networks generate different artifacts and instabilities. Range of failure to restore massive changes so there is a broad range of instabilities as regards structural changes.

Shiddieqy et al. (2017) implemented a deep convnet raspberry pi platform by using TensorFlow. Larger network size can increase model accuracy rate. This whole system is capable of classifying two cat and dog types which have several similarities. Raspberry pi computing capacity capable of deploying 5 layer CNN network but still not adequate for classification of applications in real time.

Tripathi et al. (2019) presented functional execution of image identification to use a small, CNN, offering less complexity and providing better classification precision for all sets of data analyzed. Automatic data image processing fills the distance between image files at the human vision and the pixel unit. Deep ConvNet are being expressively deployed for the analysis, detection, and classification of images for a variety of tasks. Such neural networks, analogous to the human algorithm, contain neurons with learning set of weights that are learned to recognize and categorize various objects or functionalities throughout the image.

5.2.1 PROPOSED WORK

The aim of proposed work is to classify tiny images among 200 classes with less computational power using deep residual network ResNet18 and ResNet50. Pre-trained ResNet18 and ResNet50 model are used as a feature extractor for transfer learning and their comparative analysis is done for finding the best solution based on Accuracy curve, Loss curve and confusion matrix.

5.2.2 DATASET DESCRIPTION

A dataset that includes images of around 100,000 belonging to 200 classes at several significant growth stages has been used. Dataset consist training and validation set with 90,000 and 10,000 images respectively.

5.2.3 DATA PREPROCESSING AND AUGMENTATION

Images are transformed from BGR to RGB because python expects (R, G, B) using the pre-process input function by default, and images are resized to prevent data leakage issues because most images look identical. Data augmentation is a technique that allows the data diversity available for training models to be substantially expanded, without actively collecting new data.

5.2.4 MODEL IMPLEMENTATION

ResNet is a strong backbone model which is very widely used in multiple computer vision tasks and developed by research team from Microsoft. Model design is focused on the principle of residual blocks that enable use of shortcut connections. Throughout the network architecture, these are basically connections where all the input is kept as it is (not weighted) and conveyed to a deeper layer. ResNets (Prabhu, 2018) have varying sizes depends on how broad any of the model layers are and how many layers it has, such that ResNet-18, ResNet-34, ResNet-50, ResNet-101, ResNet-110, ResNet-152, ResNet-164, ResNet-1202. A deep residual (ResNets) network consists of several stacked "Residual Units."

In a simplified form, each unit can be represented by Eqns. (1) and (2):

$$y_i = h(x_i) + F(x_i W_i) \quad (1)$$

$$x_{i+1} = f_{yi} \quad (2)$$

where; x_i and x_{i+1} are input and output of the ith unit, and F is a residual function with $h(x_i) = x_i$ is an identity mapping (Labach et al., 2019) and f is a ReLU function.

5.2.5 TRAINING AND FINE-TUNING

Architecture to random weights is initialized and trained for a number of epochs. The model recognizes attributes from data with each epoch. The training was using the stochastic momentum gradient descent (SGDM) optimizer, with an initial learning rate of 0.001, a mini batch size of 100 and 15 epochs overall. The performance of evaluated model is assess using accuracy as shown:

$$\text{Accuracy} = (TP + TN) / (TP + TN + FP + FN) \quad (3)$$

5.3 RESULTS AND EVALUATION

In ANACONDA 3.0 the method was put into practice. A standard PC fitted with 2.30 GHz Intel(R) Core(TM) i5-3567 CPU, 64-bit operating system, 8 GB RAM and 2 GB AMD Radeon R5 M330 graphics engine run on Windows 10 operating system has been configured throughout the entire training and testing phase of the model mentioned significantly. As the activation function, SoftMax, and ReLU are used and Adam (Adaptive Moment Estimation) is the optimizer. Accuracy curve, loss curve, and confusion matrix based analysis are per formed on the training and validation set of deep ResNet model.

Logarithmic loss of multiple classes also widely recognized as the categorical cross entropy is used as a metric for this work. A perfect classifier gets the log loss of 0 and is calculated by Eqn. (4).

$$logloss = 1/N\sum_{i}^{N}\sum_{j}^{M} y_{ij}\log\left(p_{ij}\right) \tag{4}$$

The first experiment on the ImageNet dataset performed with ResNet18 for 15 epochs. The training accuracy recorded 89% with logarithm loss 0.515 and the validation accuracy was 53% with logarithm loss 1.939. Figures 5.1 and 5.2 show the accuracy and loss curve for the experiment 1.

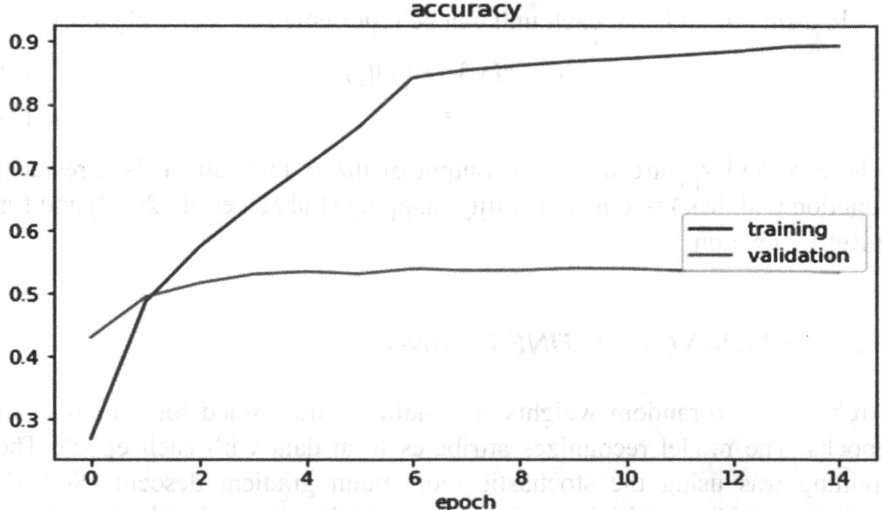

FIGURE 5.1 Training and validation accuracy of ResNet-18.

The second experiment performed with ResNet-50 for 15 epochs. The training accuracy recorded 99% with logarithm loss 0.084 and the validation accuracy was 63% with logarithm loss 1.645 (Figure 5.3). Figures 5.4 and 5.5 shows the accuracy and loss curve for the experiment 2.

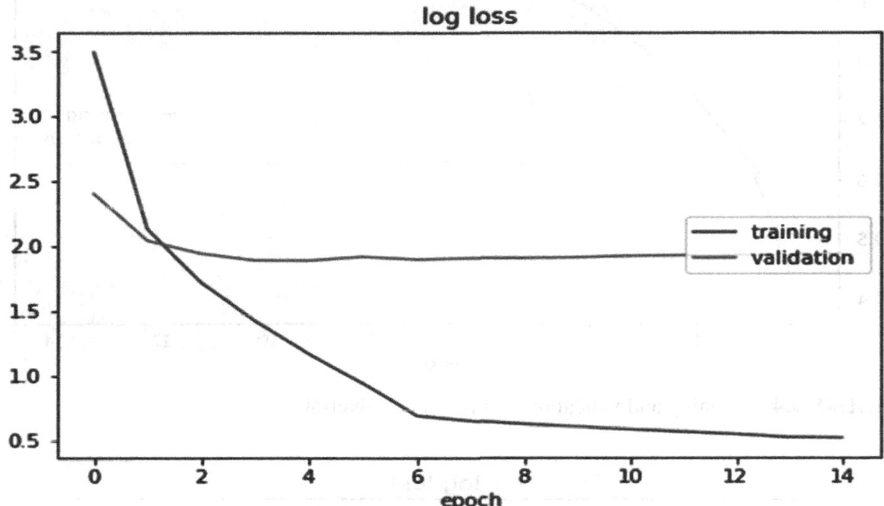

FIGURE 5.2 Training and validation loss of ResNet-18.

```
tensor([[45.,  0.,  0.,  ...,  0.,  0.,  0.],
        [ 0., 37.,  0.,  ...,  0.,  0.,  1.],
        [ 0.,  2., 31.,  ...,  0.,  0.,  0.],
        ...,
        [ 0.,  0.,  0.,  ..., 15., 10.,  0.],
        [ 0.,  0.,  0.,  ...,  5., 28.,  0.],
        [ 0.,  1.,  0.,  ...,  0.,  0., 28.]])
```

FIGURE 5.3 Confusion matrix of ResNet-18.

As per experimental results, validation accuracy and loss for ResNet50 recorded better results than ResNet18 in just only 15 epochs. The most difficult part of this work was to train such type of huge dataset on 2.30 GHz Intel(R) Core(TM) i5-3567 CPU, 8 GB RAM and 2 GB AMD Radeon R5 M330 graphics engine local machine. Due to higher computational time, less number of experiments was performed on this large dataset. However, we recorded 63% accuracy with logarithm loss 1.645 just in 15 epochs which is quite good (Figure 5.6).

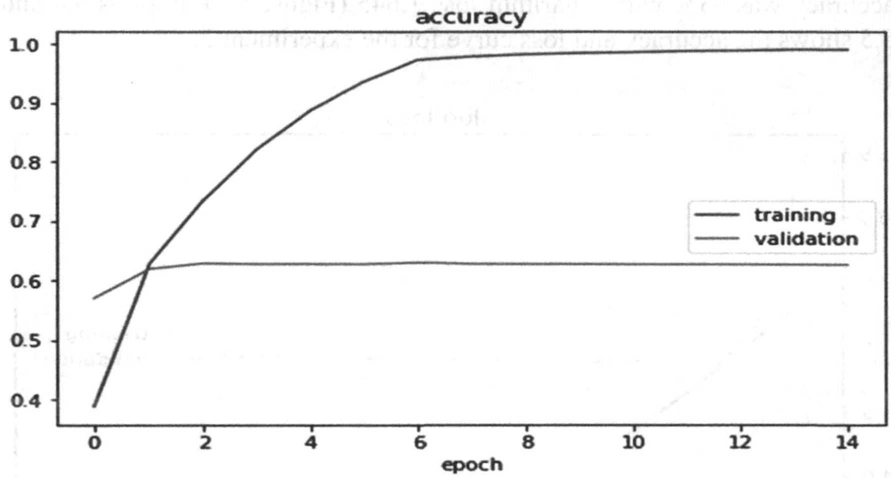

FIGURE 5.4 Training and validation accuracy of ResNet-50.

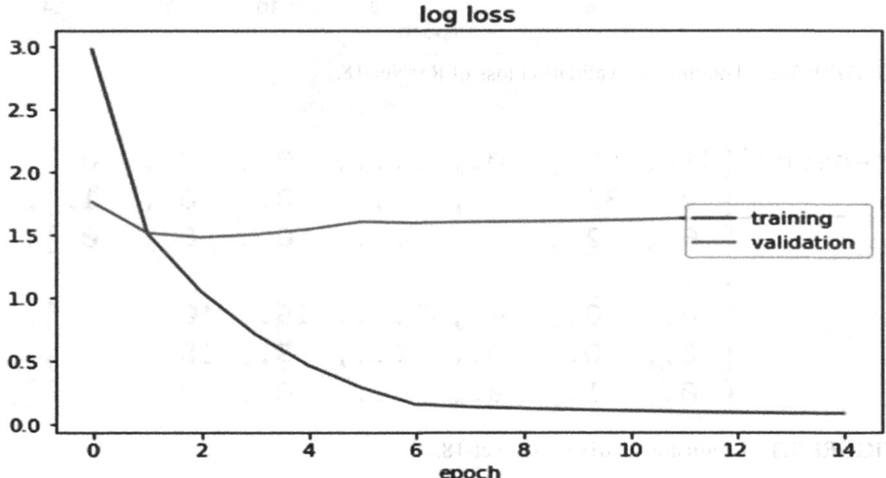

FIGURE 5.5 Training and validation loss of ResNet-50.

5.4 CONCLUSION

Classification of tiny image always be a challenging task. It depends on a detailed knowledge. In our system we developed comprehensive deep learning models for the classification of images, predicated on advanced CNN framework. A dataset has been used that includes images of around

100,280 distinct images actually belonging to 200 classes at several significant stages. Our experimental findings showed how our deep-learning model is capable of successfully categorizing various types of tiny images. While proposed work does have better results, it is possible to increase the number of epochs for better results. We hope that our proposed program will make a significant contribution to the research in image processing.

```
tensor([[47.,  0.,  0., ...,  0.,  0.,  0.],
        [ 0., 43.,  0., ...,  0.,  0.,  0.],
        [ 0.,  0., 31., ...,  1.,  0.,  0.],
        ...,
        [ 0.,  0.,  0., ..., 22.,  8.,  0.],
        [ 0.,  0.,  0., ...,  9., 30.,  0.],
        [ 0.,  0.,  0., ...,  0.,  0., 26.]])
```

FIGURE 5.6 Confusion matrix of ResNet-50.

KEYWORDS

- **convolutional neural network**
- **ResNet18**
- **ResNet50**
- **tiny images**

REFERENCES

Alom, M. Z., Taha, T. M., Yakopcic, C., Westberg, S., Sidike, P., Nasrin, M. S., & Asari, V. K., (2018). *The History Began from Alexnet: A Comprehensive Survey on Deep Learning Approaches*. arXiv preprint arXiv:1803.01164.

Antun, V., Renna, F., Poon, C., Adcock, B., & Hansen, A. C., (2019). *On Instabilities of Deep Learning in Image Reconstruction-Does AI Come at a Cost?* arXiv preprint arXiv:1902.05300.

Barbedo, J. G. A., (2018b). Impact of dataset size and variety on the effectiveness of deep learning and transfer learning for plant disease classification. *Computers and Electronics in Agriculture, 153*, 46–53.

Barbedo, J. G., (2018a). Factors influencing the use of deep learning for plant disease recognition. *Biosystems Engineering, 172*, 84–91.

Bastidas, A., (2017). *Tiny ImageNet Image Classification*. https://pdfs.semanticscholar.org/1b0c/2ba54f7e2f3f5b3a2098721d36e6079d0382.pdf (accessed on 08 December 2021).

Chandrarathne, G., Thanikasalam, K., & Pinidiyaarachchi, A., (2020). A comprehensive study on deep image classification with small datasets. In: *Advances in Electronics Engineering* (pp. 93–106). Springer, Singapore.

Labach, A., Salehinejad, H., & Valaee, S., (2019). *Survey of Dropout Methods for Deep Neural Networks.* arXiv preprint arXiv:1904.13310.

Liu, J. E., & An, F. P., (2020). Image classification algorithm based on deep learning-kernel function. *Scientific Programming, 2020.*

Prabhu, (2018). *CNN Architectures – LeNet, AlexNet, VGG, GoogLeNet and ResNet.* https://medium.com/@RaghavPrabhu/cnn-architectures-lenet-alexnet-vgg-googlenet-and-resnet-7c81c017b848 (accessed on 08 December 2021).

Rashad, M. Z., El-Desouky, B. S., & Khawasik, M. S., (2011). Plants images classification based on textural features using combined classifier. *International Journal of Computer Science and Information Technology, 3*(4), 93–100.

ResNet, AlexNet, VGGNet, Inception, (2020). *Understanding Various Architectures of Convolutional Networks.* Retrieved from https://cv-tricks.com/cnn/understand-resnet-alexnet-vgg-inception/ (accessed on 08 December 2021).

Schmidhuber, J., (2015). Deep learning in neural networks: An overview. *Neural Networks, 61*, 85–117.

Shiddieqy, H. A., Hariadi, F. I., & Adiono, T., (2017). Implementation of deep-learning based image classification on single board computer. In: *2017 International Symposium on Electronics and Smart Devices (ISESD)* (pp. 133–137). IEEE.

Srivastava, N., Hinton, G., Krizhevsky, A., Sutskever, I., & Salakhutdinov, R., (2014). Dropout: A simple way to prevent neural networks from overfitting. *The Journal of Machine Learning Research, 15*(1), 1929–1958.

Tripathi, S., & Kumar, R., (2019). Image classification using small convolutional neural network. In: *2019 9th International Conference on Cloud Computing, Data Science & Engineering (Confluence)* (pp. 483–487). IEEE.

Xin, M., & Wang, Y., (2019). Research on image classification model based on deep convolution neural network. *EURASIP Journal on Image and Video Processing, 2019*(1), 40.

CHAPTER 6

IMPROVED TDOA-BASED NODE LOCALIZATION IN WSN: AN ANN APPROACH

ARPAN CHATTERJEE, ASHISH AGAROWALA, and PARAG KUMAR GUHA THAKURTA

CSE Department, NIT Durgapur, West Bengal, India

ABSTRACT

A new node localization method depending on time difference of arrival (TDOA) concept is proposed in this chapter to obtain minimized error in the prediction of the sensor node position in a network. A multi-layer, feed-forward, back propagation neural network consisting of one input layer, three hidden layers with a (64-32-16) node structure and an output layer is used to develop the proposed model. Different cases have been introduced to train the model by varying the number of anchor nodes and their location along with the number of sensor nodes. Various results for each of the cases mentioned earlier show the effectiveness of the proposed approach in terms of the performance metrics namely root mean squared error (RMSE), average accuracy and the average range error. The proposed approach obtains an accuracy of 99.142% which outperforms the result of existing work.

6.1 INTRODUCTION

A wireless sensor network (WSN) (Patel and Kumar, 2018) is comprised of sensor devices that can communicate information, gathered from a monitored

Advances in Data Science and Computing Technology: Methodology and Applications. Suman Ghosal, Amitava Choudhury, Vikram Kumar Saxena, Arindam Biswas, & Prasenjit Chatterjee (Eds.)

field, through wireless links. The sensor is used to collect the data regarding both physical and environmental parameters, such as pressure, heat, light, etc. The output of these sensors is transmitted to the base station (BS) for further processing (Guha and Roy, 2018). In recent years, the efficient design of a WSN has become a leading area of research due to its significant nature of adaptability in diverse applications. One of the challenging applications in this field is that of node localization (Zhang et al., 2017). It is a cooperative process where nodes with known locations are used to localize nodes at unknown locations. These nodes with known locations as well as unknown locations are called anchor nodes and sensor nodes respectively.

As WSNs are penetrating into the industrial domain, many research opportunities are emerging in the area of node localization. For example, over the past decades, the safety measures adopted by the coal industry (Shrawankar and Mangulkar, 2018) has been an important factor in constraining its development around the world. It is known that coal mining accidents result in loss of life and property. In this issue, mines environment monitoring through node localization can reduce such losses and ensure miners safety as far as possible (Srikanth, Kumar, and Rao, 2018).

In this context, a time difference of arrival (TDOA) based node localization method is proposed in this chapter to obtain minimized error in the prediction of the sensor node position. The proposed approach relies on the distance between the anchor nodes and the sensor nodes. In order to obtain such distance, the time difference between the data transmission by sensor node and data reception by anchor node is considered. In this work, an artificial feed-forward neural network with three hidden layers is used. To train the model have been introduced by different cases by varying the number of anchor nodes and their location along with the number of sensor nodes.

The rest of this chapter is organized as follows: Section 6.2 introduces a literature survey for completeness of the work. Next, Section 6.3 presents the system model. The proposed approach is discussed in Section 6.4. The results are shown in Section 6.5. Finally, we conclude our work in Section 6.6.

6.2 LITERATURE SURVEY

A number of techniques have been proposed for node localization, which can be divided into two broad, categories such as range based and range free (Paul and Sato, 2017). The range based technique relies on distance measurement between the nodes. Some of those in this domain are mainly using received signal strength (RSS), TDOA (Tambe and Krishna, 2018). On the other hand, range free techniques are dependent on hop based information. In 2017, Shahra et al. in 2017 compared range free and range based localization algorithms. The result of this work shows that range based algorithms are more accurate than range free algorithms. In 2019 (Wu et al., 2019), the authors stated that TOA and TDOA, among the range based positioning methods achieve higher position accuracy to perform the measurement on minimized load requirement for the single sensor node. Samanta et al. in 2018 discussed a node localization using ANN to manipulate the location of the sensor nodes which obtains an accuracy of 98%. In order to improve the accuracy further by minimizing RMSE and Average Range Error, a new procedure for node localization is introduced here as discussed next.

6.3 SYSTEM MODEL

6.3.1 *NETWORK MODEL*

A WSN having N number of nodes in a square field of size (M × M) can be viewed as a two dimensional (2D) coordinate system. Each node is represented in a 2-axes layout as (x, y) where $0 <= x, y <= 50$. These nodes are homogeneous in nature. Let us assume that the number of anchor nodes and sensor nodes are m and $(N - m)$, respectively. The positions of the anchor nodes are assumed to be fixed while the sensor nodes are randomly deployed in the network. Let, the set of anchor node positions and the set of sensor node positions are represented by A and S, respectively. These can be represented below:

$$A = [(x_i, y_i) \,|0 \leq x_i, y_i \leq 50 \,\&\, i \in [1, m]] \tag{1}$$

$$S = [(x_j, y_j) \,|\, 0 \leq x_j, y_j \leq 50 \,\&\, j \in [1, N - m]] \tag{2}$$

In the proposed work, the values of m are considered as 3, 4, and 5 for the sake of simplicity. The network with m = 3 and the set A = [(0,0), (25,50), (50,0)] is shown in Figure 6.1. The values of A = [(0,0), (50,0), (0,50), (50,50)] and A = [(0,0), (50,0), (0,50), (50,50), (25,25)] are considered for m = 4 and m = 5, respectively.

6.3.2 *USEFUL DEFINITIONS*

In order to develop the proposed model and measure its efficiency, the following definitions are introduced next:

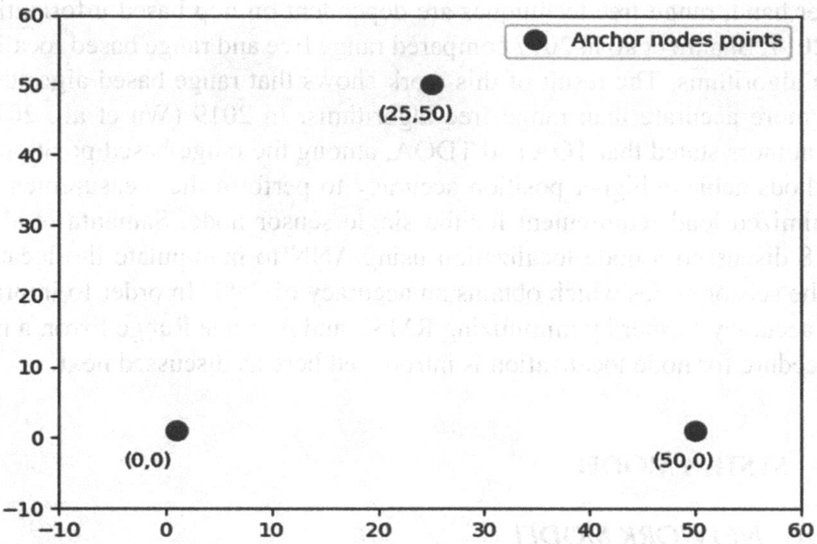

FIGURE 6.1 Network representation with set A.

- **Definition 1 (TDOA):** It is defined as the time difference between the data transmission by sensor node and data reception by anchor node. It is denoted by the following expression:

$$TDOA = d/s \tag{3}$$

where; 'd' is the distance between anchor and sensor node and 's' is the speed of data transmission by the sensor.

- **Definition 2 (Residual):** A residual is a measure of the prediction in error which is nothing but the difference between the actual and predicted distance of the sensor node from an anchor node. It is denoted by the following expression:

$$e = (d_{\text{pred}} - d_{actual}) \tag{4}$$

where; e is the residual, d_{pred} and d_{actual} are the predicted and actual distance of sensor node from anchor node respectively.

- **Definition 3 (RMSE):** It is defined as the standard deviation (SD) of the residuals. It highlights how spread out these residuals are. In the context of the proposed work, such RMSE values can be obtained by the following expression:

$$RMSE = \sqrt{\sum_{i=1}^{N-m} \sum_{j=1}^{m} e_{ij}^{2}} \tag{5}$$

where; e_{ij} is the residual of ith sensor node from the jth anchor node.

- **Definition 4 (Average Accuracy):** It defines how correctly the position of the sensor node is predicted. If the position is predicted more accurately, the error in the distance would be less and subsequently higher average accuracy can be obtained. It is calculated in percentage by using the following expression.

$$Average\ Accuracy = \left(\sum \frac{|e|}{d_{actual}} \right) \times 100 \quad (6)$$

- **Definition 5 (Average Range Error):** Let us assume, there lies a group of sensor nodes within a specific range from an anchor node. Then the average range error for this specific range is defined as the average of the errors in terms of distance for all of the sensor nodes with respect to the anchor nodes. This can be obtained as follows:

$$Average\ Range\ Error = \sum \left| \overline{\overline{d}}_i - \overline{\overline{d}}_{pred} \right| \quad (7)$$

where; $d_i \in R$ and R can be [(0,10), (10,20),(20,30), (30,40), (40,50), (50,60), (60,70)] in our proposed work.

6.3.3 MORPHOLOGICAL OPERATION BASED ENHANCEMENT

Morphological operations make use of structuring elements while enhancing retinal artifacts. Disk-shaped masks are generally used among researchers in biomedical image data enhancement. Tophat and bottom-hat transformations are represented (Digital Image Processing, Gonzalez) in Eqns. (8) and (9), respectively.

$$Tophat\ (I) = (I - (I(open)b)) \quad (8)$$

$$Bottomhat\ (I) = ((I(close)b) - I) \quad (9)$$

where; b is structuring element and morphological operation ($I(open)b$) is erosion of I by b, followed by dilation with b and $I(close)b$ is dilation of I by b, followed by erosion with b. According to Hassanpour et al. (2015), subsequent applications of tophat and bottomhat transformation lead to contrast enhancement between light and dark arenas.

$$En = I + Tophat\ (I) - Bottomhat\ (I) \quad (10)$$

When Eqn. (10) is tried with different structural elements of suitable size enhancement result varies. The output enhanced retinal image is (En).

6.4 PROPOSED MODEL

In order to obtain an improvement on average accuracy and reduce RMSE and average range error, the proposed work is carried out in two consecutive steps. Initially, an ANN model has been developed in this regard. Next, the developed model is trained with the input matrix consisting of the actual distance between anchor and sensor nodes and the coordinate (x, y) of the sensor nodes. For the purpose of testing the model, a new matrix consisting of distance is fed to the model and subsequently the model predicts the coordinate (X, Y) of the unknown sensor node.

6.4.1 ANN MODEL DEVELOPMENT

In the proposed work, a multi-layer, feed-forward, back propagation neural network is used as shown in Figure 6.2. It consists of one input layer, three hidden layers with a (64-32-16) node structure and an output layer. The output layer consists of two nodes for the coordinate X and Y. Rectified linear unit (ReLU) (Nwankpa et al., 2018) is used as the activation function to cut off all the negative values of the node output. Here, the Adam optimizer algorithm (Kingma and Ba, 2014) is used to adjust weight factors, which in turn updates the output of the nodes in order to reduce error.

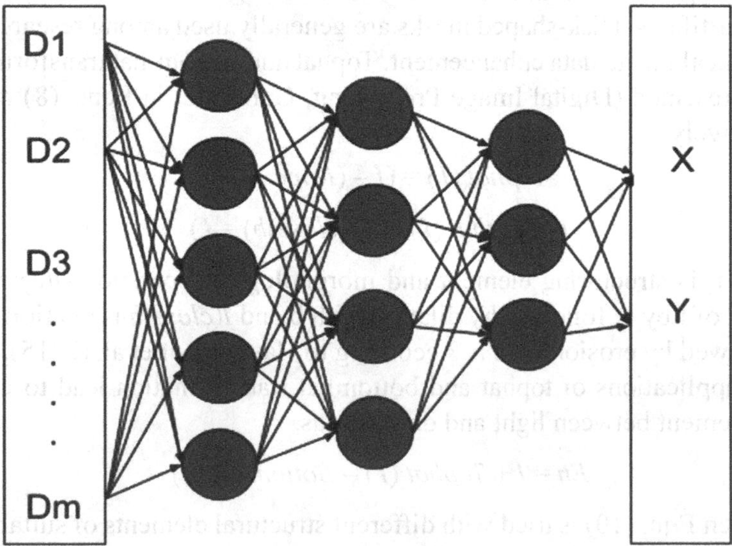

FIGURE 6.2 Feed forward neural network with one input, three hidden and two output layers.

6.4.2 MODEL TRAINING AND TESTING

There are six cases to be considered for training the model developed in the earlier step. In each of the cases, the number of anchor nodes and sensor nodes have been varied as listed in Table 6.1. The model has been trained and tested separately for each of the six different cases.

TABLE 6.1 Various Cases with the Value of m and(N-m)

Case	m	N-m
1	3	100
2	3	500
3	4	100
4	4	500
5	5	100
6	5	500

The generation of a training dataset is required for training the model. It is obtained by deploying the sensor nodes in the network area. For each of the sensor nodes, the distance from all the anchor nodes is calculated. So, a distance matrix D of size (N-m)*m is obtained. Each row in D represents a feature vector as $V_j = [D_{j1}, D_{j2}, \ldots, D_{jm}]$, where D_{ji} denotes the distance of the jth sensor node from the ith anchor node and V_j is the feature vector for the jth sensor node. The original position (x, y) of the sensor node j is the label of V_j. Each row of the input matrix consists of V_j along with its label for a particular sensor node. The input matrix is then fed to the model which is mentioned in Figure 6.3.

$$\begin{bmatrix} D_{11} & D_{12} & \cdots & D_{1m} & x_1 & y_1 \\ D_{21} & D_{22} & \cdots & D_{2m} & x_2 & y_2 \\ \vdots & \vdots & \ddots & \vdots & \vdots & \vdots \\ D_{(N-m)1} & D_{(N-m)2} & \cdots & D_{(N-m)m} & x_{(N-m)} & y_{(N-m)} \end{bmatrix}$$

FIGURE 6.3 Input matrix used.

To test the model, a random set of 30 sensor nodes is generated in the network area for each of those six cases introduced earlier. Then the feature vector for each of these test sensor nodes is computed. Similar to the input matrix used for training the model, each row of the test input matrix consists of the feature vector for a particular sensor node. This test input matrix is then fed to the trained model which in turn predicts the label (X, Y) for each of the test sensor nodes. Hence, the output of this model can be obtained as follows:

$$\begin{bmatrix} X \\ Y \end{bmatrix} = relu\ (relu\ (relu\ (D * W_1 + B_1) * W_2 + B_2) * W_3 + B_3) * W_4 + B_4 \quad (11)$$

where; D is the input row vector of length m; W_i is the weight vector at i[th] layer and B_i is the bias vector at the ith layer.

6.5 RESULTS

In this section, a comparison among the six cases is discussed in an individual and collective manner. A number of tests have been conducted using various activation functions like sigmoid, tanh, ReLU, and by changing the number of nodes in the different hidden layers. A network with three layers having (64-32-16) node structure with ReLU as the activation function provides the best result for the six cases.

It is noteworthy that 30 test sensor nodes have been generated in the network area to check the prediction accuracy of the model. The predicted position of these test sensor nodes is obtained after feeding the test input matrix to the model. The results of the original and predicted position of test sensor nodes for each of the different cases mentioned earlier are generated as shown in Figure 6.4.

To compute the RMSE value, an error matrix of size (30 ∗ *m*) is generated from the predicted position of the test sensor node. Here, an output matrix of size (30 ∗ *m*) is obtained. Each row of the output matrix is a vector of size *m* denoting the distance of the particular sensor node from the 'm' anchor nodes. The difference between the test input matrix and the corresponding output matrix provides the error matrix. The corresponding RMSE value is then computed as per case 5. On the other hand, the accuracy of a single test sensor node from a particular anchor node can be obtained by considering the ratio between error in prediction and original distance between the test sensor node and anchor node. There are (30 ∗ *m*) values obtained for all the test sensor nodes. The average of these values denotes the average accuracy of the model as per case 6. So, a comparative study on the RMSE value and the average accuracy for each of these cases has been shown in Table 6.2.

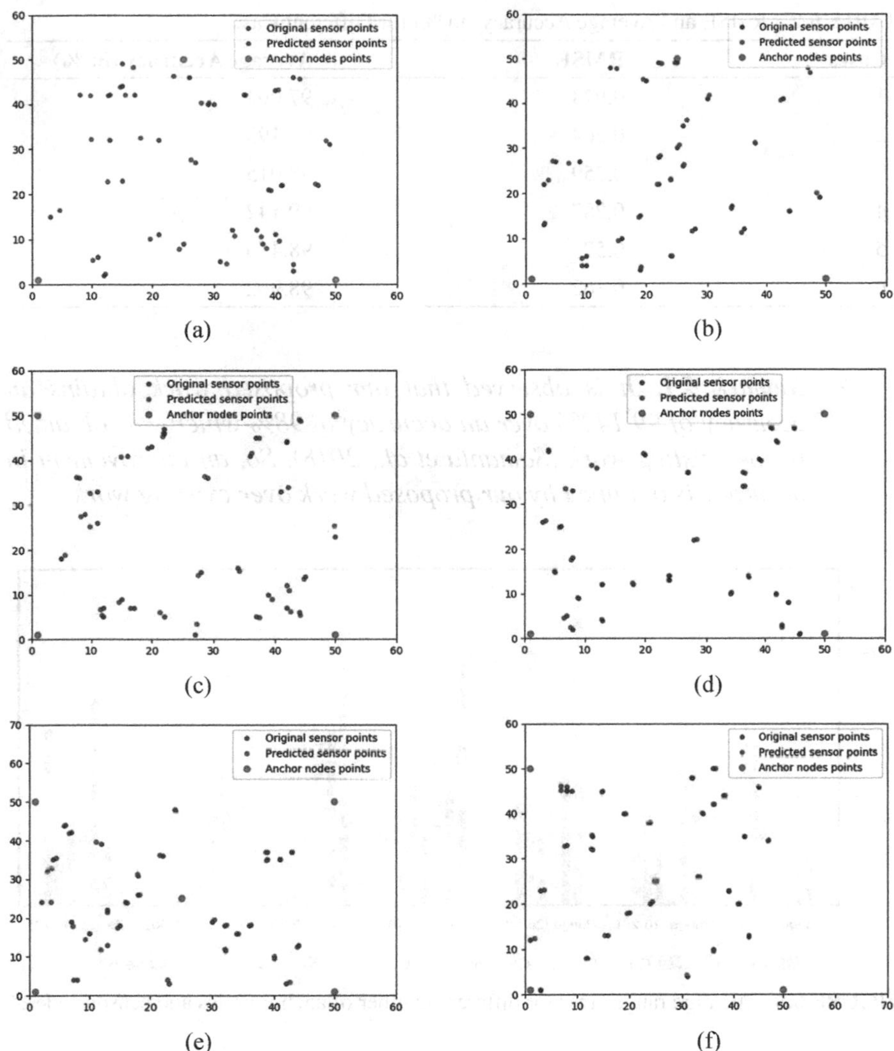

FIGURE 6.4 Original and predicted position of test sensor nodes for: (a) case 1; (b) case 2; (c) case 3; (d) case 4; (e) case 5; (f) case 6.

The range error of a test sensor node provides an estimate of the error in prediction of the position of the node when its distance from the anchor node lies within a known range. The average range error of a particular range is obtained as per case 7. In this context, from Figure 6.5, the range in which the least average error is obtained. This information can be used to minimize the error in prediction during the node localization process.

TABLE 6.2 RMSE and Average Accuracy (in %) for Different Cases

Cases	RMSE	Average Accuracy (in %)
1	0.973	97.004
2	0.564	98.193
3	0.759	98.015
4	0.287	99.142
5	0.57	98.476
6	0.362	98.902

➢ ***Remark 9.1:*** *It is observed that our proposed work obtains an accuracy of 99.142% over an accuracy of 98% which was obtained by the existing work (Samanta et al., 2018). So, an improvement in accuracy is obtained by our proposed work over existing work.*

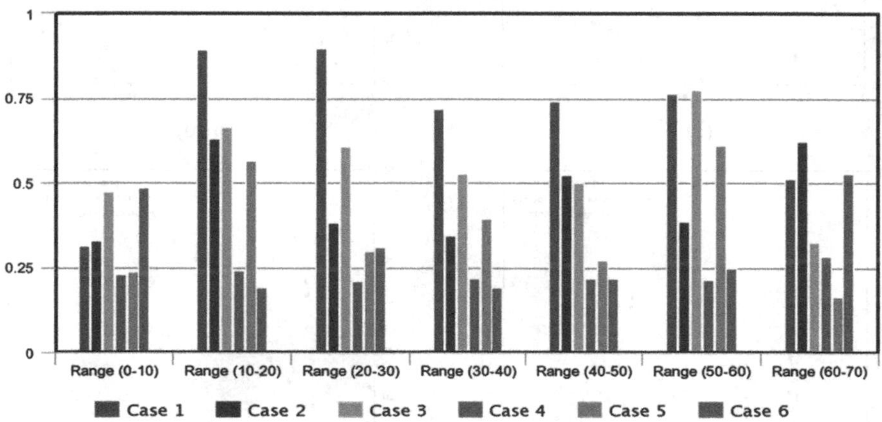

FIGURE 6.5 Average range error for different number of anchor nodes and sensor nodes.

6.6 CONCLUSIONS

A new approach of node localization in WSN is proposed in this chapter to know the position of the sensor nodes for providing better communication among the nodes. Here, an ANN model with node structure (64-32-16) is developed. The TDOA concept is used to develop the proposed model. Different cases have been considered for training and testing the model by varying the number of anchor nodes and sensor nodes in the network area. Among these, the case with 4 anchor and 500 sensor nodes provides maximum average accuracy (99.142%) and minimum RMSE (0.287). For this case, the average range error as obtained is minimum in the range of (20–30). This proposed model can be enhanced for heterogeneous nodes in the near future.

KEYWORDS

- **artificial neural networks (ANN)**
- **average accuracy**
- **average range error**
- **node localization**
- **root of mean square error (RMSE)**
- **time difference of arrival (TDOA)**
- **wireless sensor network (WSN)**

REFERENCES

Guha, T. P. K., & Roy, S., (2018). A decentralized fuzzy c-means minimal clustering protocol for energy efficient wireless sensor network. In: *2018 Fifth International Conference on Parallel, Distributed and Grid Computing (PDGC)* (pp. 24–29). Solan Himachal Pradesh, India.

Kingma, D. P., & Ba, J., (2014). *Adam: A Method for Stochastic Optimization*. https://arxiv.org/abs/1412.6980 (accessed on 08 December 2021).

Nwankpa, C., Ijomah, D., Gachagan, A., & Marshall, S., (2018). *Activation Functions: Comparison of Trends in Practice and Research for Deep Learning*. https://arxiv.org/abs/1811.03378 (accessed on 08 December 2021).

Patel, N. R., & Kumar, S., (2018). Wireless sensor networks' challenges and future prospects. In: *2018 International Conference on System Modeling & Advancement in Research Trends (SMART)* (pp. 60–65). Moradabad, India.

Paul, A. K., & Sato, T., (2017). Localization in wireless sensor networks: A survey on algorithms, measurement techniques, applications and challenges. *Journal of Sensor and Actuator Networks, 6*(4).

Samanta, R., Kumari, C., Deb, N., Bose, S., Cortesi, A., & Chaki, N., (2018). Node localization for indoor tracking using artificial neural network. *Third International Conference on Fog and Mobile Edge Computing (FMEC)* (pp. 229–233). Barcelona.

Shahra, E. Q., Sheltami, T. R., & Shakshuki, E., (2017). A comparative study of range-free and range-based localization protocols for wireless sensor network: Using COOJA simulator. *International Journal of Distributed Systems and Technologies, 8*(1).

Shrawankar, U., & Mangulkar, P., (2018). Monitoring and safety system for underground coal mines. *IEEE International Conference on Power Energy, Environment and Intelligent Control (PEEIC2018).*

Srikanth, B., Kumar, H., & Rao, K. U. M., (2018). A robust approach for WSN localization for underground coal mine monitoring using improved RSSI technique. *Mathematical Modeling of Engineering Problems, 5*, 225–231.

Tambe, K., & Krishna, M. G., (2018). An efficient localization scheme for mobile WSN. *International Journal of Innovative Technology and Exploring Engineering (IJITEE), 8*(2S).

Wu, P., Su, S., Zuo, Z., Guo, X., Sun, B., & Wen, X., (2019). Time difference of arrival (TDoA) localization combining weighted least squares and firefly algorithm. *Sensors, 19*, 2554.

Zhang, X., Tepedelenlioglu, C., Banavar, M., & Spanias, A., (2017). Node localization in wireless sensor networks. In: *Node Localization in Wireless Sensor Networks* (Vol. 12, pp. 1–62). (Synthesis Lectures on Communications). Morgan and Claypool Publishers.

PART II

Role and Impact of Big Data in E-Commerce and the Retail Sector

CHAPTER 7

IMPACT OF BIG DATA ON E-COMMERCE WEBSITES

DIVYANG BHARTIA[1] and MAUSUMI DAS NATH[2]

[1]*Student, St. Xavier's College (Autonomous), Kolkata, 30 Mother Teresa Sarani, Kolkata–700016, West Bengal, India*

[2]*Assistant Professor, St. Xavier's College (Autonomous), Kolkata, 30 Mother Teresa Sarani, Kolkata–700016, West Bengal, India, E-mail: m.dasnath@sxccal.edu*

ABSTRACT

There has been an increasing prominence in the analytics of E-Commerce business lately. Nonetheless, it remains ineffectively explored as a theory, which blocks its theoretical and useful advancement. This time dissimilar to any is confronted with touchy development in the size of information created/caught. Information development has gone through a renaissance, impacted basically by ever less expensive figuring power and the omnipresence of the web. The web has changed E-Commerce and client presently approach the wide scope of items offered through E-Commerce sites. To stay serious and protect a piece of the overall industry, E-Commerce firms figure web based showcasing systems dependent on continuous information. This has directed to a change in perspective in E-Commerce, where information is viewed as the greatest resource for the firm in understanding explicit necessities of clients, foreseeing conduct, fitting explicit requirements, and offering execution measurements to survey viability. Internet business firms are discovering approaches to remove important data from bigger datasets where information gets created at more noteworthy speed, distinctive assortment,

Advances in Data Science and Computing Technology: Methodology and Applications. Suman Ghosal, Amitava Choudhury, Vikram Kumar Saxena, Arindam Biswas, & Prasenjit Chatterjee (Eds.)

and at high volumes that are regularly alluded to Big Data. Thus, this chapter explains how E-Commerce firms are contributing gigantic to Big Data Analytics to engage them to take precise and ideal choices.

7.1 INTRODUCTION

In the world of Digital and Electronic Commerce, advertisers take each and every minute to catch the attention of the guests to fulfill their quick needs. The key segment to fulfill client needs gives them the best client experience which yields to two central points in business-brand value and client dedication. To meet with the consumer loyalty, companies develop huge information and research arrangements prompting accurately gauge and break down each and every client execution variable on the site. Big data refers to every single large volume of data that is been collected on all aspects of human life. With Big Data, every action of humans could be quantified and hence stored in the data bank which in turn assists in satisfying the needs of the customers. These data are collected from a vast array of electronic devices such as handheld devices like mobile, tablet, and even portable devices such as laptops and all kind of sensors that are connected to the internet. With Big Data, each activity of people could be evaluated and thus put away in the information bank which thus helps with fulfilling the necessities of the clients. This information is gathered from a huge swath of electronic gadgets, for example, handheld gadgets like tablets and even compact gadgets, for example, PCs, and all sort of sensors that are associated with the web. Prior to Big Data companies applied investigation to reveal patterns and increase bits of knowledge. In any case, the handling was unmistakably complex and difficult to do physically. The personalization should be possible by investigating the purchasing behavior of the client that permits advertisers to set unmistakable patterns and examples. Extra inputs permit the advertisers to adjust the endeavors and work on improving contributions for the clients. Then again advertisers may think about perusing information, buy styles and other large information sources to make customized items. Google Trends online tools help advertisers to channel down inclining points and related pursuit works permitting them to target clients locally and all around including area, district, nations, and interests to get more fascination towards the customized item or service These days utilizing Big Data, advertisers likewise set ICP (ideal customer profile) that permits them not to rely upon determined speculations but instead gives exact data about the socio-economics of the client just as their psychographic and behavioristic examples. Consequently, Big Data permits advertisers to arrive at clients at the ideal time and promote the item in a convincing manner to understand their need and increment the deal.

The chapter is coordinated as follows: Section 7.2 details overview of related work, Section 7.3 gives the exploratory study and Section 7.4 focuses on the objectives of the study. Section 7.5 gives analysis and discussions while Section 7.6 puts forward the recommendations and Section 7.7 finally includes the references.

7.2 OVERVIEW OF RELATED WORK

The role of Big Data in marketing is increasing exponentially; there is no comparison to the amount of data that is collected for the profiling of each customer. To understand the aspects of the field few empirical studies have been conducted to understand the role of big data analytics from the management point of view. The use of data is a decisive variable in the marketing decisions of the marketing managers and various related studies have been conducted to understand the various fields of big data analytics from the marketing perspective.

Shahriar and Samuel (2016) focused on more extensive conversations with respect to future examination difficulties and openings in principle and practice. Generally, the discoveries of the examination combine different BDA ideas (e.g., the meaning of large information, types, nature, business esteem, and important hypotheses) that give further experiences along with the cross-cutting investigation applications in web-based business.

Uyoyo (2014) showed us an outline of the special highlights that separate enormous information from customary datasets. Furthermore, the use of large information investigation in the E-trade and the different advances that make examination of shopper information conceivable is talked about.

Avinash and Akarsha (2017) researches how the utilization of huge information investigation is seen as worth maker that can direct E-Commerce organizations accomplish upper hand.

Moorthi, Srihari, and Karthik (2017) talks about different approaches and cycles continued in online business for business knowledge. Additionally, proposed some new strategies to improve the business knowledge in web based business field utilizing large information examination.

7.3 EXPLORATIVE STUDY

Walmart, with more than 20,000 stores in 28 nations is the biggest retailer on the planet. So, it is fitting then that the organization is building the world's biggest private cloud, sufficiently large to adapt to 2.5 petabytes of information consistently. To understand this data, Walmart has made what it calls its Data Café – a best in class investigation center point situated inside its Bentonville, Arkansas central station. The Data Café permits tremendous volumes of interior and outer information, including 40 petabytes of ongoing value-based information, to be quickly displayed, controlled, and pictured. Walmart has an expansive huge information biological system. The large information environment at Walmart forms numerous Terabytes of new information and petabytes of authentic information consistently. The examination covers a large number of items and 100's of millions clients from various sources. Walmart has comprehensive client information of near 145 million Americans of which 60% of the information is of U.S. grown-ups. Walmart accumulates data on what client's purchase, where they live and what are the items they like through in-store Wi-Fi. The enormous information group at Walmart Labs examinations each interactive activity on Walmart.com-what purchasers purchase available and, on the web, what is drifting on Twitter, nearby occasions, for example, San Francisco goliaths winning the World Series, how neighborhood climate deviations influence the purchasing behaviors, and so forth. All the occasions are caught and broke down wisely by huge information calculations to recognize important large information bits of knowledge for a huge number of clients to appreciate a customized shopping experience (Figure 7.1).

How Walmart is having a genuine effect to expand deals?

1. **Launching New Products:** Walmart is utilizing web based life information to discover about the drifting items with the goal that they can be acquainted with the Walmart stores over the world. For example, Walmart dissected internet based life information to discover the clients were distracted about "Cake Pops." Walmart reacted to this information investigation rapidly and Cake Pops hit the Walmart stores.
2. **Better Predictive Analytics:** Walmart has as of late changed its transportation approach for items dependent on large information examination. Walmart utilized prescient examination and expanded the base sum for an online request to be qualified with the expectation of complimentary delivery. As indicated by the new dispatching

strategy at Walmart, the base sum with the expectation of complimentary transportation is expanded from $45 to $50 with expansion of a few new items to improve the client shopping experience.

3. **Customized Recommendations:** Simply the way wherein Google tracks customized ads, Walmart's enormous information calculations break down MasterCard buys to give specific suggestion to its clients dependent on their buy history.

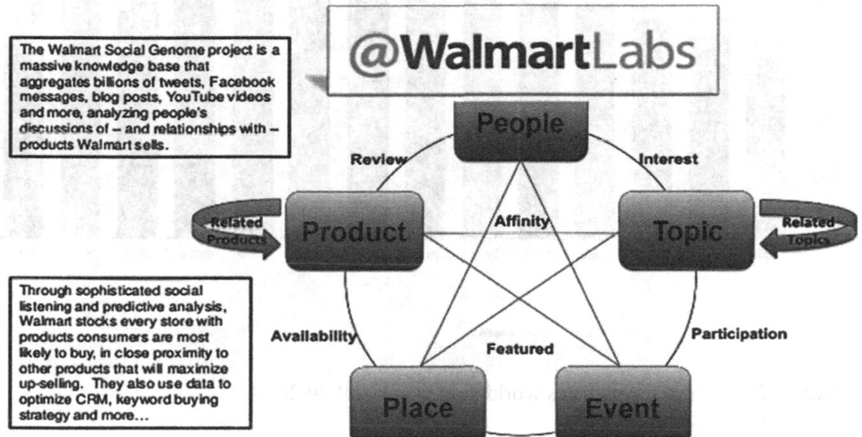

FIGURE 7.1 An overview of how Walmart Labs function.
Source: slideshare.com.

7.3.1 WAL-MART VERSUS THE OPPOSITION

Despite the fact that business development at Wal-Mart has eased back as of late, the measurements are still head and shoulders above the greater part of the business. Contrasted with the best 10 retailers internationally, just Home Depot beat Wal-Mart consequently on resources or net overall revenue. In 2013, Wal-Mart recorded a net edge of 3.5% and profit for resources of 8.6%. Costco, maybe its nearest rival, had an overall revenue of just 2% and profit for resources of 6.8% that year. On a business premise, Costco utilizes its land all the more proficiently, creating deals of $1,100 per square foot, yet its overall revenue is not even close to what Wal-Mart appreciates.

Then, Amazon is regularly observed as the greatest danger to the Wal-Mart plan of action. The web based business pioneer has been developing deals by about 20% every year, creating upper hands en route and set to top $100 billion in income this year. Incidentally, its prosperity has come from various perspectives by taking pages from the Walton playbook – putting the client first and offering the most reduced costs.

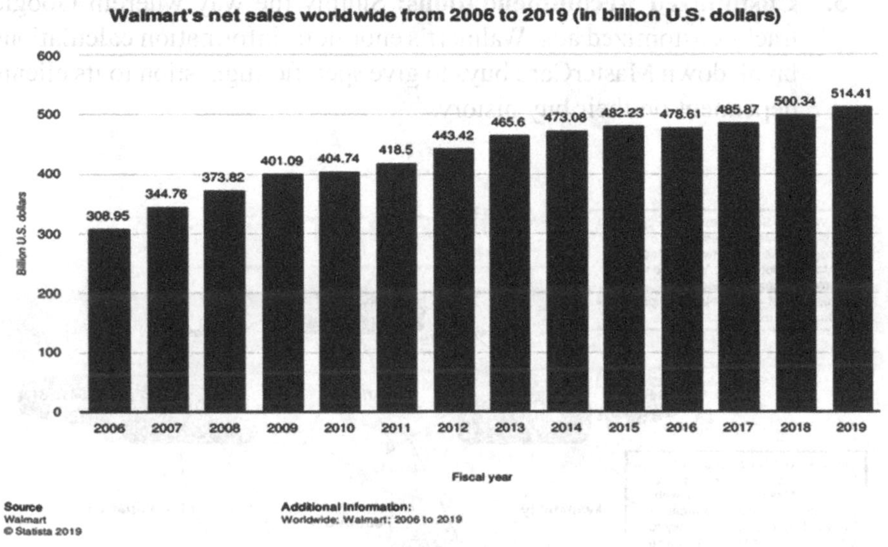

FIGURE 7.2 Walmart's net sales worldwide from 2006 to 2019.
Source: statista.com.

7.3.2 *IMAGINING THE WAL-MART OF THE FUTURE*

Wal-Mart deals have eased back as of late as U.S. superstores have all the earmarks of being arriving at an immersion point. In the course of the most recent 12 years, the organization had the option to dramatically increase income to its present sign of $488 billion, however, it will probably take any longer for it to twofold by and by to about $1 trillion. Store extension has additionally eased back altogether – all out U.S. supercenters expanded by just 3% a year ago, and a lot of that expansion originated from changing over rebate stores. Development was likewise delayed for global markets and Sam's Clubs. Conversely, the quantity of supercenters developed by 17% in 2005. Presently, the board is concentrating on two development roads: first is its neighborhood Markets. In financial 2015, the organization included 232 of these littler configuration stores, developing the base by over half. Execution at these stores has been solid with same-store deals bouncing 7.7% in the latest quarter. The subsequent open door is online business. Wal-Mart has unmistakably fallen behind Amazon in online retail, yet it needs to bend over backward to make up for lost time. The organization acquired $12 billion in online deals a year ago and expects to significantly increase that figure to $35 billion by 2018 with interests in circulation focuses and innovation. All things considered, contrasted, and almost $500 billion in complete income, that scarcely moves the needle.

7.4 OBJECTIVE OF THE STUDY

Marketers analyze Big Data to match with the expectations of customers and hence frame new product line and extend their product width. To streamline the production and manufacturing performance, marketers require huge data. But to assess the concept of Big Data from the customers perception is also one of the huge factors that should be taken into account under this scenario. Hence the objectives of the research could be categorized as under:

1. **Primary Objective:** The main objective of the chapter is to examine the perception of E-Commerce websites towards big data and personalized campaigns and analyzing its impact on buying decision.
2. **Secondary Objective:** The research chapter also tried to analyze the pros and cons regarding the recent big data analytics amalgamation.

7.5 ANALYSIS AND DISCUSSIONS

The business case for Big Data has been clear for quite a while, anyway over the latest couple of years, we have seen critical strolls in how associations are putting gigantic data to use to all the more probable fathom their own strategies and how they convey and offer to customers. From a B2B standpoint, immense data has quite recently exhibited to be a primary purpose in the movement of online strategic policies and stages. It is prepared for associations to not simply grasp the gigantic proportions of data they are gathering – yet likewise how to all the more promptly understand what their customers are expecting to get away from the modernized channels they are building.

7.5.1 EXPANDED EXAMINATION AND REVELATION OF NOTEWORTHY BITS OF KNOWLEDGE

Colossal data gets past the overall wreck of customer data that associations are gathering, zeroing in on the critical pieces of information that notice to associations what they need to think about spikes famous, customer tendencies, changes in buyer direct and that is just a glimpse of something larger. Understanding buyer lead and their specific tendencies when making a purchase are fundamental for continued with bargains and all in all accomplishment.

Big data benefits in eCommerce

Real-time, targeted promotions broadcasted directly to customers' smart phones while they shop by examining purchase history, online "travel", likes via social networks, geo-location, retailers can now create

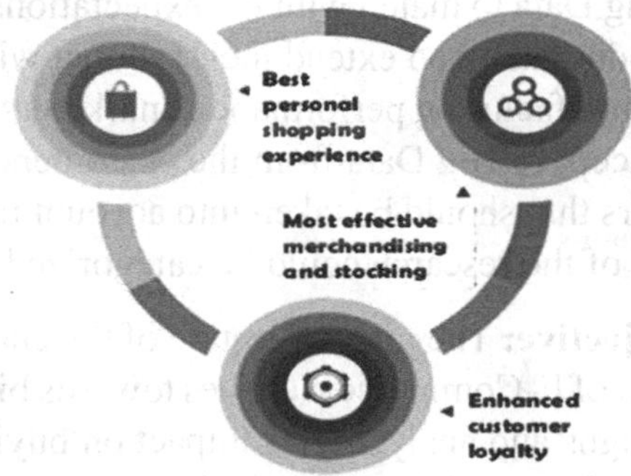
Best personal shopping experience
Most effective merchandising and stocking
Enhanced customer loyalty
Supported by data coming from online sources, retailers can now pinpoint which merchandise should be stocked at specific locations and where items should be placed throughout the store (eg: pregnant woman seeing baby products at the entrance in a shop)
By tailoring offers to each individual customer, retailers are seeing an increase in returning clients. Customers nowadays are looking for the easiest and most convenient way to shop and Big Data allows retailers to understand their customers' needs before they even enter a store
EITG

FIGURE 7.3 Big data benefits in ecommerce.
Source: forbes.com.

An association probably will not have a particular commitment that customers are searching for in their own thing set, yet there may be accessories inside their organic framework that have the significant applications and extensions that do. Colossal data empowers associations to fathom what they need to take any action on, considering what customers need and are looking for. Right when inspected dynamically, the move can be made immediately, which can be the differentiation that addresses the decision time an arrangement.

7.5.2 IMPROVED CLIENT ENCOUNTERS AND PURCHASER VENTURES

In a U.S. study done by Help Scout, it was found that most of Americans have given up a masterminded purchase on account of helpless help. This identical examination furthermore reported that associations lose more than $62 billion consistently taking into account helpless customer backing and experience. Clearly, customers are looking for positive experiences and bound together buying adventures that eliminate the scouring from purchasing on the web – coordinating this will be especially critical to a business' fundamental concern. Customers that can find what they need, when they need it are happy customers who become reiterate ones – especially when the things they need are no matter how you look at it place with a clear a path to purchase.

7.5.3 SIMPLER AND PROGRESSIVELY SECURE APPROACHES TO FIND AND PAY ON THE WEB

Associations can use colossal data for a load of improvements in how customers pay on the web and how these trades stay cautious. This separated deception for constant similarly as tax avoidance deceives that appear as though they are true trades. Since tremendous data consolidates unmistakable portion limits into one central stage it diminishes the threat of deception while moreover making trades less complex for customers.

Joining and packaging data for logically direct, quick, and brilliant assessment, tremendous data has prepared for associations to build up their customer base while furthermore scaling express procedures, like web business, for extended triumphs. As both the business and purchaser universes continue making more data, it is inflexibly huge for associations to stay before how they track, dismember, and put this data to authentic use. For B2B web business to continue creating, it will require moving past essentially looking at the data open–forefront advancements and assessment approaches ought to be shipped off totally get the prizes and reinforce long stretch accomplishment.

All things considered, the imbuement of tremendous data has given associations an intuitive strategy to locate the critical encounters from inside their instructive assortments that will advance business endeavors, for instance, online business frameworks, progressively productive. Exploiting huge data gives associations a lively and smoothed out a way to deal with focus on significant pieces of data–which would then have the option to be used to improve customer experience over the entire buying adventure, making brisk proposals and unraveling approaches to purchase, while keeping trades logically secure.

7.6 RECOMMENDATIONS

The research work is based to break down the significance of Big Data for promoting of customized items which has just been restricted with the respondents of Kolkata. The information for this exploration work just centered around not many of the areas, for example, video spilling applications, and Amazon for instance to site the information. The territory for customized promoting is very wide and need exceptional consideration as large information bolsters in surrounding the legitimate prerequisites of the clients in both on the web and disconnected markets. The part for this research is very tremendous and can be beseeched through different elements of focusing on and situating procedures utilized by advertisers to investigate the customized showcases through large information. The socioeconomics utilized right now be additionally investigated while the locational portion could be expanded for a superior extent of research and more extensive research space.

7.7 CONCLUSION

Personalization is undoubtedly an appealing vision for advertisers. A dream which is at some point tempered by unreasonable desires, wrong application approaches and security stricken laws that give clients more control on close to home requests. Huge information without a doubt assumes a significant job in alluring the vision of personalization and produce crusades remembering potential clients. It gives a feeling that brands care for the clients and henceforth clients pay off by indicating more dependability towards those brands. According to a review directed by Forbes, 64% of the respondents accept that information-driven advertising is fundamental for the development of the current hyper-serious economy. This exploration concentrated basically on to comprehend the client's recognition of the customized market and its significance through the investigation of large information. The investigation demonstrated that however, individuals have less thought on enormous information. Research yet firmly concurs that it is significant for advertisers to infer ends and comprehend the purchasing conduct of the clients, consequently concentrating on personalization. Respondents have reasonable information about Marketing, information-driven markets, and data innovation yet subliminally they are a piece of huge information and consequently follow an extraordinary example that has been offered by the advertisers. The instances of Walmart and Amazon unmistakably depicted that clients do acknowledge items which are customized by the advertisers for explicit clients at certain purpose of time. In spite of the fact that an ongoing McKinsey study of senior promoting pioneers expressed that solitary 15% of the CMOs accept that their organization has been on a correct track with personalization, however, today's pioneers have discovered demonstrated that there has been an expansion of 10–30% increment in deals. Henceforth, one might say that personalization is not only a strategy to assemble brands; it additionally presents upsell open doors for increment of deals.

KEYWORDS

- **analytics for E-commerce**
- **big data analytics**
- **decision making**
- **E-commerce**

REFERENCES

Data Analytics in E-Commerce Retail. https://towardsdatascience.com/data-analytics-in-e-commerce-retail-7ea42b561c2f (accessed on 08 December 2021).

Ecommerce Insights on the Go. https://www.bigcommerce.com/blog/ecommerce-big-data/ (accessed on 08 December 2021).

How Big Data Analysis Helped Increase Walmarts Sales Turnover. https://www.dezyre.com/article/how-big-data-analysis-helped-increase-walmarts-sales-turnover/109 (accessed on 08 December 2021).

How Big Data Analytics Has Changed Ecommerce Industry. https://www.smartdatacollective.com/how-big-data-analytics-has-changed-ecommerce-industry/ (accessed on 08 December 2021).

How Big Data Benefits E-Commerce. https://dg1.com/blog/how-big-data-benefits-e-commerce (accessed on 08 December 2021).

Three Key Benefits of Using Big Data in B2B E-Commerce. https://www.itproportal.com/features/three-key-benefits-of-using-big-data-in-b2b-e-commerce/ (accessed on 08 December 2021).

Walmart Sales Forecasting. https://medium.com/analytics-vidhya/walmart-sales-forecasting-d6bd537e4904 (accessed on 08 December 2021).

CHAPTER 8

A REVIEW OF THE REVOLUTIONIZING ROLE OF BIG DATA IN RETAIL INDUSTRY

MAUSUMI DAS NATH and MADHU AGARWAL AGNIHOTRI

Department of Commerce, St. Xavier's College (Autonomous), Kolkata, West Bengal, India, E-mails: m.dasnath@sxccal.edus (M. D. Nath), madhu.cal@sxccal.edu (M. A. Agnihotri)

ABSTRACT

Information is present all around us in different kinds of forms. Voluminous amount of information is generated daily in several forms. It has become highly appreciable as information provides benefit both to the individuals as well as to the business organizations. The existence of any business in this competitive market depends solely on acquiring accurate information at a particular time. Companies must have prior information concerning their customers, products or inventory, market, and themselves too. To carry out this task the structured and the unstructured data must be combined efficiently to improve the business transactions, analyze the past performance, and forecast the market advantages. This led to a new concept termed as Big Data which companies rely on. There are several factors that contributed to the exponential usage of today's big data. It helps them to store, process, and analyze data, monitor, and improve their performance and also generate revenue. It helps to reengineer business models and also transform the decision-making process. In this chapter, we portray some features of Big Data and how it revolutionizes the decision-making process and the overall performance of the business organization.

Advances in Data Science and Computing Technology: Methodology and Applications. Suman Ghosal, Amitava Choudhury, Vikram Kumar Saxena, Arindam Biswas, & Prasenjit Chatterjee (Eds.)

8.1 INTRODUCTION

Information is scattered everywhere in the environment and it is generated daily through machines also. It is in the form of either unstructured, semi-structured or structured form. It is to be captured, stored, processed, and analyzed for variety of activities and events that occur all around. Thus, the concept of Big data has emerged. In the last decade, giant companies like Google, Facebook, and Amazon are all using this new technology to sustain in the global market by gaining a competitive advantage over other organizations. Big data details about events, preferences, and behaviors taking place all around and finally gives access to a wider reach to the voluminous amount of data from gigantic and multiple resources in a lesser amount of time. Companies collect and store data, and then analyze to find out new revenue streams. The data are sorted so as to process them in a lesser time. Different companies take the help of big data for a reason of their own. For example, telecommunications company use big data to assess the data traffic and the volume of incoming data. Amusement park like Disney Land introduced "magic wristbands" to improve the attention of park visitors. Prediction is carried out by wholesale company Amazon which helps them to outline the customer's behavioral pattern and their buying preferences. The opportunities of big data is enormous. Companies can get a clear picture of their future business value and thus big data helps in making a better decision in their business activities. Big data finds its application in many areas like telecommunications, IT sector, retail industry, finance, health as well as crime to name a few. Hence, Big data has drawn up as a significant area of research for many researchers, it finds its enormous impact in business environment. Although it has several challenging issues, the benefits have solved many areas of concern in many sectors of business. The implementation of big data gives an efficient way to manage huge sets of data effectively in a cost effective manner in lesser amount of time. Thus, big data helps in monitoring the performance of the organization and ultimately enables in taking better decisions.

The presentation of the chapter is organized as follows. Section 8.2 focuses on the literature review and Section 8.3 states the objectives of the study. Role of big data in the retail sector is given in Section 8.4. Discussions are given in Section 8.5. Section 8.6 illustrates the future directions and concluding remarks.

8.2 LITERATURE REVIEW

Big data finds its applications and advantages in a business environment. Ajah and Nweke (2019) reviewed and discussed the upcoming trends, business opportunities and limitations of big data and how it has helped the organizations to create successful business strategies and remain competitive in the business market. Furthermore, this chapter also brought forth the different applications of big data and business analytics, how the data sources were being generated in these applications and their main features. The reviewers portrayed the challenges for a big data project and showed a path for future research too. Again, Berntzen, and Krumova in 2017 applied Porter's value chain to big data and a framework has been suggested to figure out the possible business opportunities for the business organization. The most frequent revenue models have been discussed with its application. Moreover, Banica and Alina focused (2015) on how big data helps organizations to gain a competitive advantage over others. Moreover, it highlights how big data helps to build, organize, and perform analysis from huge datasets by providing a three-layered architecture. Furthermore, Pugna et al. in their chapter (2019), outlined that there are some organizational challenges due to the use of Big Data, but its impact on the business environment is enormous, especially on performance management. Managers' views and understanding have been analyzed in terms of their performance in the company and also on the decision-making process. After adopting the grounded theory, it has been found that the key areas need special attention and understanding. Gao, Chunli, and Chuanqi (2016) have discussed the validation and quality assurance of big data, including the main concepts. Furthermore, the chapter presented a comparison among big data validation tools and several major market players were considered and discussed. Moreover, the primary issues, challenges, and needs of big data have been discussed immensely. On the other hand, Furht, and Flavio explained in the chapter (Furht and Flavio, 2016) the term big data and its role. It also highlights the main technological aspects in a big data environment. Alsghaier et al. in their research (2017) focused on some aspects of Big Data and its effect on organizations' business performance and how firms use the famous open source platform Hadoop to process the huge amount of data. Alam et al. in their chapter (2014) focused on the role of big data in the business and its various challenges. Why these challenges are not dealt in the planning stage by the organizations have also been discussed in details. Here, Alsghaier et al. (2017) illustrated on some important points on Big data and its need on business performance and how Hadoop can be used to process the data to win over others in this competitive market. Again, Kubina, Michal, and Irena (2015) showed the path as to how to reach a competitive advantage by using big data. Several research techniques like content analysis, documents study, comparative analysis, process analysis, statistical analysis, empirical research were used.

8.3 OBJECTIVES OF THE STUDY

The objective of this study is to explore the ways in which big data has played a pivotal role in business. this study focuses on finding out the revolutionary contribution of big data in retail sector. It is based on secondary data.

8.4 ROLE OF BIG DATA IN RETAIL

Various use cases of retails sector has been studied to understand the business model of traditional retail sector and the retail sector where big data concept has been implemented. The observations made are then analyzed to find out the significant contributions of the technique adopted.

8.4.1 TRADITIONAL RETAIL BUSINESS MODEL

Retail business focuses mainly on delivering good shopping experience to customers to retain them. Customer's shopping experience is influenced by various factors such as quality of the product or the service on offer, the channels of delivery, time-frame, communication experience, payment flexibility, incentives on offer, etc. Retail Business comprises of significant dimensions such as customer value proposition, revenue model and operating model. Customer value proposition is centered on the overall customer wellbeing. It is about giving a good experience to customer so that they stay with the company in long run. Customer's satisfaction would be highest if the product or services availed from a company are of utmost quality within reasonable price range. Also, the way the facilities, services, and incentives are offered plays a critical role in achieving higher customer satisfaction. Revenue model for business is also crucial for business as the survival of business depends on customer as well as the revenue line. Continuous flow of revenue is very important and hence, business needs to have a systematic cost model and corresponding earning model in place so that business can sustain in long run. The source of money for business, how the money will be distributed through various retail locations to meet up all expenses and target revenue to be generated for smooth conduct of business are of critical importance. The distribution of product and services on offer for customer, flow of funds (cost as well as earning), and customer satisfaction are the key components that builds a sound retail business system. The soundness of the system entirely depends on the availability and accessibility to right data at right time (Figure 8.1).

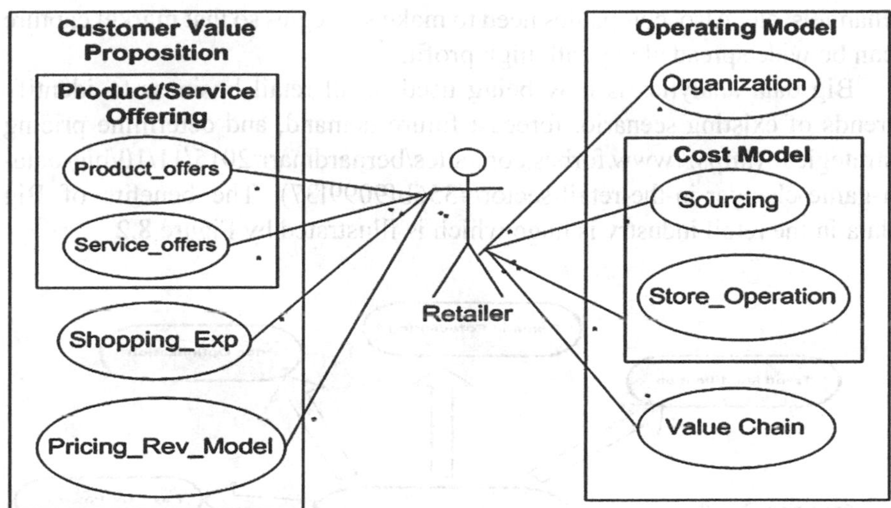

FIGURE 8.1 Retail business model portraying two significant dimensions.

In traditional mode, the data management may be manual or digitized. However, digitization does not always mean integrated systems. Digitized system usually maintains data is maintained in flat file format. In such cases, the issues of data replication, issues of multiple versions of same data, data accessibility at various locations, maintenance of user profiles for filtered data access, data status at various locations are serious concerns. Traditional retail business needs to transform their model to integrate customer requirements, inventory, distribution channel activities, and build optimized business model for smooth operations. The modern retail business model is data driven model and hence, Big Data is playing a crucial role in providing upgraded and integrated system.

8.4.2 RETAIL BUSINESS MODEL IMPLEMENTING BIG DATA

Due to ubiquitous nature of digital technology, customers can take informed decisions about the products they want to buy or the services they want to avail. As an amazing pool of updated information has become very handy for online users, it has become very easy to compare the offers made by various companies, price comparisons, reviews given by existing customers, and the list is endless. Therefore, it is very essential for business to adapt technologies that can provide deep insights into data to understand customer preferences for the products, the manner in which those are offered, preferred delivery channels, etc. Also, companies need to make strategies so that market capture can be wide spread along with high profit.

Big data analytics is now being used in all retail business to identify trends of existing scenario, forecast future demand, and determine pricing strategies (https://www.forbes.com/sites/bernardmarr/2015/11/10/big-data-a-game-changer-in-the-retail-sector/#356bf9099f37). The benefits of Big data in the retail industry is huge which is illustrated by Figure 8.2.

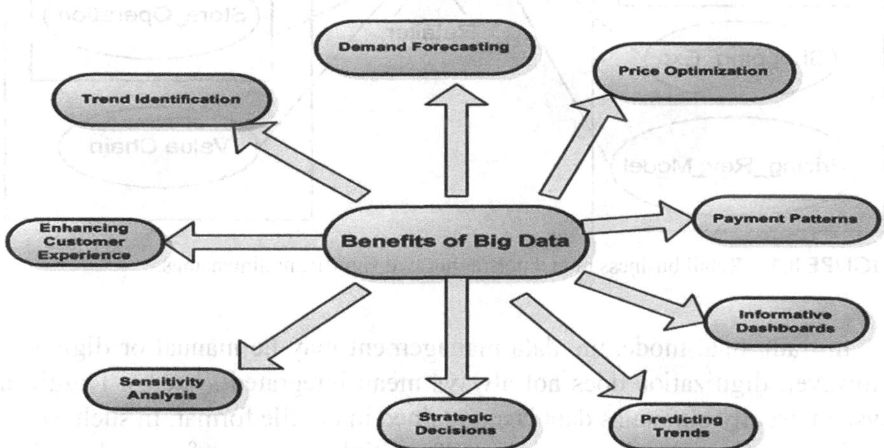

FIGURE 8.2 Benefits of big data in the retail industry.

1. **Trend Identification:** This is achieved on the basis of the historical data available with the company about customers. Existing pattern of customers can help to find out the purchase pattern of customers throughout the year. Also, region-wise list as well as season-wise trends can also be identified with in depth analysis. Also, special machine learning algorithm can be developed to filter out the data available on social media, communities, forums, and blogs to understand the interest of potential customers. It helps to understand, the purchase preference of existing customer. Browsing habit of people also tells a lot about their thought process which may be highly affected due to hype in social media or in non-digital platforms also. Retailers must deploy tools to drag out such minute details about customers to read their mind for strategic planning.
2. **Predicting Trends:** Nowadays, retailers use a variety of tools available to them to find out the season's trends. They can choose

what items they should have in their shelves so as to attract the customers, both old and new. Trend forecasting algorithms extract the information from social media websites and web browsing history to analyze out what is causing a murmur, and ad-buying data is analyzed to find out what marketing departments will be focusing on.

3. **Demand Forecasting:** Once trends have been identified, through Machine Learning tool, future demand to be raised by customers for the products and the services can be predicted precisely. Demographic detail of customer reflects lot of information. Country's economic condition and the related parameters also help to understand the scenario of the market and accordingly gives a good idea about future demand. Market Basket Analysis helps retailers to identify the customer preferences very clearly. Proper future assessment helps retailer to assess the requirements well in advance and accordingly enables them to tap the opportunity of high sales in a specific season or region.
4. **Price Optimization:** Trend and demand helps to understand clearly the market scenario and customer response. Accordingly, to tap the potential customers and to retain the existing ones, retailer needs to provide products and services in the right combination of price and quality keeping in mind several factors such as season, region, competitor's offerings, economic indicators. Pricing strategies plays a crucial role in the survival of the business. Big data analytics helps business to understand the price scenario and determine the price fluctuation scenario for their own company. It can give clear indications for the right time for price hike or price drop.
5. **Market Identification:** Big data analytics helps tremendously in finding out which customers will be buying which type of product in a specified time period in a specified region. This helps the retailer to be ready with the required stocks, as demand has been assessed accurately. It also enables retailers to be ready with the delivery channel and all the logistics in place so that timely order fulfillment can be achieved.
6. **Market Segmentation:** Big data enable the retail industry to segment the market. They narrow down their segments as "young minds,' "digital people," Senior citizens," etc. Then they again sub categorize as sport lovers, movie viewers, etc. Based on these categories and

sub-categories, the success or the failure rate depends (https://indata-labs.com/blog/impact-of-big-data-on-business). In the traditional retail business model, marketing was mainly a one-way communication. Instant Customer communication or feedback was not prevalent. But, with the advent of big data and application of various algorithms, market segmentation has become more easy and effective. One-way communication have been transformed into a bi-directional communication. Applications of this concept of big data can remove certain segmentation barriers and help the retail industry on a micro level with personalized, customized messages to which the potential customers can relate and identify.

7. **Payment Patterns:** It has been observed that even though many online visitors visits the retailer website and add product to their cart, some do not complete the final payment process due to various reasons. Data analytics has reflected that the hesitation in making online payment, not being convinced about the safety measures on the retailer website, required card is not available with the customer at that moment, inability to complete the online transaction within valid session time frame and few more such reasons prohibits the ultimate conversion of the visitor into customer (https://www.simplilearn.com/big-data-transforming-retail-industry-article). Thus, multiple payment mode facility must be deployed on the retailer website for having high conversion rate of visitors into customers.
8. **Informative Dashboards:** Highly informative and attractive dashboards on retailer website puts a very good impression on the minds of customers. The matrices about the product performance, market penetration, customer feedback, can be accurately calculated and put up on the dashboard with the help of analytics tools.
9. **Making Strategic Decisions:** Companies are highly benefitted by the enormous amount of information extracted, gathered, stored, and analyzed. The analysis helps them to take a strategic marketing decision in all their activities in future.
10. **Sensitivity Analysis:** It is a mathematical model which marketers often use to find out the uncertainty in the market. By applying sophisticated machine learning based algorithms, companies can quantify the uncertainties prevailing in the market and figure out the optimal parameter settings required in their model. This data can be used to accurately predict what the top selling products or brands in a category would be.

8.5 ANALYSIS AND DISCUSSIONS

Modern retail business can be highly benefitted with the utilities big data can bring into. Retail business has always been working with lot of data to understand customer scenario and accordingly designing the strategies to tap the market. However, big data analytics helps them to get hold of unstructured data also so that better understanding of customers can be gained. This ultimately helps in managerial as well as operational efficiency in retail sector.

As compared to the traditional retail model, big data help companies aim at offering improved customer services, better product, and pricing services, which thereby can help to quantify their profit. As retail sector focuses primarily on customers, big data enables the companies to retain their old customers and extend their customer base too. Other facts include better target marketing, cost reduction, and improved efficiency of existing processes which leads to better performance and helps them to sustain in this competitive market.

Thus, Big data help business organizations to analyze information and improve on taking better and effecting strategic decisions. Moreover, the retail sector collects a large amount of data through RFID, POS scanners (https://indatalabs.com/blog/impact-of-big-data-on-business), various customer loyalty programs, etc. Application of big data enables in detecting and reducing frauds and enables the timely analysis of inventory. Earlier, big data was mainly implemented by large business houses that could afford the technologies but, recently, both large and small business enterprises are highly relying on this game changing concept of big data for intelligent business purposes. Hence, the demand for big data proliferates in the market domain and specifically in the retail segment. Enterprises from across different industries consider ways of how big data can be used in business. Therefore, the benefit of big data are to improve productivity, identify customer needs and demands, offer a competitive advantage, and chalks out ways for sustainable economic development. Even business owners are increasingly investing in big data solutions to optimize their operations and manage data traffic. Vendors are adopting big data solutions for better procurement methods, hassle-free supply chain management, etc., to work on data efficiently and effectively. The integration of the different business processes has enabled the strategic managers to take decisions accurately and timely. The benefits and its impact is so huge in the retail industry, that companies who had not implemented this concept are thinking of shifting to this technological advancement (Figures 8.3 and 8.4).

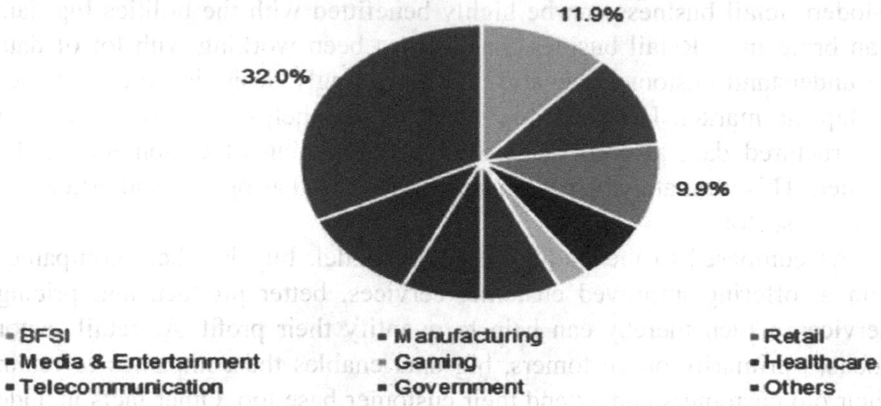

FIGURE 8.3 Big data scenario (in %) in 2018.
Source: https://indatalabs.com/blog/impact-of-big-data-on-business.

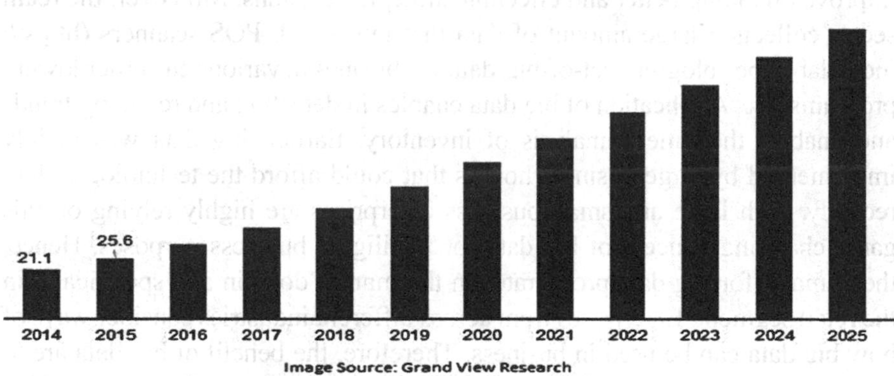

FIGURE 8.4 Rising trend of big data market by 2025.
Source: https://indatalabs.com/blog/impact-of-big-data-on-business.

8.6 CONCLUSION AND THE FUTURE SCOPE

From Figure 8.4, it is seen that the big data market is expected to witness remarkable growth by the year 2025. An important reason is an exponential growth in the volume of both structured and unstructured data. Apart from other factors increased technological penetration in the market and the wide usage of smart phones has led to the generation of larger amounts of data. The escalating need for analyzing data will lead to the rise of demand for big data over the forecast period. Furthermore, the number of online businesses in the industry is also growing, owing to enhanced profit margins. Other industries, such as healthcare, utilities, oil, and gas, logistics, manufacturing, and banking, will widely use online platforms to provide customized d services to customers. These have led to the rise of global big data market growth. Even, socio-economic benefits are also associated with big data. Therefore, several government agencies have laid down policies for promoting the development of big data. But, usage of big data in several industries, such as healthcare, oil, and gas, and so on, has been growing in a snail's speed. Adopting new technology, implementing certain sophisticated algorithms based on data analytics, AI, and quantum computing has been a challenging task. Moreover, it is expensive to adopt and integrating different operational units is cumbersome. Furthermore, there should be strong and efficient security measures to combat the data breaches if it occurs.

KEYWORDS

- **big data**
- **machine learning tool**
- **retail industry**
- **strategic decision-making**

REFERENCES

Ajah, I. A., & Henry, F. N., (2019). Big data and business analytics: Trends, platforms, success factors and applications. *Big Data and Cognitive Computing, 3*(2), 32.

Alam, J. R., Asma, S., Ramzan, T., & Muneeb, N., (2014). A review on the role of big data in business. *International Journal of Computer Science and Mobile Computing, 3*(4), 446–453.

Alsghaier, H., Mohammed, A., Issa, S., & Samah, A., (2017). The impact of big data analytics on business competitiveness. *Proceedings of the New Trends in Information Technology (NTIT)*.

Alsghaier, H., Mohammed, A., Issa, S., & Samah, A., (2017). The importance of Big Data Analytics in business: A Case study. *American Journal of Software Engineering and Applications, 6*(4), 111–115.

Banica, L., & Alina, H., (2015). Big data in business environment. *Scientific Bulletin-Economic Sciences, 14*(1), 79–86.

Berntzen, L., & Krumova, M., (2017). *Big Data from a Business Perspective, Conference Paper.* doi: 10.1007/978-3-319-65930-5_10.

Furht, B., & Flavio, V., (2016). Introduction to big data. In: *Big Data Technologies and Applications* (pp. 3–11). Springer, Cham.

Gao, J., Chunli, X., & Chuanqi, T., (2016). Big data validation and quality assurance--issuses, challenges, and needs. In: *2016 IEEE Symposium on Service-Oriented System Engineering (SOSE)* (pp. 433–441). IEEE.

https://indatalabs.com/blog/impact-of-big-data-on-business (accessed on 08 December 2021).

https://www.digimarc.com/resources/retail-tech-spendingreport?utm_source=google&utm_campaign=planet%20retail&utm_medium=ppc&gclid=EAIaIQobChMIosXP88Pz5wIVSA4rCh3TxAP-EAAYASAAEgLsRPD_BwE (accessed on 08 December 2021).

https://www.forbes.com/sites/bernardmarr/2015/11/10/big-data-a-game-changer-in-the-retail-sector/#356bf9099f37 (accessed on 08 December 2021).

https://www.simplilearn.com/big-data-transforming-retail-industry-article (accessed on 08 December 2021).

Kubina, M., Michal, V., & Irena, K., (2015). Use of big data for competitive advantage of company. *Procedia Economics and Finance, 26*, 561–565.

Pugna, I. B., Adriana, D., & Oana, G. S., (2019). Corporate attitudes towards Big Data and its impact on performance management: A qualitative study. *Sustainability, 11*(3), 684.

PART III

Algorithm for Load Balancing in Cloud Computing

CHAPTER 9

LOAD-BALANCING IMPLEMENTATION WITH AN ALGORITHM OF MATRIX BASED ON THE ENTIRE JOBS

PAYEL RAY, ENAKSHMI NANDI, RANJAN KUMAR MONDAL, and DEBABRATA SARDDAR

Department of Computer Science and Engineering, University of Kalyani, West Bengal, India, E-mails: payelray009@gmail.com (P. Ray), nandienakshmi@gmail.com (E. Nandi), ranjangcett@gmail.com (R. K. Mondal), dsarddar1@gmail.com (D. Sarddar)

ABSTRACT

Cloud computing is the on-demand delivery of IT resources over the internet with pay-as-you-go pricing. Instead of purchasing, retaining, and managing personal data centers and servers, you can access technology services, such as computing capability, storage, and databases, on as-necessary support from a cloud provider. Cloud computing is internet-based computing. There are a more significant number of servers associated with the system to provide many kinds of network services to give cloud clients. Deficient quantities of servers related to the cloud have to produce extra per task at a time. So, it is a complex issue to perform all tasks at a peculiar time. Some systems carry out all tasks, so there is a need to readjust all amount of loads at a time. Load balancing adjusts the closure time and presents all tasks, respectively. There is not available always to remain a corresponding amount of servers to carry out the same tasks. Tasks to be dealt with in cloud computing would be higher than the linked servers. Inadequate servers have to go on a million volumes of tasks at a time. We propose algorithms where some machines

Advances in Data Science and Computing Technology: Methodology and Applications. Suman Ghosal, Amitava Choudhury, Vikram Kumar Saxena, Arindam Biswas, & Prasenjit Chatterjee (Eds.)

perform the jobs here. The jobs are higher than the machines and balance all machines to exploit the quality of services in cloud computing.

9.1 INTRODUCTION

Cloud system is a just now progressing procedure to give online sources, storage, and enable consumers to manage applications with improved scalability, ease of use, and fault tolerance (Chan, Jiang, and Huang, 2012). The Cloud system is about storing the substance on isolated workstations instead of on individual physical systems or new machines (Jeevitha, Chandrasekar, and Karthik, 2015). This instruction can turn back accepting the network on any machine, quite where in the world, if that machine can support cloud system systems (Chen et al., 2013). The cloud system contains a front-end, that is the user side, and a back-end that is an association of the workstations and physical machines held by an agent keeping the data (Krakowiak, 2009). A central workstation that is a splinter of the back-end following protocols and uses middleware to be in contact between networked physical systems (Dinh, Lee, Niyato, and Wang, 2013). The Cloud system collects all the computing resources and deals with them mechanically (Pareek, 2013). Its features characterize a cloud system: on-need self-service, pooling of sources, perfect entry to the network, the flexibility of service convenience, and efficiency of services utilized by individual clients (Mansouri and Buyya, 2016; Sibiya, Venter, and Fogwill, 2015). Cloud system has hugely been with devices like Google Drive replacing Microsoft Office, Amazon Web Services re-establish traditional business data storage, websites replacing branch agencies, and Dropbox storing all our data and files (Kabakus and Kara, 2017).

9.2 CLOUD COMPUTING FEATURES

Cloud computing features are as following (Kaur and Luthra, 2012):

1. **Service Provides on-Demand:** When any applicant demands services and resources, thus the cloud deliver services on demand.
2. **Rapid Flexibility:** The number of resources in the cloud can be raised and deteriorate smoothly.
3. **Resource Pooling:** Resources are designated at different positions corresponding to the buyer's demands.

4. **Pay per Use:** Conforming to the consumer's application of computing resources is asked to be paid off.

9.3 LOAD BALANCING

In the cloud computing background, load-balancing is a method to share workloads amongst their multiple machines to avoid overload problems (Ray and Sarddar, 2018). It assists in controlling the resources and services (Chaudhari and Kapadia, 2013). It makes better utilization of resources and assists in improving the act of the system. Load balancing is the technique to determine the overloaded machine (cloud workstations) and help transfer the extra load to other machines. The following Figure 9.2 step is as:

1. First, a client sends a demand for service or resource;
2. Then though, web demand goes to load balancing (CLB);
3. From CLB, the demand goes to the Database workstation through a cloud workstation. After that Database workstation full field, the demand in a proper manner;
4. In this way, incoming, and outgoing demand and the reply will continue to carry out with any failure. It divides load balancing algorithms into two groups: static load balancing and dynamic load balancing algorithm:
   i. **Static Load Balancing Algorithms:** The loads do not depend on the system's present status, but it needs facts concerning the nodes' uniqueness. This type of algorithm partitions the traffic uniform among the workstations. By this approach, the traffic on the workstations will be handling simple, and consequently, it will make the situation better.
   ii. **Dynamic Load Balancing Algorithms:** Dynamic algorithms are more flexible than the static algorithms, and they do not rely on prior learning but depends on the present status of the system. A dynamic algorithms algorithm search throughout the whole network and selects the appropriate weights on the workstation, and it refers to the lightest workstation to balance the traffic. However, selecting a suitable workstation requires secure communication within networks, leading to better traffic attached to the system.

9.4 RELATED WORKS

Cloud computing presents a variety of utilities to the end-user, such as media sharing, online software, gaming, and online storage. In a cloud environment, all nodes represent all tasks or subtasks (Ritchie and Levine, 2005). The opportunistic load balancing algorithm (OLB) plans to stay all nodes active despite the current workloads of all nodes (Braun et al., 2001). The OLB algorithm allocates tasks to produce nodes randomly. The minimum completion time (MCT) algorithm allows all tasks to the nodes, considering the scheduled MCT of this task over other nodes (Wang et al., 2010). The Min-Min scheduling algorithm (MM) considers the same scheduling approach as the MCT algorithm to allot each task, the node to end this task with MCT over other nodes (Che-Lun, Hsiao-His, and Yu-Chen, 2012). The load balance Min-Min (LBMM) algorithm (Kokilavani and George, 2011) enforces the MM scheduling technique and load balancing program. It can escape the unnecessarily duplicated assignment (Min-You, Wei, and Hong, 2000). Load Balancing with Job Switching (Ray and Sarddar, 2018) minimize loads of oppressive loaded system to under loaded machine by replacing a specific task way. Load Balancing of an Unbalanced Matrix with Hungarian method (Ranjan et al., 2017) utilizes the Hungarian algorithm, where tasks are more extensive than all systems. The load balancing of the unbalanced cost matrix (Ranjan et al., 2017) is the same as the other algorithm.

9.5 PROPOSED METHOD

To determine the matrix cost as well as a combination of tasks(s) vs machine(s) of an unbalanced matrix, to concentrate on a problem consisting of 'm' machines M={M1, M2, …, Mm}. A set of 'n' jobs J={J1, J2, …, Jn} is considered to be assigned for execution on the 'm' available machines and the execution cost Cij, where i = 1, 2, …, m, and j = 1, 2, …, n are mentioned in the cost matrix where m>n. First of all, we obtain the figure for each row and each column of the matrix; keep the results in the array, i.e., Row-sum, and Column-sum. Then we select the first n rows by Row-sum, i.e., starting with most minimums to next minimum to the array Row-sum and deleting rows corresponding to the left behind (r) jobs. Accumulate the results in the new array that should be the array for the first sub-problem. Repeat this process until remaining jobs become less than a machine when remaining jobs are less than n, then, deleting (c) columns by Column-sum, i.e., the corresponding value(s) maximum to the next maximum to form the last sub-problem. Save the results in the new array that shall be the array for the last sub-problem.

9.5.1 HUNGARIAN METHOD

Steps of Hungarian method (Kuhn, 1955) as follows:

- **Step 1:** The input of this algorithm is an n×n square.
- **Step 2:** Find the smallest element from each row and subtract it from each element in the corresponding row.
- **Step 3:** Similarly, for each column, find the smallest element and subtract it from each element in the corresponding column.
- **Step 4:** Cover all zeros in the subtracted matrix with a minimum number of horizontal and vertical lines. If lines numbers are n, then an optimal assignment exists. The algorithm stops. Otherwise, if lines numbers are less than n then, go to the next step.
- **Step 5:** Get the smallest element that is not covered by a line in Step 4, deduct with the lowest uncovered element from each uncovered element, and include the lowest uncovered element to each element covered two times.

9.6 PROPOSED ALGORITHM

To present an algorithmic representation of the method, Consider a problem that consists of a set of 'm' Machines $M = \{M_1, M_2, \ldots, M_m\}$. A set of 'n' jobs $J = \{J_1, J_2, \ldots, J_n\}$ is considered to be assigned for execution on 'm' available machines and the execution cost C_{ij}, where $i = 1, 2, \ldots, m$, and $j = 1, 2, \ldots, n$, where m>n, i.e., the number of jobs is more than a number of machines.

- **Step 1:** m×n matrix.
- **Step 2:** Add all elements of each column of the matrix, namely, Column-sum in each column, respectively.
- **Step 3:** Select the first k column by Column-sum, i.e., starting with most minimums to next minimum to the array Column-sum until some rows and number of columns are equal. Make a new balanced problem.
- **Step 4:** Concern Hungarian method to solve the problem.
- **Step 5:** List the element costs of identical machines.
- **Step 6:** Stop.

9.7 ILLUSTRATION OF AN EXAMPLE

Consider a problem in which a set of 5 machines M={M_1, M_2, M_3, M_4, M_5}, and a set of 8 jobs J={J_1, J_2, J_3, J_4, J_5, J_6, J_7, J_8}. The assignment matrix contains the execution costs of every job to each machine.

➢ **Steps 1 to 2:** Input: 5×8 matrix.

J_n/M_m	M_1	M_2	M_3	M_4	M_5
J_1	151	277	185	276	321
J_2	245	286	256	264	402
J_3	246	245	412	423	257
J_4	269	175	145	125	156
J_5	421	178	185	425	235
J_6	257	257	125	325	362
J_7	159	268	412	256	286
J_8	365	286	236	314	279

➢ **Step 3:** To obtain the sum of a column of the matrix, i.e., the sum of each column is as follows:

J_n/M_m	M_1	M_2	M_3	M_4	M_5
J_1	151	277	185	276	321
J_2	245	286	256	264	402
J_3	246	245	412	423	257
J_4	269	175	145	125	156
J_5	421	178	185	425	235
J_6	257	257	125	325	362
J_7	159	268	412	256	286
J_8	365	286	236	314	279
Column-Sum	**2,113**	**1,972**	**1,956**	**2,408**	**2,298**

➢ **Step 4:** To equal rows and columns, we add some columns as dummy based on minimum column-sum until a number of rows and number of columns will be equal.

J_n/M_m	M_1	M_2	M_3	M_4	M_5
J_1	151	277	185	276	321
J_2	245	286	256	264	402
J_3	246	245	412	423	257
J_4	269	175	145	125	156
J_5	421	178	185	425	235
J_6	257	257	125	325	362
J_7	159	268	412	256	286
J_8	365	286	236	314	279
Column-Sum	**2,113**	**1,972**	**1,956**	**2,408**	**2,298**

- **Step 5:** This corresponds to the following optimal cost in the original cost matrix using Hungarian method.

J_n/M_m	M_1	M_2	M_3	M_4	M_5
J_1	151	277	185	276	321
J_2	245	286	256	264	402
J_3	246	245	412	423	257
J_4	269	175	145	125	156
J_5	421	178	185	425	235
J_6	257	257	125	325	362
J_7	159	268	412	256	286
J_8	365	286	236	314	279

- **Step 6:** Assign all jobs to corresponding machines.

J_n/M_m	M_1	M_2	M_3	M_4	M_5
Jn	310	423	361	264	156

Final Result

$M_1 \rightarrow J1*J7 \rightarrow 151 + 159 = 310$

$M_2 \rightarrow J3*J5 \rightarrow 245 + 178 = 423$

$M_3 \rightarrow J6*J8 \rightarrow 125 + 236 = 361$

$M_4 \rightarrow J2 \rightarrow 264$

$M_5 \rightarrow J4 \rightarrow 156$

9.8 RESULT ANALYSIS

Figure 9.1 shows our proposed work result. Figure 9.1 explains the execution time for each task at each node. To analyze the performance of our approach is judged by other methods shown in Figure 9.1. Figure 9.1 displays the comparison of the execution time of each computing node among our approach. The total completion time for completing each task by using the proposed work, LBMM, and MM, are 17, 27, and 42 ms, respectively. Our approach attains the least completion time with enhanced load balancing than available other algorithms.

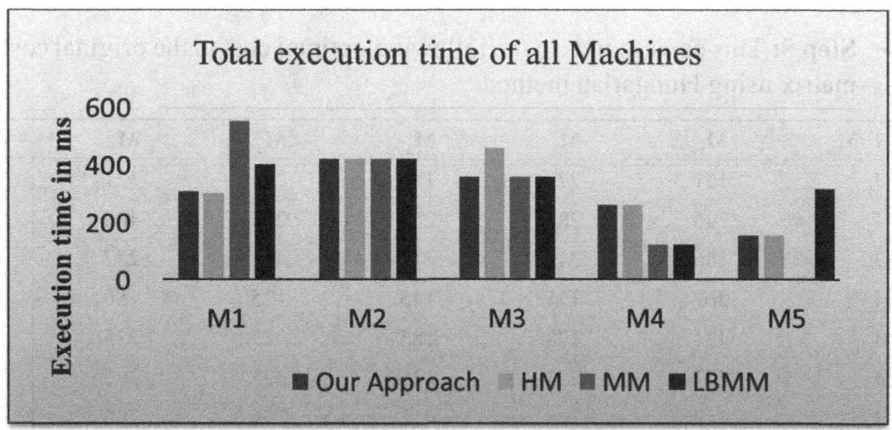

FIGURE 9.1 Execution time (ms) of each task at each computing node.

9.9 CONCLUSION

The result of this problem, the matrix, can be solved with the support of the Hungarian method. It gives each task or subtask with minimal completion time to its corresponding machine so that our result is excellent for the system. Though some machines execute more than one task or subtask completion time is excellent, and loads are being balanced with our proposed work. Without some alternative consideration, our approach has continuously been satisfied in various cases and prospects; we figure out a further way of dealing with our difficulties sometimes.

KEYWORDS

- **load balancing**
- **minimum completion time**
- **unbalanced matrix**

REFERENCES

Braun, T. D., Siegel, H. J., Beck, N., Bölöni, L. L., Maheswaran, M., Reuther, A. I., Robertson, J. P., et al., (2001). A comparison of eleven static heuristics for mapping a class of independent tasks onto heterogeneous distributed computing systems. *Journal of Parallel and Distributed Computing, 61,* 810–837.

Chan, M. C., Jiang, J. R., & Huang, S. T., (2012). Fault-tolerant and secure networked storage. In: *Digital Information Management (ICDIM), 2012 Seventh International Conference on* (pp. 186–191). IEEE.

Chaudhari, A., & Kapadia, A., (2013). Load balancing algorithm for azure virtualization with specialized VM. *Algorithms, 1,* 2.

Che-Lun, H., Hsiao-His, W., & Yu-Chen, H., (2012). Efficient load balancing algorithm for the cloud computing network. In: *International Conference on Information Science and Technology (IST 2012)* (pp. 28–30).

Chen, C. H., Lin, H. F., Chang, H. C., Ho, P. H., & Lo, C. C., (2013). An analytical framework of a deployment strategy for cloud computing services: A case study of academic websites. *Mathematical Problems in Engineering, 2013.*

Dinh, H. T., Lee, C., Niyato, D., & Wang, P., (2013). A survey of mobile cloud computing: Architecture, applications, and approaches. *Wireless Communications and Mobile Computing, 13*(18), 1587–1611.

Jeevitha, M., Chandrasekar, A., & Karthik, S., (2015). Survey on verification of storage correctness in cloud computing. *International Journal of Engineering and Computer Science, 4*(09).

Kabakus, A. T., & Kara, R., (2017). A performance evaluation of drop box in the light of personal cloud storage systems. *International Journal of Computer Applications, 163*(5).

Kaur, R., & Luthra, P., (2012). Load balancing in cloud computing. In: *Proceedings of International Conference on Recent Trends in Information, Telecommunication and Computing, ITC.*

Kokilavani, T., & George, A. D. I., (2011). Load balanced min-min algorithm for static meta-task scheduling in grid computing. *International Journal of Computer Applications, 20*(2), 43–49.

Krakowiak, S., (2009). Distributed under a Creative Commons license. *Middleware Architecture with Patterns and Frameworks.* http://creativecommons.org/licenses/by-nc-nd/3.0/ (accessed on 10 January 2022).

Kuhn, H. W., (1955). The Hungarian method for the assignment problem. *Naval Research Logistics Quarterly, 2*(1, 2), 83–97.

Mansouri, Y., & Buyya, R., (2016). To move or not to move: Cost optimization in dual cloud-based storage architecture. *Journal of Network and Computer Applications, 75*, 223–235.

Min-You, W., Wei, S., & Hong, Z., (2000). Segmented min-min: A static mapping algorithm for meta-tasks on heterogeneous computing systems. In: *Heterogeneous Computing Workshop, 2000, (HCW 2000) Proceedings, 9*[th] (pp. 375–385). IEEE.

Pareek, P., (2013). Cloud computing security from single to multi clouds using secret sharing algorithm. *International Journal of Advanced Research in Computer Engineering & Technology, 2*(12), 3261–3264.

Ranjan, K. M., Payel, R., Enakshmi, N., Biswajit, B., Manas, K. S., & Debabrata, S., (2017). Load balancing of unbalanced matrix with Hungarian method. In: *International Conference on Computational Intelligence, Communications, and Business Analytics* (pp. 256–270). Springer, Singapore.

Ranjan, K. M., Payel, R., Enakshmi, N., Priyajit, S., & Debabrata, S., (2017). Load balancing of the unbalanced cost matrix in a cloud computing network. In: *Computer, Communication and Electrical Technology: Proceedings of the International Conference on Advancement of Computer Communication and Electrical Technology (ACCET 2016)* (p. 81). West Bengal, India. CRC Press.

Ray, P., & Sarddar, D., (2018). Load balancing with inadequate machines in cloud computing networks. In: Mandal, J., & Sinha, D., (eds.), *Social Transformation – Digital Way. CSI 2018, Communications in Computer and Information Science* (Vol. 836). Springer, Singapore.

Ritchie, G., & Levine, J. A., (2005). Fast, effective local search for scheduling independent jobs in heterogeneous computing environments. *Journal of Computer Applications, 25*, 1190–1192.

Sibiya, G., Venter, H. S., & Fogwill, T., (2015). Digital forensics in the Cloud: The state of the art. In: *IST-Africa Conference, 2015* (pp. 1–9). IEEE.

Wang, S. C., Yan, K. Q., Liao, W. P., & Wang, S. S., (2010). Towards a Load Balancing in a three-level cloud computing network. In: *CSIT* (pp. 108–113).

CHAPTER 10

LOAD-BALANCING ALGORITHM WITH OVERALL TASKS

ENAKSHMI NANDI, PAYEL RAY, RANJAN KUMAR MONDAL, and DEBABRATA SARDDAR

Department of Computer Science and Engineering, University of Kalyani, West Bengal, India, E-mail: nandienakshmi@gmail.com (E. Nandi)

ABSTRACT

The innovative system cloud is a model that is based on virtualization. It is a new name of existing technology. Cloud computing mainly serves computing resources as a service to cloud clients. This new system is used to store and access hard drives. This new computing paradigm provides marvelous opportunities to solve the large-scale scientific issues. To utilize the applications of cloud in different field, several challenges have to face, in that case task scheduling is one of the important factor. In a cloud system, optimum utilization of computing resources is always challenging. A cloud service giver ensures to serve computing resources efficiently to cloud client at optimum cost. Load balancing techniques play a vital role in the competent deployment of computing entities. Different algorithm related to load balancing migrates an overloaded virtual machine task to an under the loaded machine, with disturbing the existing background. Various load balancing techniques are suggested by cloud researchers. Hence, a resourceful load balancing process has required. This research chapter depicts an assessment of the various loads balancing method in the cloud system. Also, a comparative analysis has been presented based on various performance measuring parameters. Cloud

Advances in Data Science and Computing Technology: Methodology and Applications. Suman Ghosal, Amitava Choudhury, Vikram Kumar Saxena, Arindam Biswas, & Prasenjit Chatterjee (Eds.)

computing is the convenient aptitude application on the Internetworking. Cloud service providers deal with applications and resources. It faces different challenges. Load balancing related problem is great issue. Load balancing is nothing but proper distribution of the tasks among different nodes for the best utilization of resources. The major target of a variety of type load balancing algorithms is to decrease the waiting time and turnaround time. Within this chapter, we are proposing a new scheduling technique in a distributed system is choosing suitable entity with their total task. It is a very easy way to decide on a suitable node. This idea would provide the effective usage of computing entities and preserve cloud load balancing.

10.1 INTRODUCTION

Cloud computing is an example that interiors a collection of clients with admittance to scalable, virtual nature of resources over the network (Chan, Jiang, and Huang, 2012). Cloud computing is a form to extract and run IT resources (Jeevitha, Chandrasekar, and Karthik, 2015). In data centers, applications are run in physical workstations that are often provisioned with a high workload (Chen et al., 2013). This configuration makes data centers costly to maintain with significant overhead (Krakowiak, 2009). Data centers are more flexible, secure, and give excellent support for on-demand resource allocations (Dinh, Lee, Niyato, and Wang, 2013). It abstracts workstation diversity, performs workstation consolidation, and enhances workstation utilization (Pareek, 2013). A host is able to run multiple virtual machines with various resource specifications and varied workloads (Mansouri and Buyya, 2016). Physical workstations hosting in homogeneous virtual machines with unpredictable workloads may cause unbalanced resource usage (Sibiya, Venter, and Fogwill, 2015).

Cloud data centers are extremely unpredictable due to (Kabakus and Kara, 2017; Youseff, Butrico, and Da Silva, 2008; Jain and Paul, 2013):

- Irregular buyer demand pattern;
- Wavering resource usage by virtual machines;
- Various rates of arrivals and departure of customers; and
- The performance of the host may differ when handling various load levels (Kumar and Lu, 2010). These reasons are enough to trigger unbalanced loads in the data center, which needs a load balancing mechanism to keep away from presentation degradation and service level agreement destructions (Wang et al., 2010).

10.2 CLOUD COMPUTING AND LOAD BALANCING

10.2.1 CLOUD SYSTEM

Cloud computing is an on-demand set of connections admittance to a team of organize able entities that is able to be hastily prerequisite and unconfined with nominal management endeavor (Moura and Hutchison, 2016). The Cloud technology is based on virtualization concept which is the prime aspect of this system. Virtual machine is a software completion of the physical entity (Zhang et al., 2011). A hypervisor, which is called a virtual machine monitor, is accountable for distribution a single physical occurrence of cloud entities among diverse occupants (Sabahi, 2011).

The main aim of this computing is to progress the presentation of the system and to get optimal resource exploitation, greatest throughput, highest response time, and evading overload (Mishra and Mishra, 2015).

10.2.2 CLASSIFICATION OF LOAD BALANCING TECHNIQUE

Load balancing technique have been divided based on the present state of the system and who instigated the process (Gupta and Deshpande, 2014).

1. **Depending on which Client Initiates the Method:**
   i. **Sender-Initiated:** Sender or client begins the completion of the load balancing technique helps to identify necessitate for load balancing (Eager, Lazowska, and Zahorjan, 1985).
   ii. **Receiver-Initiated:** Workstation initiates the completion of the load balancing process to detect the requirement for load balancing (Adhikari and Patil, 2012).
   iii. **Symmetric:** This category of the algorithm is a blend of the sender-beginner category and receiver-start up category methods (Eager, Lazowska, and Zahorjan, 1985).
2. **Depending on the Existing Static System Algorithm:** In the static algorithm, there is a uniform distribution of traffic among the workstations (Vakkalanka, 2012). This algorithm needs a prior understanding of system entities such as the judgment of shifting the load does not rely on the running state of the system (Mani, Suresh, and Kim, 2005). The static method is perfect in the system, which has fewer inequalities in load (Kumar et al., 2001).
3. **Dynamic Algorithm:** In case of dynamic method for balancing heavy load, the lightest workstation in the whole system is looked

upon and preferred (Patil and Shedge, 2013). For statement with the set of connections is essential, which is able to enlarge the traffic in the system (Akyildiz et al., 1999).

10.3 LOAD BALANCING

Load Balancing specifically signifies to transfer loads from overloaded process to other under-loaded processes (Ansari and Sachin, 2015). The main difficulty is how to choose the best node to execute a scrupulous job and transfer a load to a suitable node (Bittencourt, Miyazawa, and Vignatti, 2012). Based on the implementation method, load balancing algorithms can be divided in two ways (Willebeek-LeMair and Reeves, 1993):

- Static algorithms; and
- Dynamic algorithms.

Obtainable load balancing methods in cloud computing regard as a range of parameters like (Rastogi, Bansal, and Hasteer, 2013; Bhargava and Goyal, 2013; Sran and Navdeep, n.d.):

1. **Throughput:** To estimate the number of jobs, those executions have been accomplished.
2. **Fault Tolerance:** It signifies capability of an algorithm to achieve consistent load balancing despite of random node or link letdown.
3. **Migration Time:** It is the time to migrate the task or entities from one node to another node.
4. **Response Time:** The quantity of time required to reply to a meticulous load balancing method in a scattered system.
5. **Resource Utilization:** It is the method to ensure the deployment of resources.
6. **Scalability:** It is the metrics measures capability of an algorithm to achieve load balancing for a scheme with some fixed number of nodes.
7. **Performance:** It is the process to test out the competence of the system.
8. **Energy Consumption:** It is the process to help in evades the overheating condition by balancing the workload transversely every nodes.

10.4 RELATED WORKS

Load balancing is a process which is utilized for distributing processing loads to slighter dealing out nodes for developing the entire presentation of a scheme

(Begum and Prashanth, 2013). In a distributed system situation, it is the development of distributing load among a range of other nodes of an arrangement to acquire superior entity operation and task response time (Alakeel, 2010). An ultimate load balancing technique should keep away from overfilling or under loading of any definite node (Moharana, Ramesh, and Powar, 2013). But, in the case of a cloud atmosphere, the assortment of a load balancing is not purely method due to its engagement of additional restriction like throughput, security, dependability, etc. (Mangla and Goyal, 2017). Hence, goal of an algorithm in cloud computing is to obtain better the response time of the task by distributing a largely load of the method. The algorithm must also make certain that it is not overloading any explicit node (Haryani and Jagli, 2014).

Load balancers can work in two ways: one is cooperative and non-cooperative. In a cooperative way, all nodes work concurrently to reach the familiar goal of optimizing the on the whole response time. In the non-cooperative way, jobs run separately to recover the response time of local jobs (Moharana, Ramesh, and Powar, 2013).

Cloud computing supplies a collection of duties to the customer, such as multi-media division, online office software, online storage, etc. (Shilpa, Kulkarni, and Kulkarni, 2014). The opportunistic load balancing algorithm (OLB) proposes to remain distinct node busy despite the existing workloads of every node (Saranya and Maheswari, 2015). OLB algorithm allotted schedules to reflect nodes in arbitrary order. The minimum completion time algorithm (MCT) disperses schedules to the nodes containing the anticipated MCT of this schedule over other nodes (Patel, Mehta, and Bhoi, 2015). The Min-Min scheduling algorithm (MM) presumes the identical scheduling come near as the MCT (Kaur and Singh, 2014) algorithm to consigning a schedule, the node to conclude this schedule with MCT over other nodes (Kaur and Patra, 2013). The load balance min-min (LBMM) scheduling method (Rahm, 1996) implement the MM scheduling technique and load balancing stratagem and this also evaded the unreasonably replicate project (Shidik, Azhari, and Mustofa, 2015).

10.5 PROPOSED WORK

Diversity of nodes exists in a cloud system. Each nodes cannot perform similar tasks; therefore, just reflect on the processing of the entities is insufficient at the time of an entity is preferred to accomplish a particular assignment. Thus, to pick a suitable entity to perform an assignment is awfully significant in cloud computing.

The different schedule has a diverse attributes to disburse implementation. Thus, its requirements a little of the entities of definite; for case in point, as soon as implementing individual sequence assemblage, it has to necessitate toward memory enduring. To obtain the best result in the completing of each schedule, so we will intend schedule property to imagine a dissimilar circumstance inconsistent in which it is according to an entity of schedule requirement to deposit resolution variable.

The process of our approach is presented as follows:

- **Step 1:** Respective process to calculate the total execution time of every schedule for all entities, correspondingly.
- **Step 2:** It is to acquire the schedule having the greatest execution time.
- **Step 3:** It is to locate the entity which is not dispersed and has the least execution time for the schedule preferred in Step 2. Then, respective schedule is stimulated to the elected entity for computation purpose.
- **Step 4:** If, in step 2, no unallocated entity can be selected, then that all entity, together with unallocated and allocated entities, should be reassessed. The smallest amount execution time of an allocated entity is the summation of the smallest amount completion time of the consign schedule on this entity and the smallest amount execution time of the existing schedule. The lowest amount execution time of an unallocated entity is the existing smallest amount execution time for the schedule. It is to determine the unallocated entity or allocated entity having the lowest amount execution time for the schedule selected in Step 2 afterward, this schedule is stimulated to the elected entity for computation.
- **Step 5:** Perform once more Step 2 to Step 4 up to all schedules have been accomplished absolutely.

10.6 ILLUSTRATION

In the subsequent part, an example to be computed by means of the projected algorithm is going behind.

Schedule Entity	N_{11}	N_{12}	N_{13}	N_{14}
T_1	13	14	11	15
T_2	17	25	14	26
T_3	27	32	13	34
T_4	18	25	19	32

- **Step 1:** It is to estimate the total execution time of distinct schedule for all entities, individually.

Schedule Entity	N_{11}	N_{12}	N_{13}	N_{14}	Total Schedule
T_1	13	14	11	15	53
T_2	17	25	14	26	82
T_3	27	32	13	34	106
T_4	18	25	19	32	94

- **Step 2:** Then locate the schedule that has the greatest execution time.

Schedule Entity	N_{11}	N_{12}	N_{13}	N_{14}	Total Schedule
T_1	13	14	11	15	53
T_2	17	25	14	26	82
T_3	27	32	13	34	106
T_4	18	25	19	32	94

- **Step 3:** Next to locate the unallocated entity which has the least execution time for the schedule preferred in Step 2. Then, this schedule is stimulated to the elected entity for computation.

Schedule Entity	N_{11}	N_{12}	N_{13}	N_{14}	Total Schedule
T_1	13	14	11	15	53
T_2	17	25	14	26	82
T_3	27	32	13	34	106
T_4	18	25	19	32	94

- **Step 5:** Perform over again Step 2 to Step 4 up to all schedules have been executed totally.

Schedule Entity	C_{11}	C_{12}	C_{13}	C_{14}	Total Schedule
T_1	13	14	11	15	53
T_2	17	25	14	26	82
T_3	27	32	13	34	106
T_4	18	25	19	32	94

Final Result

Schedule Entity	C_{11}	C_{12}	C_{13}	C_{14}
T_1	13	14	11	15
T_2	17	25	14	26
T_3	27	32	13	34
T_4	18	25	19	32

10.7 COMPARISON

Figure 10.1 illustrates the execution time for distinct schedule at several computing entities. To analyze the performance of our projected method is evaluated with a new method by the case shown in Figure 10.1. Figure 10.1 demonstrates the assessment of the execution time of distinct computing entity among our method. Respective completion times for ultimate all schedules by means of the proposed technique, LBMM, and MM, are 24, 33, and 35 ms, respectively. Our technique attains the MCT and enhanced load balancing than AK algorithms.

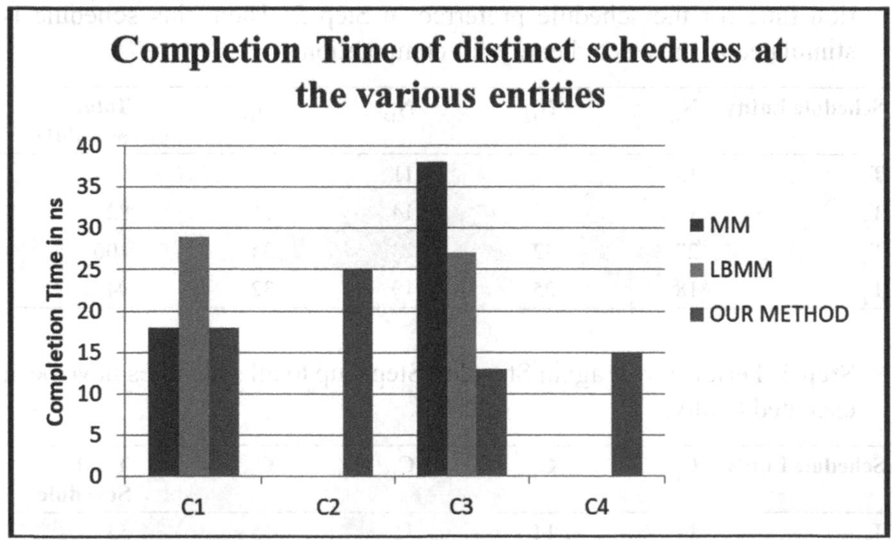

FIGURE 10.1 The assessment of the completion time of distinct schedule at the several entities.

10.8 CONCLUSION

In our chapter, we projected a competent scheduling technique, LBOT, for the cloud computing environment to dispense schedule to work out entities according to their respective resource competence correspondingly, our method can attain superior load balancing and performance than other existing method, such as LB3M, LBMM, and MM, from the case history.

KEYWORDS

- **cloud computing**
- **distributed system**
- **load balancing**

REFERENCES

Adhikari, J., & Patil, S., (2012). Load balancing the essential factor in cloud computing. *International Journal of Engineering Research & Technology (IJERT).*

Akyildiz, I. F., Janise, M., Loren, C. M, Ramon, P., & Yelena, Y., (1999). Medium access control protocols for multimedia traffic in wireless networks. *IEEE Network, 13*(4), 39–47.

Alakeel, A. M., (2010). A fuzzy dynamic load balancing algorithm for homogenous distributed systems. *International Journal of Computer Science and Network Security, 10*(6), 153–160.

Alnusairi, T. S., Shahin, A. A., & Daadaa, Y., (2018). *Binary PSOGSA for Load Balancing Task Scheduling in Cloud Environment.* arXiv preprint arXiv:1806.00329.

Ansari, M. B., & Sachin, P., (2015). Load balancing and issue monitoring of cloud nodes. *International Journal, 5*(7).

Begum, S., & Prashanth, C. S. R., (2013). Review of load balancing in cloud computing. *International Journal of Computer Science Issues (IJCSI), 10*(1), 343.

Bhargava, S., & Goyal, S., (2013). Dynamic load balancing in cloud using live migration of virtual machines. *Computing, 3*(7), 8.

Bittencourt, L. F., Miyazawa, F. K., & Vignatti, A. L., (2012). Distributed load balancing algorithms for heterogeneous players in asynchronous networks. *Journal of Universal Computer Science.*

Chan, M. C., Jiang, J. R., & Huang, S. T., (2012). Fault-tolerant and secure networked storage. In: *Digital Information Management (ICDIM), 2012 Seventh International Conference* (pp. 186–191). IEEE.

Chen, C. H., Lin, H. F., Chang, H. C., Ho, P. H., & Lo, C. C., (2013). An analytical framework of a deployment strategy for cloud computing services: A case study of academic websites. *Mathematical Problems in Engineering, 2013.*

Dinh, H. T., Lee, C., Niyato, D., & Wang, P., (2013). A survey of mobile cloud computing: Architecture, applications, and approaches. *Wireless Communications and Mobile Computing, 13*(18), 1587–1611.

Eager, D. L., Lazowska, E. D., & Zahorjan, J., (1985). A comparison of receiver-initiated and sender-initiated adaptive load sharing. *ACM SIGMETRICS Performance Evaluation Review, 13*(2), 1–3.

Gupta, E., & Deshpande, V., (2014). A technique based on ant colony optimization for load balancing in cloud data center. In: *Information Technology (ICIT), 2014 International Conference on* (pp. 12–17). IEEE.

Haryani, N., & Jagli, D., (2014). A dynamic method for load balancing in cloud computing. *IOSR Journal of Computer Engineering (IOSR-JCE), 16*(4), 23–28.

Jain, R., & Paul, S., (2013). *Network Virtualization and Software Defined Networking for Cloud Computing: A Survey* (Vol. 51, No. 11, pp. 24–31). IEEE Communications Magazine.

Jeevitha, M., Chandrasekar, A., & Karthik, S., (2015). Survey on verification of storage correctness in cloud computing. *International Journal of Engineering and Computer Science, 4*(09).

Kabakus, A. T., & Kara, R., (2017). A Performance evaluation of drop box in the light of personal cloud storage systems. *International Journal of Computer Applications, 163*(5).

Kaur, D., & Singh, S., (2014). An efficient job scheduling algorithm using min-min and ant colony concept for grid computing. *International Journal of Engineering and Computer Science, 3*(07).

Kaur, R., & Patra, P. K., (2013). Resource allocation with improved min-min algorithm. *International Journal of Computer Applications, 76*(15).

Krakowiak, S., (2009). Distributed under a Creative Commons license. *Middleware Architecture with Patterns and Frameworks*. http://creativecommons.org/licenses/by-nc-nd/3.0/ (accessed on 10 January 2022).

Kumar, K., & Lu, Y. H., (2010). Cloud computing for mobile users: Can offloading computation save energy?. *Computer, 43*(4), 51–56.

Kumar, R., Agrawal, G., Theobald, K., Zoppetti, G. M., & Gao, G. R., (2001). Compiling several classes of reductions on a multithreaded architecture. In: *Proceedings of Mid-Atlantic Student Workshop on Programming Languages and Systems.*

Mangla, P., & Goyal, S. K., (2017). Study of various heuristic approaches in cloud computing. *International Journal of Advanced Research in Computer Science, 8*(3).

Mani, V., Suresh, S., & Kim, H. J., (2005). Real-coded genetic algorithms for optimal static load balancing in distributed computing systems with communication delays. In: *International Conference on Computational Science and its Applications* (pp. 269–279). Springer, Berlin, Heidelberg.

Mansouri, Y., & Buyya, R., (2016). To move or not to move: Cost optimization in a dual cloud-based storage architecture. *Journal of Network and Computer Applications, 75*, 223–235.

Mishra, N. K., & Mishra, N., (2015). Load balancing techniques: Need, objectives, and major challenges in cloud computing-a systematic review. *International Journal of Computer Applications, 131*(18).

Moharana, S. S., Ramesh, R. D., & Powar, D., (2013). Analysis of load balancers in cloud computing. *International Journal of Computer Science and Engineering, 2*(2), 101–108.

Moharana, S. S., Ramesh, R. D., & Powar, D., (2013). Analysis of load balancers in cloud computing. *International Journal of Computer Science and Engineering, 2*(2), 101–108.

Moura, J., & Hutchison, D., (2016). Review and analysis of networking challenges in cloud computing. *Journal of Network and Computer Applications, 60,* 113–129.

Padhy, R. P., (2012). Virtualization techniques & technologies: State-of-the-art. *International Journal of Global Research in Computer Science (UGC Approved Journal), 2*(12), 29–43.

Pareek, P., (2013). Cloud computing security from single to multi clouds using secret sharing algorithm. *International Journal of Advanced Research in Computer Engineering & Technology, 2*(12), 3261–3264.

Patel, G., Mehta, R., & Bhoi, U., (2015). Enhanced load balanced min-min algorithm for static meta task scheduling in cloud computing. *Procedia Computer Science, 57*, 545–553.

Patil, U., & Shedge, R., (2013). Improved hybrid dynamic load balancing algorithm for distributed environment. *International Journal of Scientific and Research Publications, 3*(3), 1.

Pearce, M., Zeadally, S., & Hunt, R., (2013). Virtualization: Issues, security, threats, and solutions. *ACM Computing Surveys (CSUR), 45*(2), 17.

Rahm, E., (1996). Dynamic load balancing in parallel database systems. In: *European Conference on Parallel Processing* (pp. 37–52), Springer, Berlin, Heidelberg.

Rastogi, D., Bansal, A., & Hasteer, N., (2013). Techniques of load balancing in cloud computing: A survey, In: *International Conference on Computer Science and Engineering (CSE)* (pp. 140–143).

Sabahi, F., (2011). Security of virtualization level in cloud computing. In: *Proc. 4th Intl. Conf. on Computer Science and Information Technology* (pp. 197–201). Chengdu, China.

Saranya, D., & Maheswari, L. S., (2015). Load balancing algorithms in cloud computing: A review. *International Journal of Advanced Research in Computer Science and Software Engineering Research Paper.*

Shidik, G. F., Azhari, M. K., & Mustofa, K., (2015). Evaluation of selection policy with various virtual machine instances in dynamic VM consolidation for energy efficient at cloud data centers. *JNW, 10*(7), 397–406.

Shilpa, S., Kulkarni, S., & Kulkarni, S., (2014). Analysis of load balancing algorithms in cloud computing and study of game theory [J]. *Analysis, 3*(4), 1818–1823.

Sibiya, G., Venter, H. S., & Fogwill, T., (2015). Digital forensics in the cloud: The state of the art. In: *IST-Africa Conference, 2015* (pp. 1–9). IEEE.

Sran, N., & Navdeep, K., (2013). Comparative Analysis of Existing Load balancing techniques in cloud computing. *International Journal of Engineering Science Invention, 2*(1), 60–63.

Vakkalanka, S., (2012). Classification of job scheduling algorithms for balancing load on web servers. *International Journal of Modern Engineering Research (IJMER), 2*(5), 3679–3683.

Wang, L., Gregor, V. L., Andre, Y., Xi, H., Marcel, K., Jie, T., & Cheng, F., (2010). Cloud computing: A perspective study. *New Generation Computing, 28*(2), 137–146.

Willebeek-LeMair, M. H., & Reeves, A. P., (1993). Strategies for dynamic load balancing on highly parallel computers. *IEEE Transactions on Parallel and Distributed Systems, 4*(9), 979–993.

Youseff, L., Butrico, M., & Da Silva, D., (2008). Toward a unified ontology of cloud computing. In: *Grid Computing Environments Workshop, 2008, GCE'08* (pp. 1–10). IEEE.

Zhang, F., Chen, J., Chen, H., & Zang, B., (2011). CloudVisor: Retrofitting protection of virtual machines in multi-tenant cloud with nested virtualization. In: *Proceedings of the Twenty-Third ACM Symposium on Operating Systems Principles* (pp. 203–216). ACM.

CHAPTER 11

IMPLEMENTATION OF LOAD BALANCING OF MATRIX PROBLEM WITH THE ROW SUM METHOD

PAYEL RAY, ENAKSHMI NANDI, RANJAN KUMAR MONDAL, and DEBABRATA SARDDAR

Computer Science and Engineering, University of Kalyani, West Bengal, India, E-mail: payelray009@gmail.com (P. Ray)

ABSTRACT

Cloud computing is turning into one of the most emerging technologies based on the internet. In the simplest terms, cloud computing means storing and gaining access to data and application over the internet as an alternative to your computer's hard drive. Cloud computing is the delivery of on-demand computing services from applications to storage and deals with power typically over the internet and on a pay-as-you-go basis. There is a wide range of network servers joined to online computing to provide different types of online services to cloud clients. There are some fewer figures of network servers having joined the cloud networks that have to perform greater tasks at a limited time. So, it is certainly easy to accomplish the entire works at a specific time. Few systems complete each task, so there is a demand for balancing workloads at a limited time. Here Load balancing reduces the time limit of each task in an exacting process.

There are invariably no chances to continue an equal quantity of network servers to accomplish the same quantity of jobs in the same period. All tasks to be performed would be higher than the connection to the servers. There are inadequate servers that have to hand over a million jobs in an instant.

Advances in Data Science and Computing Technology: Methodology and Applications. Suman Ghosal, Amitava Choudhury, Vikram Kumar Saxena, Arindam Biswas, & Prasenjit Chatterjee (Eds.)

We will propose an algorithm that insufficient nodes perform the jobs here; some jobs are much more than the nodes and balance each node to improve the quality of services in cloud computing.

11.1 INTRODUCTION

Cloud computing is the exploitation of online sources accepted as a service over an arrangement (Chan et al., 2012). The name emerges from the usual make use of a cloud-shaped sign as an abstraction for the system communications containing in system figures (Jeevitha, Chandrasekar, and Karthik, 2015). The Cloud system entrusts distant services with a client's data, program, and computation. Cloud computing comprises resources made last on the web as provided by third-party services. These services serve the perfection of the entrance to current software functions and high-end networks of physical workstation machines (Sarddar and Ray, 2017). The purpose of cloud computing is to treat traditional device or high-performance computing practice, again, and repeatedly made use by research resources, to execute trillions of calculations per unit time, in consumer-oriented applications for example financial portfolios, to deliver personalized guidance, to allow data storage or to power extensive, immersive physical machine games. The cloud computing uses methods of comprehensive collections of workstations, running low-cost customer PC technology with unique connections to extend data-handling chores across them. This distributed infrastructure encloses enormous pools of systems connected. It requires virtualization techniques to handle the effect of cloud computing.

11.2 CLOUD COMPUTING AND LOAD BALANCING

11.2.1 CLOUD SYSTEM

Cloud system (cloud computing) is a method for enabling expedient, on-demand network right off the entrance to a shared pool of configurable resources competent to be quickly stipulation and issued with the slightest management attempt (Moura and Hutchison, 2016). It bases the cloud system on virtualization. Virtualization is the major feature of the cloud system. The cloud system virtualizes an individual system into several virtual systems (Ray and Sarddar, 2018). A virtual machine is a software implementation of the physical machine. A low-level program called virtual machine monitor is reliable for the sharing of an individual physical machine among different resources.

Cloud computing aims to enhance the function of the system and to perform the most essential resource expenditure, greatest throughout, least response time and avoiding extra load.

11.3 CLASSIFICATION OF ALGORITHMS

It has divided the load balancing algorithms based on the recent state of a system and who started the process:

1. Depending on which client initiates the process (Eager, Lazowska, and Zahorjan, 1985):
   i. **Sender-Initiated:** Sender or client starts the completion of the load balancing algorithm on analyzing the need for load balancing.
   ii. **Receiver-Initiated:** Receiver or workstation starts the finalization of a load balancing algorithm on diagnosing the demand for load balancing.
   iii. **Symmetric:** This category of the algorithm is a merging of the sender-initiated category and receiver-initiated category algorithms.
2. Depending on the present state of the static system algorithm: In the static algorithm, there is a uniform distribution of traffic among the workstations. This algorithm requires a prior recognition of the resources of the system so that the judgment of moving the loads not depend on the presented situation of the system.
3. **Dynamic Algorithm:** In the dynamic algorithm, for balancing the load, the lightest workstation in the entire process is considered and selected. For this, real-time connection with the Internet is required that can expand the data transfer in the system. Here to determine to manage the load, it capitalizes on the living shape of the system.

11.4 LOAD BALANCING IN CLOUD

Load balancing is a process of reassigning the entire load to the dissimilar nodes of the distributed system to form resource use proper-formed and to gain an enhanced response time of the job, alongside removing circumstances where a few of the nodes are extra loaded whereas several other nodes are loaded easily. A load-balancing method strives to build up the application of resources with a light load or idle resources efficiently by freeing the resources with an extra load. The process seeks to handle the load amongst each obtainable resource. It strives to reduce the make-span of the successful exploitation of sources.

In distributed processes, comprised of homogeneous and distributed resources, it has researched load balancing processes extensively. These processes will not be living in cloud architecture because of its heterogeneity, scalability, and freedom. This creates load-balanced scheduling for cloud computing extra high and a provocative issue for many analysts.

The Non-traditional methods differ from the conventional process because it brings about the most beneficial results in a concise hour. There is no most excellent scheduling process for a cloud system. The alternative opportunity is to determine a convenient approach to apply in known cloud surroundings owing to the factors of the tasks, servers, and network heterogeneity.

11.5 RELATED WORKS

1. **LBMM (Kokilavani and Amalarethinam, 2011):** It is a static balancing algorithm. This algorithm uses load balancing amongst machines considering it. The Min-Min algorithm demonstrates the lowest completion time for all jobs. Then it upholds the tasks with the least completion time amongst all the tasks. This algorithm follows by permitting the positions to the process design the lowest makespan. Min – it schedules Min states similar procedures in the prospect of all jobs. The algorithm demonstrates the Min-Min schedule and creates an opinion of the node with the drawing makespan. Consequent to the system requires the task with the least execution time. It is referring to the execution time of the chosen job for each source. It considers the peak execution time of the elected a job with the make-span. If the execution time is smallest, therefore, it shares the preferred job to the machine, holding the maximum makespan. Else, the latest maximum makespan of the work is preferred, and the steps are consistent. The program tests if it assigns each system with each task. In these conditions where the figure of undersized jobs is higher than the number of abundant jobs in meta-job, these algorithms have improved appearance. This algorithm does not recognize light and upgrade systems, heterogeneity, and jobs.

2. **Two-Phase Scheduling (Mishra, Sahoo, and Parida, 2018):** It is the formation of OLB and LBMM algorithms to make superior completion competency and moves on with the process load balancing. OLB scheduling keeps on each process in managing situations to have a hold of the purpose of load balancing with the LBMM algorithm is maintained to diminish the completion of the time of all jobs on the system by this means diminishing the complete execution time. This algorithm was working on having rescued from the process of sources and bought the effort competency.
3. **Min-Min (Shaw and Singh, 2014):** It assumes MM with a load balancing strategy. In this approach, it splits all tasks into subtasks. The completion times of each subtask on every server, and the threshold, are measured. All subtasks, along with the server to raises the shortest volume of execution time for the subtask goes into Min time. The subtask having the least amount of execution time amongst all subtasks is selected and gave away to the associated server.
4. **LB3M (Madni et al., 2017):** This scheme determines the average completion time of all subtasks on all servers. The subtask has the largest average completion time, and the server is making the least completion time determined. At present, the assigned server will deal with the chosen subtask. If the server is before assigned, formerly the procedure evaluates the completion time of the server. For the designated server, make-span is the summation of the makespan of appointed tasks and completion time for the current subtask. For the unallocated server, make-span is the execution time of the remaining task. This proceeding is continuous in the subsequently unassigned subtasks.

11.6 PROPOSED METHOD

To determine the matrix cost in addition to a combination of the job(s) vs. machine(s) of an unbalanced matrix, to concentrate on a problem consisting of machines $M=\{M_1, M_2, \ldots, M_m\}$. The jobs $J = \{J_1, J_2, \ldots, J_n\}$, is considered to be assigned for execution on the 'm' obtainable machines with the implementation cost C_{ij}, where $i = 1, 2, \ldots, m$, and $j = 1, 2, \ldots, n$ are mentioned in the matrix in which m is greater than n. First of all, we get the sum of all rows and all columns of the matrix, store the results in the array, row-sum, and column-sum. Then to choose the first n rows by row-sum, i.e., starting with smallest to next smallest to the array row-sum and deleting rows related to the remaining (r) jobs. Store the results in the new array that should be the array for the first sub-problem. Do this process again until remaining jobs become less than a machine when remaining jobs are less than n, then, deleting (c) columns by column-sum, i.e., corresponding to the value(s) most higher to next higher to form the last sub-problem. Accumulate the results in the new array that will be the array for the last sub-problem. Apply the Hungarian method (Dive, Ingale, and Shahade, 2013) to get the best possible solution of all sub-problems, which is now becoming a balanced problem. Lastly, rearrange all the sub-problems to acquire the best possible cost.

11.6.1 ALGORITHM

To present an algorithmic illustration of the method, consider a problem that consists of 'm' machines $M = \{M_1, M_2, \ldots, M_m\}$. Moreover, 'n' jobs $J = \{J_1, J_2, \ldots, J_n\}$ is considered to be assigned for execution on 'm' obtainable machines with the implementation cost C_{ij}, where $i = 1, 2, \ldots, m$, and $j = 1, 2, \ldots, n$, where $m>n$, i.e., the jobs is greater than machines.

- **Step 1:** Input: m×n matrix.
- **Step 2:** To calculate the row sum of all tasks, correspondingly.
- **Step 3:** To divide the problem into two components. The first component will be the n×n matrix based on the highest row-sum, and the remaining components would be the n×m format where n>m. To calculate easy, we follow this rule (First, the component will be followed by a single machine with a single job with executes shortest execution time based on highest row-sum, and the second component will be followed by a device that performs minimum execution time based on maximum row-sum.).
- **Step 4:** In the second component, find least cost of unassigned jobs J_i of all already assigned machines from highest row-sum, and it is to be attached to the related machine and remove from the assigned job and related machine from the matrix and find next least cost of unassigned jobs from next maximum row-sum and assign the task to its matching machine and so on in anticipation of each task assigned to their consequent device.
- **Step 5:** So, each job assigned as a minimum to its related machine and more than two jobs are not allocated to any device.
- **Step 8:** Stop.

11.7 ILLUSTRATION OF AN EXAMPLE

Let us a set that has 5 machines M = {M_1, M_2, M_3, M_4, M_5}, and 8 jobs J = {J_1, J_2, J_3, J_4, J_5, J_6, J_7, J_8}. The set contains the execution costs of every job to each machine.

➢ **Step 1:** Input: 5×8 matrix.

J_n/M_m	M_1	M_2	M_3	M_4	M_5
J_1	151	277	185	276	321
J_2	245	286	256	264	402
J_3	246	245	412	423	257
J_4	269	175	145	125	156
J_5	421	178	185	425	235
J_6	257	257	125	325	362
J_7	159	268	412	256	286
J_8	365	286	236	314	279

➢ **Step 2:** It is to calculate the Row Sum of each task, respectively.

J_n/M_m	M_1	M_2	M_3	M_4	M_5	Row-Sum
J_1	151	277	185	276	321	1,210
J_2	245	286	256	264	402	1,453
J_3	246	245	412	423	257	1,583
J_4	269	175	145	125	156	870
J_5	421	178	185	425	235	1,444
J_6	257	257	125	325	362	1,326
J_7	159	268	412	256	286	1,381
J_8	365	286	236	314	279	1,480

➢ **Step 3:** We split the matrix into two parts. The first part will be the n×n matrix based on the maximum row-sum, and the remaining parts would be the n×m format. Easy to calculate, we follow this rule. First, the part will be followed by one machine with one job with executes minimum execution time based on maximum row-sum, and the second part will be followed by a device that performs minimum execution time based on maximum row-sum.

J_n/M_m	M_1	M_2	M_3	M_4	M_5	Row-Sum
J_2	245	286	256	264	402	1,453
J_3	246	245	412	423	257	1,583
J_5	421	178	185	425	235	1,444
J_7	159	268	412	256	286	1,381
J_8	365	286	236	314	279	1,480

➢ **Step 4:** Like all machines assigned by jobs by now, thus we want to remain the lowest cost of unassigned jobs in the employed system. Find the minimum value of unassigned jobs J_i of all already assigned devices from maximum row-sum, and it is to be attached to the corresponding machine and remove from the assigned job and similar machine from the list and find next minimum cost of unassigned jobs from next maximum row-sum and allocate the task to its corresponding machine and so on until all tasks assigned to their corresponding device.

J_n/M_m	M_1	M_2	M_3	M_4	M_5	Row-Sum
J_1	151	277	185	276	321	1,210
J_4	269	175	145	125	156	870
J_6	257	257	125	3255	362	1,326

➢ **Step 5:** So, all jobs assigned at least to its corresponding machine, and more the two jobs are not allocated to any device.

J_n/M_m	M_1	M_2	M_3	M_4	M_5
J_1	151	277	185	276	321
J_2	245	286	256	264	402
J_3	246	245	412	423	257
J_4	269	175	145	125	156
J_5	421	178	185	425	235
J_6	257	257	125	325	362
J_7	159	268	412	256	286
J_8	365	286	236	314	279

Final Result

$M_1 \rightarrow J_1 * J_2 \rightarrow 151 + 245 = 396$

$M_2 \rightarrow J_3 \rightarrow 245$

$M_3 \rightarrow J_6{*}J_8 \rightarrow 125 + 236 = 361$

$M_4 \rightarrow J_4{*}J_7 \rightarrow 125 + 256 = 381$

$M_5 \rightarrow J_5 \rightarrow 235$

Our approach: 1,618

M_m	M_1	M_2	M_3	M_4	M_5
Costs	396	245	361	381	235

11.8 RESULT ANALYSIS

The experiments were performed by following table execution time (ms) of each task at all nodes. The version of the system is Intel Core i3, 4th Generation processor, 3.4 GHz CPU and 4 GB RAM running on Windows 7 platform. We have considered makespan and we have conducted two sets of simulation scenarios as follows.

Figure 11.1 explains the implementation time for each task at several computing nodes. To compute the presentation of our proposed work is evaluated with other processes shown in Figure 11.1. Figure 11.1 displays the evaluation of the execution time of all nodes among in our approach. For completing each task by using the proposed algorithm, LBMM, HM, and MM, the makespan is 396, 423, 461, and 555 ms, correspondingly. Our approach attained the lowest completion time and improved balancing of the load than other algorithms.

11.9 CONCLUSION

It assigns None of the works to dummy machines. The result of the unbalanced matrix problem, the matrix obtained with the back of the Hungarian method, is pointed out. It appoints all subtasks with makespan to its related system so that the result is farthest desirable for the system. Although some devices perform over one job, makespan is most convenient, and loads are adjusting with our proposed work.

The technique is presented in an algorithmic shape and implemented on the many sets of input data to test the act and usefulness of the algorithm. Applying to load balancing is straightforward.

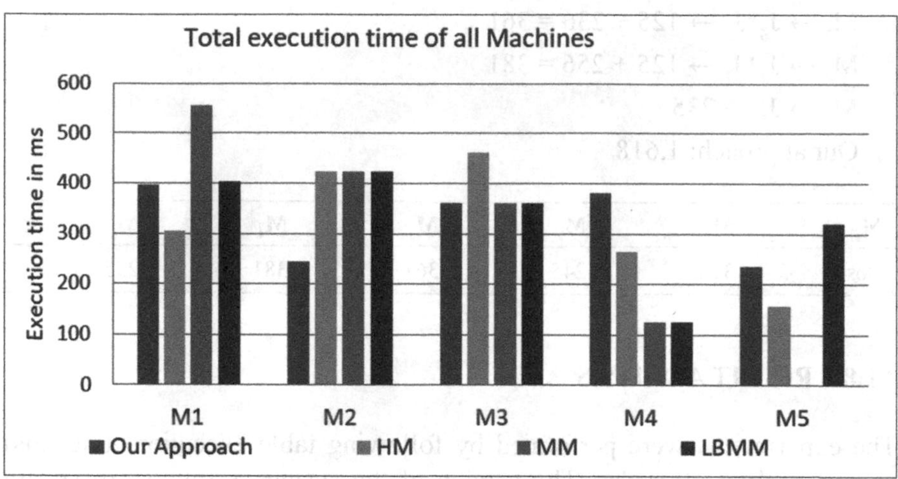

FIGURE 11.1 Execution time (ms) of each task at different computing nodes.

KEYWORDS

- **cloud computing**
- **load balancing**
- **minimum completion time**

REFERENCES

Chan, M. C., Jiang, J. R., & Huang, S. T., (2012). Fault-tolerant and secure networked storage. In: *2012 Seventh International Conference on Digital Information Management (ICDIM)* (pp. 186–191). IEEE.

Dive, S. R., Ingale, A. D., & Shahade, M. R., (2013). Virtual machine and network-performance, study, advantages, and virtualization option. *International Journal of Advanced Research in Computer Science, 4*(6).

Eager, D. L., Lazowska, E. D., & Zahorjan, J., (1985). A comparison of receiver-initiated and sender-initiated adaptive load sharing. *ACM SIGMETRICS Performance Evaluation Review, 13*(2), 1–3.

Jeevitha, M., Chandrasekar, A., & Karthik, S., (2015). Survey on verification of storage correctness in cloud computing. *International Journal of Engineering and Computer Science, 4*(9).

Kokilavani, T., & Amalarethinam, D. G., (2011). Load balanced min-min algorithm for static meta-task scheduling in grid computing. *International Journal of Computer Applications, 20*(2), 43–49.

Madni, S. H. H., Latiff, M. S. A., Abdullahi, M., & Usman, M. J., (2017). Performance comparison of heuristic algorithms for task scheduling in the IaaS cloud computing environment. *PloS One, 12*(5), e0176321.

Mishra, S. K., Sahoo, B., & Parida, P. P., (2018). Load balancing in cloud computing: A big picture. *Journal of King Saud University-Computer and Information Sciences.*

Moura, J., & Hutchison, D., (2016). Review and analysis of networking challenges in cloud computing. *Journal of Network and Computer Applications, 60*, 113–129.

Ray, P., & Sarddar, D., (2018). Load balancing with inadequate machines in cloud computing networks. In: Mandal, J., & Sinha, D., (eds.), *Social Transformation – Digital Way* (Vol. 836). CSI 2018. Communications in computer and information science. Springer, Singapore.

Sarddar, D., & Ray, P., (2017). A new effort for distributed load balancing. *International Journal of Grid and Distributed Computing, 10*(8), 1–9.

Shaw, S. B., & Singh, A. K., (2014). A survey on scheduling and load balancing techniques in a cloud computing environment. In: *Computer and Communication Technology (ICCCT), 2014 International Conference* (pp. 87–95). IEEE.

PART IV

Advances in Embedded System-Based Applications

CHAPTER 12

MICROCONTROLLER-BASED WEARABLE LOCATION TRACKER SYSTEM WITH IMMEDIATE SUPPORT FACILITY

SUMANTA CHATTERJEE, DWAIPAYAN SAHA, INDRANI MUKHERJEE, and JESMIN ROY

Department of Computer Science and Engineering, JIS College of Engineering, Kalyani, West Bengal, India, E-mail: Sumanta.chatterjee@jiscollege.ac.in (S. Chatterjee)

ABSTRACT

The work commitments are commendable but along with this security also comes together and as much dedications are there; their lives are also at risk and so it is our responsibilities to take care of them. So, a location tracker device is proposed which is named as "wearable location tracker" and it is basically a device in any form such as pins, wearable watches or even customized shirts which gives the real time location of the person and there are some special features also added to it such as the pulse rate count and the body temperature (BT) of the people. This chapter is mainly based on how one can save people from danger and provide them immediate help to evacuate them from danger. This device not only gives location but it has the facility of SOS system for sending message to near one's and emergency contact to provide immediate help.

Advances in Data Science and Computing Technology: Methodology and Applications. Suman Ghosal, Amitava Choudhury, Vikram Kumar Saxena, Arindam Biswas, & Prasenjit Chatterjee (Eds.)

12.1 INTRODUCTION

Security is one of the most important and highly concerned aspects of every one's life. As the overall circumstances are not in a good scenario along with security, safety for women and children are also gaining concern as the days are passing. Along with safety, tracking is also as major concern as more number of crimes on children and women are reported nowadays.

So, every near and dear one is of them want to monitor their activities and want to keep track of their loved ones and over the phone is not always possible. Taking precautions and preventions is always better than meeting with an accident. So, our proposed idea will serve the best for this purpose which is built using microcontroller. A system in a form of a chip that does a job can simply be called a microcontroller. It contains a small amount of RAM, an integrated processor, which are very useful while interacting with things connected to the microprocessor. And so, as our product too have. Various parameters are used along with the microcontroller for making this wearable tracker and advanced technologies are used to their uttermost to serve the best to the mankind.

Our proposed idea mainly focuses on making a wearable tracker device when people will be facing any danger, they can take help by using our proposed device just by tapping the SOS button to request for immediate help and in case of people working in any fields the device will monitor their pulse rate and their vital body parameters and accordingly if anything goes wrong the device will send signals for immediate arrangements for rescue. This proposed device can be of great help for the safety measurements of women and kids as well. This device will be of great help for people to provide them immediate help. Thus, the device is of great importance for every aspect people.

12.2 RELATED WORK

There are various existing plans regarding location tracker device. Some of them are as in subsections.

12.2.1 SMART IOT DEVICE FOR CHILD SAFETY AND TRACKING

This chapter deals with proposal of system or device which will track children's activity and will update constantly with the help of internet of things (IoT). IoT devices are smart and advanced devices, which are able to take decisions by sensing the environment surrounding the device. IoT connects billions of devices together with the help of the internet connections and it has the ability to transfer large number of data over a network without requiring human-to-human or human-to-computer interaction. IoT is an emerging field now and it is gaining a lot of popularity also.

The proposed system has the facility of camera, GPS system, hear beat sensor and touch sensor which can monitor and track their children's activity and they can keep a record of them (Figure 12.1).

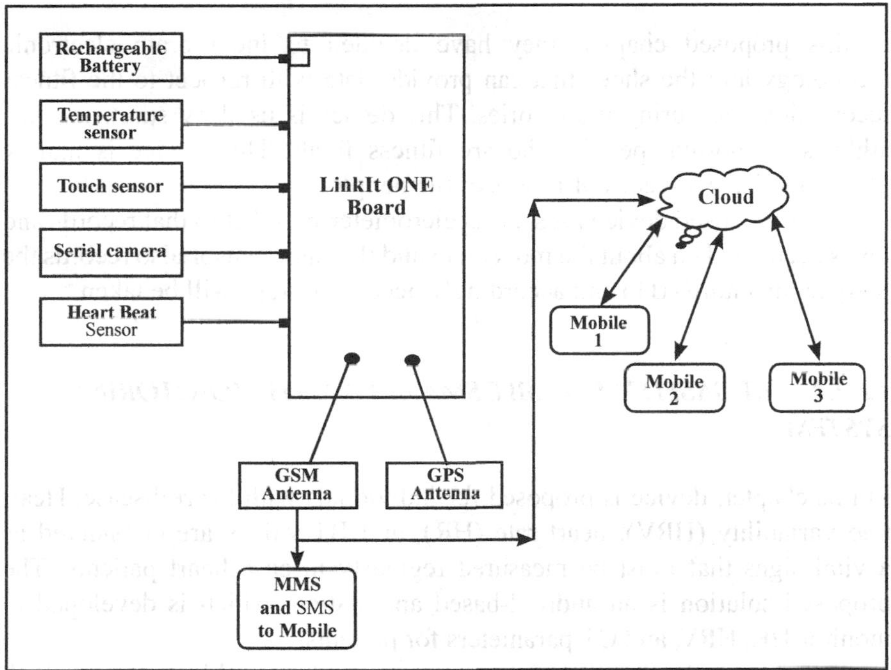

FIGURE 12.1 The workflow diagram of the IoT device for child safety and tracking.

12.2.2 GPS AND SMS-BASED CHILD TRACKING SYSTEM USING SMART PHONE

This chapter mainly focusses on safety of child. The authors have here proposed a solution based on Android which will help them to track the real time location of their wards. Recently, the smart phones are well equipped with sensors that are capable in doing this task of giving the location of the user.

Since the services are provided by smart phones, so the solution presented, takes the advantage of that to provide the required services. Hence real time location is obtained using the sensors like GPS and GSM. The system has to parts parents' part and children part. The parent's side will send a request to the child's device to send the location of the child, and the child device will acknowledge the same and send a reply.

12.2.3 SMART WEARABLE BLUETOOTH FITNESS TRACKER

In this proposed chapter, they have decided to incorporate electronic technology into the shoes that can provide data with respect to the fitness record thus measuring the calories. This device is used by sportspersons, athletes, or normal people who are fitness freak. This device is mainly planned to have a record of fitness of the people.

Their proposed device uses an accelerometer-based chip that records and gives accurate data about the movement and the same sensor also records the body temperature (BT) and accordingly necessary steps will be taken.

12.2.4 IoT BASED WEARABLE SMART HEALTH MONITORING SYSTEM

In this chapter, device is proposed for individuals with heart disease. Heart rate variability (HRV), heart rate (HR), and BT values are considered as a vital signs that must be measured regularly in case heart patients. The proposed solution is an android-based application, which is developed to monitor HR, HRV, and CT parameters for patients.

The proposed measuring system consists of wearable sensors which constantly measures patient signs. Then immediately send the measured signals to android interface via wireless connection. If the earlier recorded critical values for the patient have exceeded, the HR, HRV, CT values and along the real time location of patient is sent both to family members and doctor as e-mail and twitter notification. This particular wearable system helps the users to move around without any fear. It also ensures that e very citizen feels safe wherever they go.

12.2.5 TEMPERATURE REGULATING TIMEPIECES

This chapter worked with body parameters to track people in need. This device uses BT sensor to record the vital body parameters and provide immediate help to the victim.

And lots of other tracking devices are available which used GPS and GSM module to track the person who needs help in their workplace or in personal area of interest. And with the help of GSM module, they can send help message to the contacts registered to the GSM module and easily immediate help can be given to them.

12.2.6 PEDESTRIAN TRACKING WITH SHOE-MOUNTED INERTIAL SENSORS

In this chapter, authors mainly focused on delivering a system to rescue firefighters or other emergency responders. They developed a new system named as NavShoe, which uses a new and advanced approach for position tracking based on inertial sensing. The wireless inertial sensor is very small enough to easily tuck into the shoelaces, and sufficiently have low power to run all day on a small battery. It cannot be used alone with precise registration for close-range objects, but in applications in outdoor augmenting distant objects can be used and a user would rarely notice the NavShoe's meter-level error which is combined with any of the error in the head's assumed location is relative to the foot. NavShoe can hugely reduce the database search space for computer vision, which makes it much simpler and more robust. The NavShoe device not only provides robust approximate position, but it also provides an extremely accurate orientation tracker on the foot (Figures 12.2 and 12.3).

There are various tracking devices which are used for tracking logistic and tracking various work field people. Thus, they are of great importance for helping mankind.

12.3 PROPOSED IDEA

Leena, living in a metropolitan city, is walking an along a lonely street, it is dark, late at night, the street has dim light. Her parents are tensed, waiting, and praying for her to come home safely. Her father is bed ridden, her brother studies in a school, and she is working to earn for her family. She reaches her home safely, at 10.30 p.m. Despite living in a metropolitan city, her parents are praying so that their daughter reaches home safely. This concern pinpoints and tells us that her daughter is not safe in her own city.

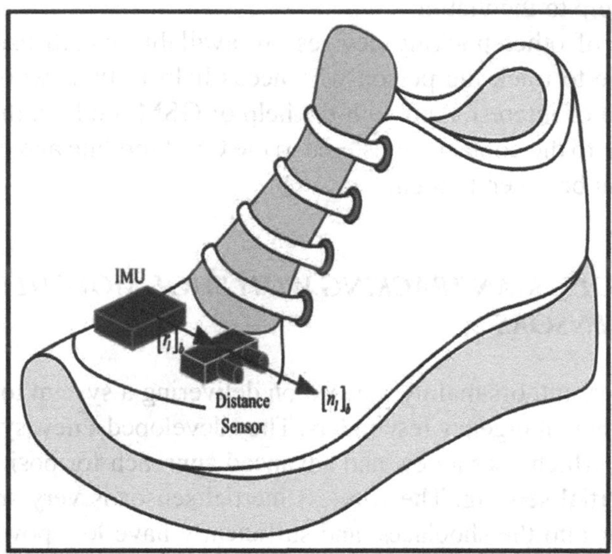

FIGURE 12.2 Proposed diagram of the pedestrian tracking with shoe-mounted inertial sensors.

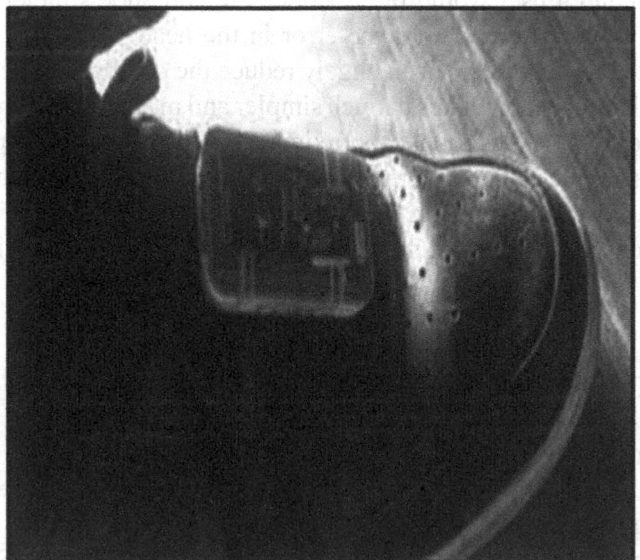

FIGURE 12.3 A proposed idea in the form of a demo product.

Shivu comes home alone from his school. As long as he does not come back, there are a lot of tension and negative thoughts regarding his safety that come across the mind of his mom. An innocent child is not safe either.

In today's world, security, and safety are the biggest concerns for an individual, especially for women and children. They are kidnapped, trafficked, physically abused, harassed, and what not. There are a lot of instances when our near and dear ones want keep a track of our location so that they can monitor us and feel assured about our safety. But is it only enough to get a location of an individual? How will someone get to know whether an individual is in a danger or need any help?

For this, a product has been devised which is the solution to all these problems. The product is a wearable device which can be in any form of a wearable device, pins, or even custom shirts. The system will not only monitor the real time location but also will have the facilities to send alert at the time of emergency and need. This system will also help in mitigating tracking problem even if a fisherman loses his boat in the sea (Figure 12.4).

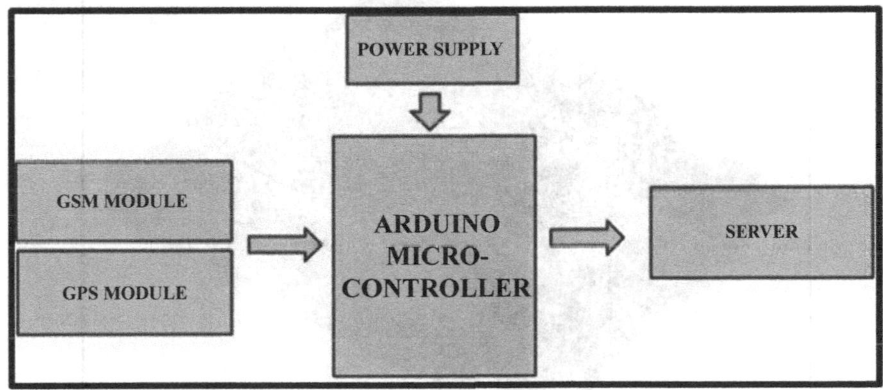

FIGURE 12.4 Block diagram of the proposed system.

12.4 HARDWARE REQUIREMENT

In order to make the proposed idea feasible and make such a device which has been proposed above, a few necessary components are needed which will give life to the proposed device in reality. Below a list of required hardware has been given along with their specification and utility:

12.4.1 ARDUINO UNO MICROCONTROLLER

The Arduino Uno is a microcontroller board based on ATmega328P micro-controller. This micro-controller board is made of Digital and Analog input/output (I/O) pins which can be even interconnected with other expansion boards and electrical circuits. The board contains 14 digital I/O pins (six for PWM output), 6 analog I/O pins, and which can be programed with the Arduino IDE, with a type B USB cable. USB cable or any external 9-volt battery provides power to the Arduino micro-controller board, where the power is between the range of 7–9 volts. It is similar to Arduino Nano and Leonardo. The Arduino-Uno board is first in a series of USB-based Arduino boards. The version 1.0 of the Arduino IDE was one of the reference versions of Arduino, which are reinvented incorporating various other changes. The ATmega328 on the board is pre-programed with a bootloader which allows us in uploading new code without any use of external programmer (Figure 12.5).

FIGURE 12.5 Arduino Uno microcontroller.

12.4.2 NEO 6M GPS MODULE

The NEO-6M GPS module is a well-known name while considering any GPS receiver which is built with 25 × 25 × 4 mm ceramic antenna, providing a very strong satellite search capability. The power and signal indicators are very useful, which can monitor the status of the module. It also has the efficient facility of data backup, in case accidental power failure (Figure 12.6).

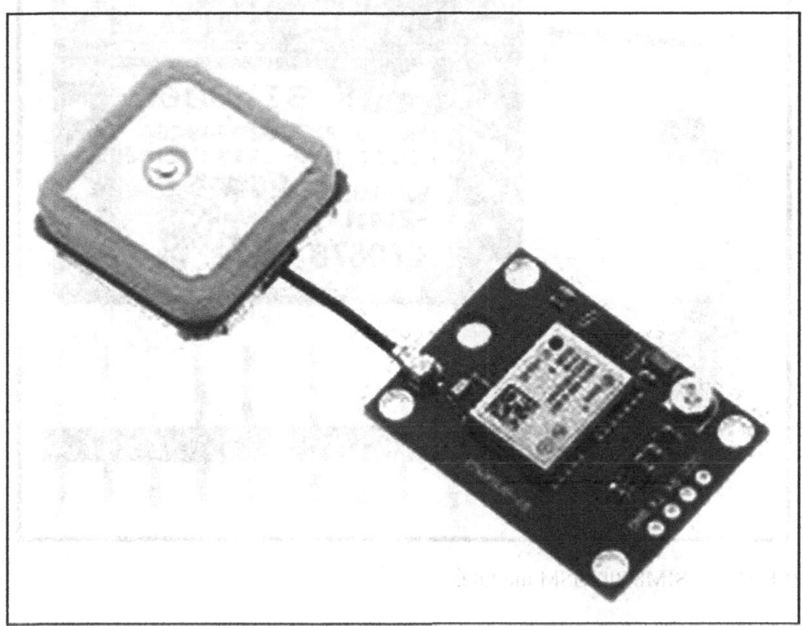

FIGURE 12.6 Arduino Uno microcontroller.

12.4.3 SIM800L GPRS/GSM MODULE WITH ANTENNAE

SIM800 is a quad-band GSM/GPRS module that works on frequencies 900 MHz EGSM, 850 MHz GSM, 1,800 MHz, and 1,900 MHz PCS. It is also featured with GPRS multi-slot class which consists of 12/class 10 (optional), and also it supports the sequences of CS-1, CS-2, CS-3, and CS-4 GPRS coding schemes. It also has one UART port. It is also featured with one USB port which can mostly be used for updating the firmware and useful for debugging. Receiver output and microphone input are also possible through audio channels. SIM800L has been configured with one SIM card interface, which integrates the TCP/IP protocol. The operation of SIM800L is performed on a supply within the range of 3.4 to 4.4 V. Making calls, sending, or receiving texts, audio over the internet, are some of the utilities of SIM800L. Hence, making it possible and useful when applied in projects, such as home and agriculture automation (Figures 12.7 and 12.8).

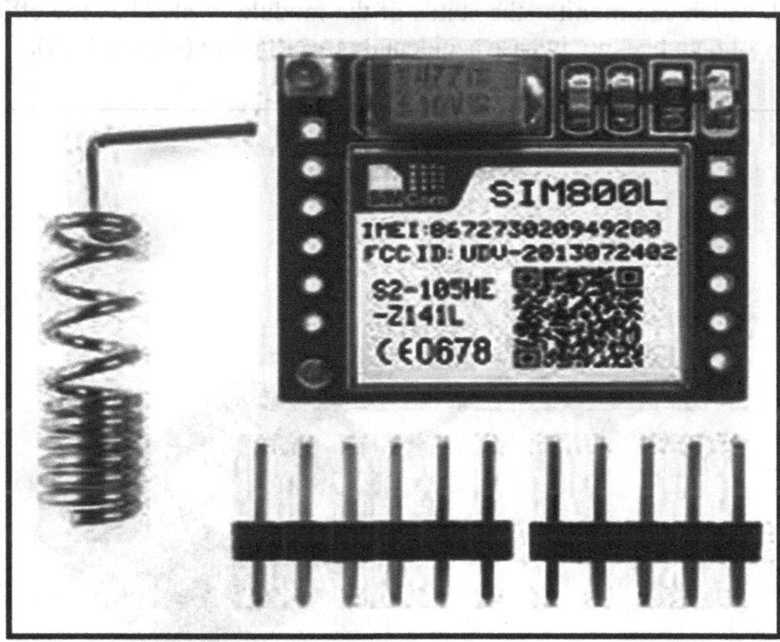

FIGURE 12.7 SIM800L GSM module.

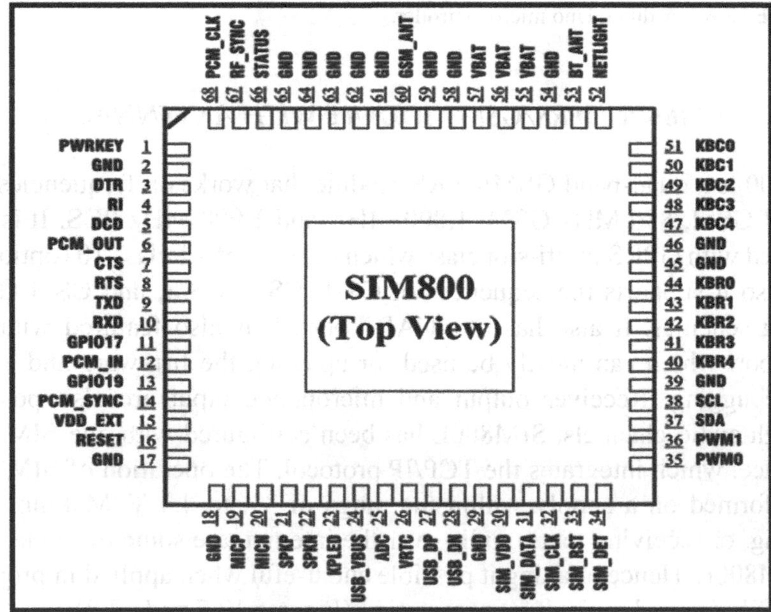

FIGURE 12.8 Pin-diagram of SIM800L GSM module.

12.4.4 *JUMPER WIRES*

Jumper wires are simply different colored wires that are used to connect components without soldering with the help of the pins that are present at the end of the wires. They are very useful in forming the circuit. And their different color codes make it easy to differentiate one from the other, thus easily distinguishable. These are usually used along with breadboards in order to connect the circuit and change the circuit as and when needed (Figure 12.9).

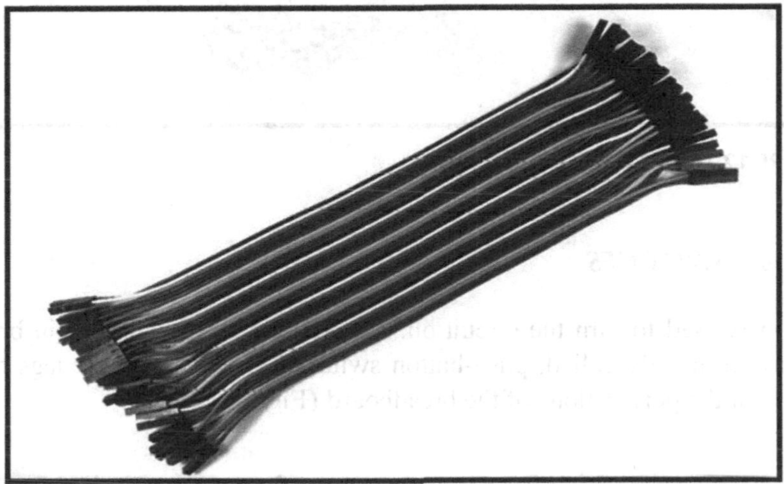

FIGURE 12.9 Jumper wires.

12.4.5 *BREADBOARD*

A tiny holed rectangular box usually made using plastic is generally referred to as a breadboard. These holes being very useful help one to insert any electronic components with ease, in the process of making a prototype or a product. It is a base solemnly used in the procedure of the very basic construction for making realistic and miniature prototype of an electronic model. Nowadays, we have breadboards that need no soldering and has perforated plastic blocks with a number of tin or alloy made clips, and sockets for better insertion of the electronic components. The clips are many a times referred to as contact points or tie points. A breadboard can be specified and segregated with the help of tie points (Figure 12.10).

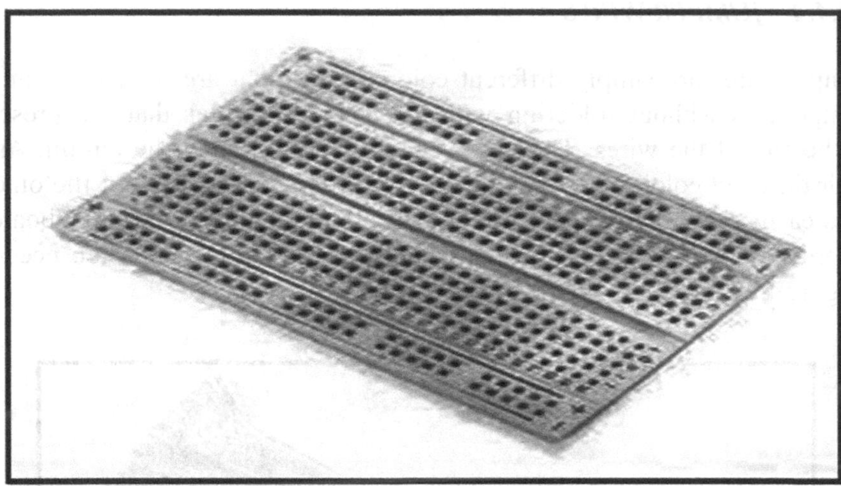

FIGURE 12.10 Bread board.

12.4.6 SWITCHES

Switch is used to turn the circuit on. The switches that are used in breadboards are mostly called, push-button switches. They have four legs to fit well with the perforations of the breadboard (Figure 12.11).

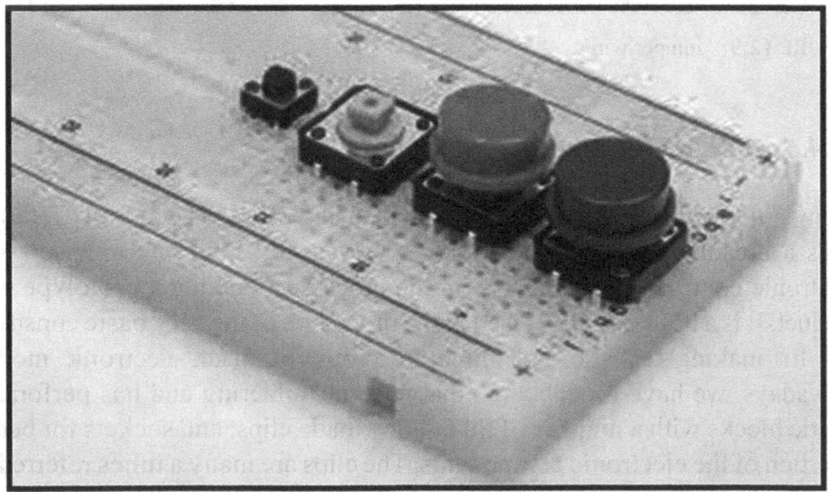

FIGURE 12.11 Switches.

12.4.7 9V BATTERY SOURCE

A 9volt battery source is used a source of power in the circuit. The 9V battery format is often accessible in its basic carbon-zinc and alkaline chemistry, in basic lithium iron di-sulfide, and also present in the form of nickel-cadmium, nickel-metal hydride and lithium-ion, which is rechargeable (Figure 12.12).

FIGURE 12.12 9V battery source.

12.5 METHODOLOGY

The system will be in a form of a wearable device. And each device will be having a registration key. With each key two mobile numbers will be registered, in which regular updates of the location of the user will be sent. The user can change these registered numbers if needed. For changing the registered numbers, the user will have to log in to a website and make the necessary changes. Since all the devices have a unique registration key, all the record of a particular device will be maintained. The information won't be disclosed, and will be kept under proper secured supervision.

The user will be able to turn the device on. Once the device is turned on, the real time location and the time of the user will get logged in the local database. This information then will be periodically get logged into the cloud. Similarly, location of the user will be sent to the two registered mobile number. The location of the user will be sent to the two registered mobile numbers after a fixed interval of time.

There are many other systems, in the market which gives the location of the user but there is a specialty of this device. But only tracking of the location isn't enough, is it? We are getting the location of the user, but what if the person is danger? What if the user is unable to make a call, and ask for help? In that case what is the need of this location tracker? There has to be something which can give an edge over these simple trackers.

The specialty of this system gives this device an edge over all the existing devices. It has an exclusive feature of SOS facility. Well, this SOS facility is not the only specialty. This device can send alerts to the alert center, when the SOS button is pressed. As soon the alert center receives the alert message it checks the immediate locality of the distressed user and alerts the nearest helps available at that point of time, so that the victim, the distressed can be rescued from that danger. And this unique system makes this system more special, and the only one of its kind. This system has a lot of battery backup (Figure 12.13).

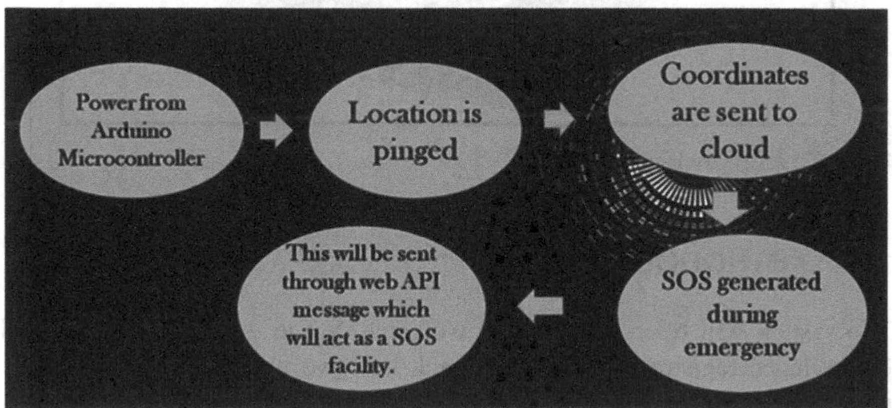

FIGURE 12.13 Working principle of this system.

For instance, when a person wearing the device, is in danger, at that time he/she can press the SOS button, and immediately an alert message will go to the registered mobile numbers. As this system is unique, as soon as the SOS button is hit, the recent last location is sent to the immediate local supports available like fire brigade, police, and ambulance.

The system is smart. As the local area changes, the local supports change too and accordingly this system works. The working of will be clearer with the next few instances.

For example, a woman wearing the device is walking on the street, and she is in danger. There are a group of eve teasers following her, or harassing her. She immediately when presses the SOS button, so an alert message goes to the alert center. Once the alert center receives this alert message it checks the present location of the user, and checks for the immediate help available. The immediate help available locally is the police station, hospitals, so these places will get an alert message along with the location of the user, and a rescue operation can be arranged for. Also, another alert message is sent to the two registered mobile number.

Similarly, when a fisherman loses his boat in the mid sea, and he presses the SOS button. The SOS message reaches the alert center. The alert center checks the locality of the user, checking he is the mid sea, the local support available will be the coast guards; police ambulance in this case will not be the effective help. Thus, the SOS button on being pressed, the alert message along with the location of the user is sent to the coast guards, who are the immediate support avail then (Figures 12.14 and 12.15).

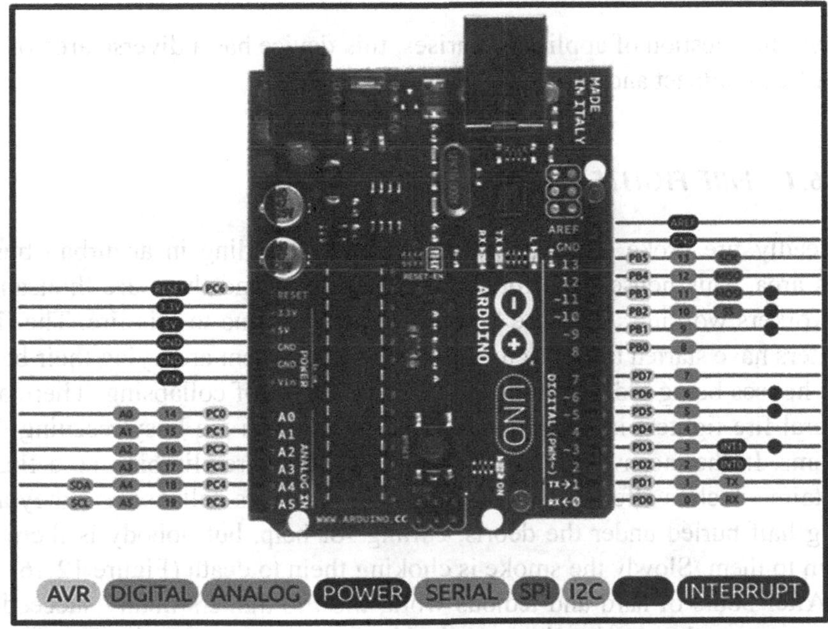

FIGURE 12.14 Pin diagram of Arduino Uno.

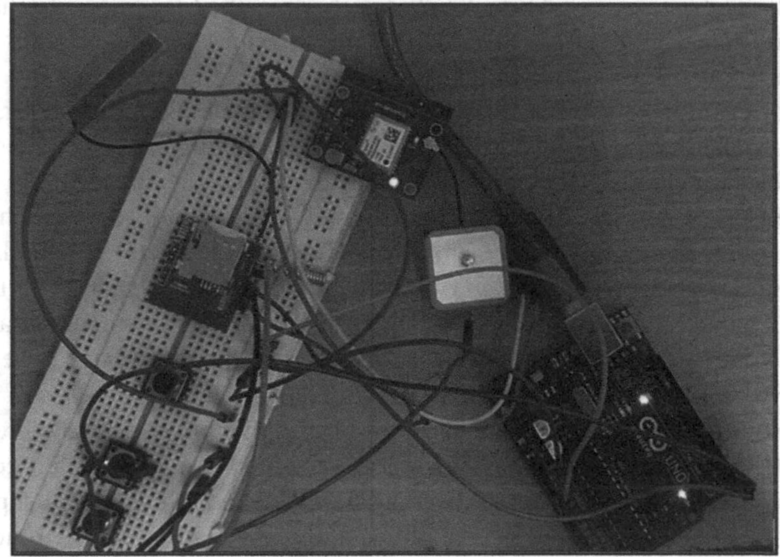

FIGURE 12.15 Circuit of the system.

12.6 USED CASES

When the question of application arises, this device has a diverse area of its significant impact and application:

12.6.1 FIRE FIGHTER SAFETY

A deadly fire broke out in an old multi-story building in an urban business area. This house is surrounded other houses and there are thousands of workers working. All these houses are affected due to this fire. The fire fighters have started their rescue operation. All of them are trying their best. The houses being old, the building is on the verge of collapsing. There are several fire fighters inside the house and all of them are busy rescuing the victims. In the meanwhile, two of the fire fighters are climbing up a flight of stairs which is affected by fire. Suddenly the stairs collapse and they are lying half buried under the debris, calling for help, but nobody is there to listen to them. Slowly the smoke is choking them to death (Figure 12.16).

After hours of hard and tedious work, the fire fighters finally succeeded in evacuating the people who were inside the houses. Suddenly, they notice that two of their brave officers are nowhere to be seen. It is evident that those brave officers have got stuck somewhere inside or have been lying under the debris. But how will the others know the location of their brave colleague? This device will help them. This device will be worn by the fire fighters. As soon as they will enter particular premises, their location will be sent to the database, and each fire fighter's location will be tracked. When the fire fighters face such a situation like the above, they can press the SOS button, and the alert message will help to locate the victims. Thus, if a fire fighter faces such a situation like both of them had then with the help of this device, they will be able to survive. The brave fighters who give fight the fire to save our lives, will not lose their precious lives due to lack of proper information.

FIGURE 12.16 This device is very beneficial for the fire fighters.

12.6.2 FISHERMEN SAFETY

While we all are in our resting time, at the time fishermen gets ready for their daily routine. For them only we people are able to have fishes as our meal but we do not have any idea through how much hard work fishermen goes for catching those fishes. Every day at the dawn, they prepare themselves for going at the middle of the sea for fulfilling their duty.

Suddenly, a storm breaks out at the mid shore and the weather condition around the sea or any other water body drastically changed within a minute and respectively high tides occurred and the waves are irregular and on tenterhooks and the fisherman is in the midst of the sea. Can you imagine the condition of the fisherman? Their life is under threat but we need to assure them a danger free life. So, our proposed device is a life savior for them. Our proposed device is of wearable type and fisherman can wear the device and then they can carry out their work. As soon as they move towards the midst of the sea or any other water bodies their locations will be recorded and sent to the database and the each and every fishermen's location will be tracked.

When fishermen face situation like tempest, they can press the SOS button and immediately an alert message will be generated and will be sent to the registered contacts so that immediate rescue operations are arranged. And if they are not in the condition to press the SOS button, their location can be tracked and immediate supports could be sent to rescue them. They risk their lives for us and it is our duty to save and provide them security.

12.6.3 CHILD SAFETY

Nowadays, child safety is a big concern for the urban parents, who mostly stays away from their children due to work for the entire day almost. At this situation, the wearable location tracker can be designed and implemented to be used in their lifestyle. It will definitely prove very effective in ensuring the safety to their children, even kids. This device will also prove efficient when the children are out of home, i.e., in schools, coaching classes or in any other place. The parents are always in a worry of the safety of their children, where our proposed product can be a boon for them. It is very easy to monitor for the parents and keep a track of their most loved ones. It is also very user-friendly and thus, children wearing it, will not face any sort of problem. It can be used for age group 3 years and above.

As for example, when a child goes to school by a pool car or school bus, there may be such situations of child trafficking, in that case, the child just needs to press the SOS button. Immediately, emergency support of police will be gathered and an alert message will also be sent to the parents' mobile number which had been registered initially in the device. Thus, it will take care of the children, even when they are away from their parents.

12.6.4 WOMEN SAFETY

Women are another important part of our society. They are often heard of becoming victims of bullying and harassment in public. In recent times, the security of women is a high point of concern for all, even the governing bodies. In this situation, this wearable location tracker can mitigate the problem up to a certain stage. Nowadays, rape has become a common issue, which is the call of the hour for us and needs to be put an end to this, as early as possible. Thus, the women get victimized the most in the society. For them, the above mentioned device will rescue them, if caught in such horrendous incidents.

Now, for example, if a woman is traveling alone late night, she definitely needs security. At this situation, she can wear this device and make sure that the power button is ON. Thus, in any chance if she faces some weird or unnatural circumstances, or if she senses danger, can simply hit the SOS button. As soon as possible, a nearby help, like police can be provided; and also, an alert message will be delivered to the closed ones of the victim, and she can be rescued. Thus, it will reduce the cases of women harassment and will also help in building a better place for them to live.

Thus, we can see that this device is not only for the fishermen, but also for the fire fighters, women, children, for each and every section of the society. This device is for the benefit, for the betterment for the good and security of the entire society at large.

12.7 RESULTS

The wearable location tracker during disaster, will not only be able to detect the location of the user, but also will provide support with nearest assistance in case of any danger suspected. The device will also be helpful to the user to get rescued by the local supports available. For example, if a person in a building, caught in dangerous fire breakout, then the assisting body will be the fire-brigade and police. But, if a fisherman loses his boat in mid water body or falls in any type of danger, then in this case the rescuing body will be the coast-guards and not the police. Thus, it is clear from the above instance that the support or the help that will be generated, entirely depends on the last tracked location of the user, i.e., the last updated location of the user stored in the cloud. The person wearing this device will not feel any extra weight, as it is extremely light in weight, which is one of the key features of the device.

Another striking feature is that, it has an exclusive SOS facility inbuilt in the device itself, so that at times of any emergency, whenever the button is hit, local support will be arranged automatically by determining the recent location of the user. Our device is facilitated with two buttons, first one is the power button, having the power symbol and green in color and second one is the most important, the SOS button. Since, it will sense the danger; it has a vibrant red color to make it more noticeable at times of emergency.

It also has an extra and highlighting facility which proves beneficial in terms of power saving and also accounts for being more user friendly. Suppose, the user instead of pressing the power button hit the SOS button or vice-versa; then in that case, it has the technology that unless and until the power had not been turned ON, the SOS button will not work and similarly, if the user hit the SOS button mistakenly instead of power button; then no alert message will be generated because the device will not start operating before turning ON the power button. The device is also made water-proof and even heat-resistant, thus making it fit for use even underwater to assist the fishermen, whenever he senses danger or loses his boat in sea.

12.8 CONCLUSION

By designing this wearable location tracker, it is ensured that in near future it can reduce the number of lost cases, also the cases relating to women and child trafficking, since their location can be tracked using this device. And not only the location, but also nearest local help is assured to the victim; and along with this, an alert message will also be delivered to the registered mobile number of the close ones. Furthermore, the device can also solve problems related to tracking of fishermen, if caught in some danger in the mid sea or loses his boat during extreme weather conditions. Hence, reducing the life risks faced by them. Superior GPS module, GPRS/GSM module are used which reduces the chances of network failures in most of the remote areas than expected. In this particular model, Arduino-NANO is used to design the device, which is better compared to its UNO version based on the size. Because of using Arduino-NANO, the size of the device is reduced which is very easy to wear and carry. This location tracker if used efficiently, i.e., if the SOS button is pressed on time, can really save precious life. It will act as a savior to the human being.

12.9 FUTURE SCOPE

In future, this location tracker can be given a new shape with some added features, which will make this more efficient. We are continuously working to increase the efficiency of our product by applying various available latest technologies. In this device, Arduino-Uno microcontroller has been used, but in near future Flora micro-controller can also be used to scale it down to a much smaller size than present. The wearable location tracker, is formulated with a SOS button, which needs to be hit at times of emergency; but in future this can also be made fully automatic, i.e., the device will not require any sort of human intervention to press the SOS button during emergency. The button will be then automatically activated and start performing the needful by determining the vital body parameters and also some advanced technologies which will determine the type of hazard the person is in. This device can also be applied in cases of child safety and women security in recent times and in near future. It can also find its application in the field of defense, by serving our country's backbone, our defense sector. Thus, serving the mankind at its best.

12.10 BUSINESS PLAN

This device will not only secure a lot of safety, but also has a potential to give a way of living to a lot of people and secure their lives. This device will be manufactured by a particular company and the business setup will ensure monopoly regarding the selling, maintenance of the service. This company will also deal with the other aspects related to the device such as security of information and the data hiding part. This company will also have provisions for the emergency alert systems, which will be very active and quick to response when there is a time of emergency.

As mentioned earlier each device will be registered onc, so the entire data will be logged into the company's log, where with each device corresponding details of the user will be maintained. Since the company will have the policy of monopoly, there will a lesser threat of duplicate products. Moreover, there will be no compromise with the quality of the device and especially the safety and security of our users.

As mentioned above, the information will not be disclosed, and will be kept under proper secured supervision, for that special and efficient team will be present, who will look into the matter of safety and security of the data, and will be supervising the entire event.

All these work will need a lot of people to work together. There will be offices of this company in various parts of the world. Thus, this company will be a home to thousands of employees. This company will bring smiles in the faces of millions of people by offering them jobs, and ensuring the safety and security of the users.

Each device will cost approximately Rs. 2,250/-. And as the number of production increases, and the production is done on a large scale, there will be a considerable drop in the price of each device.

KEYWORDS

- **Arduino microcontroller**
- **cloud**
- **global positioning system (GPS) module**
- **global system for mobile communications (GSM) module**
- **SOS button**

REFERENCES

An end-to-end location and regression tracker with attention-based fused features. In: *2019 International Joint Conference on Neural Networks (IJCNN)Date of Conference: 14–19 July 2019*. Date Added to IEEE Xplore: 30 September 2019 INSPEC Accession Number: 19028209. doi: 10.1109/IJCNN.2019.8852288Publisher: IEEE.

Debajyoti Pal, Anuchart Tassanaviboon, Chonlameth Arpnikanondt, & Borworn Papasratorn (2020). *Quality of Experience of Smart-Wearables: From Fitness-Bands to Smartwatches* (Vol. 9, No. 1, pp. 49–53). In: IEEE Consumer Electronics Magazine Date of Publication: 04 December 2019 INSPEC Accession Number: 19194203. doi: 10.1109/MCE.2019.2941462. Publisher: IEEE.

Design and development of wearable GPS tracking device by applying the design for wearability approach. In: *2018 4th International Conference on Science and Technology (ICST), Date of Conference: 7–8 Aug. 2018*. Date Added to IEEE Xplore: 12 November 2018 INSPEC Accession Number: 18233756. doi: 10.1109/ICSTC.2018.8528636 Publisher: IEEE.

Design of handheld positioning tracker based on GPS/GSM. In: *2017 IEEE 3rd Information Technology and Mechatronics Engineering Conference (ITOEC), Date of Conference: 3–5 Oct. 2017*. Date Added to IEEE Xplore: 01 December 2017INSPEC Accession Number: 17411810. doi: 10.1109/ITOEC.2017.8122477 Publisher: IEEE.

Development of real time life secure and tracking system for swimmers. In: *2019 3rd International Conference on Computing Methodologies and Communication (ICCMC), Date of Conference: 27–29 March 2019*. Date Added to IEEE Xplore: 29 August 2019INSPEC Accession Number: 18958280. doi: 10.1109/ICCMC.2019.8819859 Publisher: IEEE.

Development of real time life secure and tracking system for swimmers. In: *2019 3rd International Conference on Computing Methodologies and Communication (ICCMC), Date of Conference: 27–29 March 2019*. Date Added to IEEE Xplore: 29 August 2019INSPEC Accession Number: 18958280. doi: 10.1109/ICCMC.2019.8819859 Publisher: IEEE.

Energy efficient assisted GPS measurement and path reconstruction for people tracking. In: *2010 IEEE Global Telecommunications Conference GLOBECOM 2010, Date of Conference: 6–10 Dec. 2010*. Date added to IEEE Xplore: 10 January 2011INSPEC Accession Number: 11743331. doi: 10.1109/GLOCOM.2010.5684306 Publisher: IEEE.

Foxlin, E. (2005). Pedestrian tracking with shoe-mounted inertial sensors. In: *IEEE Computer Graphics and Applications* (Vol. 25, No. 6, pp. 38–46). Date of Publication: 07 November 2005. INSPEC Accession Number: 8662444. doi: 10.1109/MCG.2005.140. Publisher: IEEE.

Instantaneous feedback pedometer with emergency GPS tracker. In: *2018 2nd International Conference on I-SMAC (IoT in Social, Mobile, Analytics and Cloud) (I-SMAC) I-SMAC (IoT in Social, Mobile, Analytics and Cloud) (I-SMAC), 2018 2nd International Conference on Date of Conference: 30–31 Aug. 2018*. Date Added to IEEE Xplore: 28 February 2019INSPEC Accession Number: 18490586. doi: 10.1109/I-SMAC.2018.8653718 Publisher: IEEE.

Internet of Things: Solar array tracker. In: *2017 IEEE 60th International Midwest Symposium on Circuits and Systems (MWSCAS), Date of Conference: 6–9 Aug. 2017*. Date Added to IEEE Xplore: 02 October 2017Electronic ISSN: 1558–3899INSPEC Accession Number: 17214744. doi: 10.1109/MWSCAS.2017.8053109Publisher: IEEE.

Optimization of laser trackers locations for position measurement. In: *2018 IEEE International Instrumentation and Measurement Technology Conference (I2MTC), Date of Conference: 14–17 May 2018*. Date Added to IEEE Xplore: 12 July 2018 INSPEC Accession Number: 17917295. doi: 10.1109/I2MTC.2018.8409835 Publisher: IEEE.

User localization using wearable electromagnetic tracker and orientation sensor. In: *2006 10th IEEE International Symposium on Wearable Computers Date of Conference: 11–14 Oct. 2006*. Date Added to IEEE Xplore: 22 January 2007Print ISBN: 1–4244–0597–1 INSPEC Accession Number: 10236845. doi: 10.1109/ISWC.2006.286343Publisher: IEEE.

CHAPTER 13

DESIGN AND MODIFICATIONS OF EYEDROP BOTTLE HOLDER DEVICE: A SIMPLEST WAY TO INSTALL OCULAR DRUG INDEPENDENTLY

SIRAJUM MONIRA, VISHAL BISWAS, and SATYAM JHA

Department of Bachelor in Optometry, Pailan College of Management and Technology, Joka, Kolkata–700104, West Bengal, India,
E-mail: srisikhsa@gmail.com (Sirajum Monira)

ABSTRACT

To design a simple eye, drop bottle holder device that can be used to put ocular drugs independently even the patients with severe visual impairment. We have worked in three phases; in Phase 1, designing of the non-electronic device (eye dropper) it is a non-electronic device and we are calling it "Easy eye dropper." It involves less effort for the patient. The leverage which is present, it is specially designed with extended arms, so that it can deliver eye drops using with the lightest finger press and drop will be fallen inside the eyes. The eyepiece and locks position is constructed in such way that is ideal for the user. Now as the size of the drops varies from length to diameter (5 ml to 10 ml), all the diameter and length of all the drops are not same which are available in the market, to get rid of it we a have a way out here (as we used the screw tightened mechanism of a compass and a pencil). Like the same way, here we have placed a screw system, where you can put the eye drop inside the device and then according to the diameter you just tighten the screw. Hence it will be fit inside, then place the device on the eye, by the help

Advances in Data Science and Computing Technology: Methodology and Applications. Suman Ghosal, Amitava Choudhury, Vikram Kumar Saxena, Arindam Biswas, & Prasenjit Chatterjee (Eds.)

of leverage press it automatically, with one gentle press a drop will come out and will go directly inside the eye.

In Phase 2, designing of the electronic device (electronic eye dropper). We have designed another electronic model with some additional features which can be considered as the upgraded version of the non-electronic model. It includes an eye detector sensor and some audio output along with a micro vibrator. This technology will reduce the effort of the patients in terms of his/her attentiveness as with this electronic technology a sensor will detect that whether the tip of the bottle is placed on its proper place or not. A beep electrical sound will confirm the number of drops fallen in the eye. Also, we have incorporated a micro vibrator in it keeping the deaf and mute patients who may not detect the sound sensor.

One press = one beep sound = one drop will get out

In Phase 3, practical approach (pilot project) we have a plan to conduct a pilot project with this newly invented device to see the durability of one eye drop bottle manually and by these electronic and non-electronic devices.

The construction of "Easy eye dropper" is very simple in use for all the subjects whenever they try to install the eye drops in their eyes without anyone's help. It also saves the number of attempts required to install the eye drops but the effectiveness of this device yet not tried practically. This phase will be done in the coming days. A new modified eye drop bottle holder may be an effective device and useful invention in terms of saving the time, effort, wastage of drug quantity and will create independence of self-installing drugs into eyes.

13.1 INTRODUCTION

There are a majority of ocular diseases where the mainstay of treatment of the patients includes topical therapy. If we overview the various ocular conditions, starting from a very minor ocular disturbance to a severe chronic problem is treated with the help of topical therapy along with its other additional treatment. As there are many kinds of eye drops, it is hard to generalize about their usage area, but the more commonly the areas we are always dependent on topical drugs mostly: Dilating drops during eye examinations, lubricating drops for dry eye symptoms, redness-relieving drops, anti-allergy drops for itchiness-relieving, pre-operative, and post-operative drugs, commonly, and frequently used antibiotic drugs and the pressure-lowering drugs used for long term treatment like glaucoma. Consider a disease like glaucoma where we have to make sure to put in the right amount of eye drops at the right time. This way the drug will help to keep the eye pressure at the correct level for the patients, which decrease the risk of their eyesight becoming poor in future. Many people find it difficult to put eye drops in initially, also many of the patients cannot even put drugs independently and it is easy to miss the regular prescribed doses. Glaucoma is a slow progressive ocular condition that can ultimately result in impaired vision or even person may reach to the condition of blindness, so it is important to take all the drops as prescribed. Also, for glaucoma sufferers, most of them have to use drops even for their lifetime.

There are various ways to install eye drops in the eye and many people take many ways for this task. Some patients try to put independently but some people take others' assistance. Before to start putting a drop one has to make sure that the hands are properly cleaned. Few drops need to be shaken. Remember. One of the simplest ways of installing drops in is to sit comfortably in front of a mirror or can be instilled drops whilst lying down, need to gently pull down the lower eyelid with a finger of any hand, with the other hand which is free, need to press the bottle according to the proper instructions and allow the drop fall into the space between eye and the lower lid. After putting the drop in the eye, the patient needs to close the eye lid gently and press on the inside corner of the eye, near the nasal area, with a finger for one minutes. This will help to reduce the rate at which the drop drains out through the tear duct into the throat cavity, rather than staying in the eye where it is essential. A little amount may still drain through the tear duct and be swallowed in our throat, that is why it felt bitter taste in throat. This small amount which we are swallowing is not harmful but, the amount of eye drops entering body through the tear duct is minimized. This is why there is a higher chance of unnecessary wastage of drugs and also the absorption of drug through the ocular surface reduced. Many times, patients also get suspected whether the right quantity of drugs has gone inside the eye or not, hence they put another drop of the drug from this suspicion which may lead to get the bottle finishes early. For this reason, most of the topical drug user takes someone's help always. To be dependent always for this task on family members, friends or other people may reduce the wastage of drugs but it leads to untimely putting of drugs due to the unavailability of helping hands on time always. If it is a glaucoma patients or post-operative cases then there is always a need for frequent eye drop installation to preserve the eyesight. There are some other situations also where one has to install eye drops independently. Mainly, the time when people are there in their work field or traveling alone in outdoors.

Keeping all those points in mind we have made “Easy Eye Dropper” which may help to get rid of both the problems in terms of drug wastage and also a dependency on drug installation. There are few aids already commercially available for this task but they have a few disadvantages which are leading to its non-popularity among the ocular drug users. Mostly all these aids are manual or non-electronics. Few designs are not suitable for all sizes of bottles, few aids not proper for gripping in hands and some are creating problem in positioning proper places.

New modified eye drop bottle holder “Easy Eye Dropper” will be an effective device and useful invention in terms of the longevity of drug quantity, wastage of drug quantity and will create independence of putting the drug into eyes and will overcome all the limitations of previously made aids.

13.2 OBJECTIVES

The main aim of our work is to design a better technology eye drop bottle holder device. Our goal is to help all the ocular drug users including the severely visually impaired people, hand tremors, arthritis, and old-aged who cannot put eye drops without any ones help and always missed the drug schedule due to lack of helping hands on time. Also, this technology may reduce the wastage of drugs which may indirectly save the health budget of any drug user. The specific features of lightweight, easy usage technique and low cost gives ocular drug user a new lease of life, a new dimension of independence and enables him to save some budget on ocular drugs purchasing.

13.3 REVIEW OF LITERATURE

Taneja et al. group published a paper on Innovative eye drop applicator namely bull’s eye applicator. This applicator is helpful for self-installing of eye drops independently. They reported the effectiveness of this eye drop device in preventing the wastage of medicine. In this device here they have invented a small pocket sized device where there is an easy handy attachment which can be attached to any kind of eye drop bottle. They also performed a study where they got a result that the use of this device reduced the wastage of drugs. The dispenser was used in few subjects divided into two groups; they were asked to apply the lubricating eye drops of (specific size bottle) in any one eye between the two without the help of device and in fellow eye with the help of this device. A count was kept on the attempts made by the users for the application of eye drops and the residual eye drops which were not used in the returned bottles were measured for quantitative assessment as planned in the objective of their study.

They concluded that the number of attempts of instillation which was made with the help of eye drop applicator was reduced significantly. This device also reduced the wastage of drugs.

But they found few limitations also like severe arthritis patients and low vision patients cannot use this. The mirror which is placed in the applicator can break easily (Mukesh, Chappidi, and Naga, 2016).

Jungueira et al. made a study on the efficacy of drugs and also to check the safety of a new device (eye drop R), which had been designed for eye drop instillation in patients having or not having glaucoma.

They have done this study on both types of patients (on glaucoma patients and healthy participants) on applying the eye drop. This study included patients with glaucoma and normal healthy participants after doing all comprehensive eye examination and measurement of the baseline intra-ocular pressure, the topical hypotensive medication had been given in both eyes and the Eye drop delivery device was made available to all subjects (with video and written instructions in them for proper evaluation) for use in one eye, which was randomly chosen by the participants. In the second phase, all the participants were evaluated by an examiner for IOP, investigation of possible related side effects, and the ease of putting eye drops. Conclusion made was that Eye drop gained a better subjective response regarding the ease of installation as compared to traditional instillation.

But this device also has some limitations like it is an invasive technique: which can cause harm to the ocular surface. It does not give a proper idea that how much amount of drop got instill (Junqueira, Lopes, and Souza, 2015).

George C. Bauer invented a device in the year 1992 by which eye drop applications became easy. The Eye drop applicator comes for an eye dropper vial comprising of a cup with a lip portion to get fitted well around the eyeball and helps in letting the eye lid remain opened. A sleeve has an internally threaded passage extending from the top of the cup. The internally threaded passageway can be removed and it is connected to an externally threaded neck of the eye dropper vial. The one who uses it will have to squeeze the eye dropper vial for directing an eye drop from the nozzle of the eye dropper vial onto the eyeball center.

This applicator also has some disadvantages like this was just a theory no working model was produced according to this theory. If it would have produced then this device would have been an invasive device which could have caused harm to the eye during instillation of drug (www.maddak.com/autodrop-eye-drop-guide-pr-28115.html?page=2).

Shai Recanati invented another type of eye drop applicator device in the year 2004. Here they invented an eye-drop applicator for putting eye-drop into a user's eye without the blinking one's eye. Eye drop applicator is inclusive of a container which has a neck and also has an opening end for storing the solution; it has a lid too; and further includes a tubular eyedropper member being attached to the lid and through which eye-drops come out from the container. This device also has limitations like the above one (https://www.guldenophthalmics.com/products/index.php/opticare-eye-drop-dispenser.html).

Another applicator called auto drop which was manufactured by the British Royal National Institute for the Blind. There is an arrangement, which holds the eye open while keeping the bottle over the eye and also it consists of a very small pinpoint like hole that allows light to enter in; as per the manufacturer, when the patient looks at the pin hole, the eye is rightly positioned to receive the drop and drop does not get waste while putting. But this device is also not suitable for the severe arthritis patients as it involves the patient's own press for instilling the drop. Also, this device is not at all available in the Indian market (https://magic-touch-eye.myshopify.com/).

Cameron Graham made an eye drop dispenser which is called Opticare Eye drop dispenser. This device is designed in such a way that it make easier for the patient to press the bottle. But there is no such studies done till yet, which tells us the effectiveness of this dispenser among the user. Also, this device is not at all available in the Indian market (https://www.contactlenskit.com/optiaide/).

Another technology of installing eye drop is by using capillary action technique. It involves putting a single drop into a holder and then bringing the holder very close to the eye, allowing the drop to fall onto the eye from the device. This technique had several advantages, including increasing the chances that only one drop will be used at a time and no more drops will be wasted; no need for the patient to tilt his head back during the installation procedure (https://www.amazon.com/Cress-Dropper-EyeDrops-Professional-Model/dp/B004R8TOGM).

The Magic Touch—The Eye Drop Helper is invented by Julius Shulman. This device, has a small rubber bell about the size of a thimble, is placed on a fingertip; with the finger positioned vertically, a drop is placed into a small groove at the tip of the device. The user then moves forward, and when the device gets close to the user eye, the drop falls onto the eye. But it has some drawbacks also like always needs to keep the device clean as it is an invasive technique. Low vision patients cannot use it. There is no study regarding the effectiveness of this procedure on the patients (Sharma et al., 2016).

OptiAid is another type of device which looks more high-tech than the other available devices, it can hold up to five drops in its five chambers. Unlike the other devices, this one is positioned in front of the eye and allows the drop to fall onto the eye, theoretically preventing the device from touching the cornea. However, it needs moving the eyelids and eye lashes out of its way to avoid the touch. Negative factors are: Not available in the Indian market. It is a complete invasive technique to install drug. No studies have still done on this innovation regarding its effictiveness (Efron, 2013).

Another aid created by Jonathan Cress an ophthalmologist. This dropper is an arched tube. It marks the users nose as a reference point so the tip can be steadily placed in the medial canthi area and the area will be properly identified.

Once the tip is properly placed in its position, the bottle is squeezed; the drop leaves the aperture, slides down properly and off the little ramp onto the delivery tip and into the user eye. The user releases the bottle when they feel the drop fallen into the eye. The current version of the device is with no measuring component, but mostly the patients gain an understanding of squeezing the bottle to release just one drop of medicine and extra drops will not be wasted. Still this device also has some limitations. Patients who have weak hands find it difficult or impossible to press the bottle by squeezing accurately (Kahook, 2007).

13.4 METHODOLOGY

The study was carried out into three phases. But in this chapter, we have described only two phases and the third phase will be done in our next work. Broadly, phase I involved in the design and implementation of the eye drop bottle holders which will be working manually. We called the aids "Easy Eye Dropper"(non-electronic model).

In phase II, we used Arduino and Ultrasonic based sensor techniques. This is an electronic model of "Easy Eye Dropper."

In phase III we have to assess the practical and functional approach of the newly made "Easy Eye Dropper" aids with both the electronic and non-electronic models.

13.4.1 PHASE-I (EASY EYE DROPPER: NON-ELECTRONIC MODEL)

This model is beneficial particularly for those who have severe arthritis, spondylitis (mainly cervical and for that they cannot move back their neck to the backside) or difficulty in lifting their hand up and reach close to the eye due to any type of systemic illness. This type of people mainly faces a lot of challenges while they are trying to install eye drops in their eyes independently. Hence, they have to be always dependable on others. The leverage that one can get from this specially designed eye dropper holder–two extended arms (14 cms) which to deliver eye drops just by using even the gentle finger touch. The eye piece arrangements and locks in the position that is perfect for you during the drug installation procedure.

Here we have implemented a screw in the dropper – now as the size of the eye drop bottles varies from length to diameter based on its commercial companies, along with the quantity of drugs also varies from company to company and brand to brand, e.g., 2.5 ml–10 ml, diameter varies from 3 cm–3.5 cm. The shape of the eye drop bottle also varies from round to oval (Table 13.1).

TABLE 13.1 Explaining the Procedure of Fitting of Bottle Into "Easy Eye Dropper" and a Simple Instructions for the Users

Fitting of Bottle Into "Easy Eyedropper"	Instructions for the Users
If the bottle is new remove the protective seal (Figure 13.1(a))	Open eyepiece and remove the bottle cap. Tight the screw or lose the screw according to the size of the bottle (Figure 13.2(a))
Open eye piece towards arms/leverage of the dropper (Figure 13.1(b))	Close the eyepiece. Rotate the eye piece towards left or right as per in which eye you are using for the most comfortable administration of the drug (Figure 13.2(b))
Place the dispenser on the flat surface and spread the two arms of the dispenser open, push eye drop bottle inside dispenser until the base of the eye drop is fully depressed. Tight/lose the screw according to the size/diameter of the eye drop bottle (Figure 13.1(c))	While sitting back or lying down, the user has to tilt his head back and place eye piece securely over the eye. Eye piece has finger space for pulling down lower eyelid with free hand (Figure 13.2(b))
Then close the eyepiece	Gently press the arms of device to install eye drops one per squeeze (Figure 13.2(c))

So, to get rid of these we have a way out here (as we use the compass during geometry in mathematics, we tightened the pencil in the compass in the other arms of the compass for drawing) like the same way we have a screw system in this device. The user needs to put the eye drop bottle inside the dropper and then according to the diameter one can tighten the screw, this way the bottle will be fit inside. After that the device needs to be placed over the eye, by the help of the leverage the extended arms need to be pressed and automatically by this gentle pressure a drop will come out and will be fallen inside the eye.

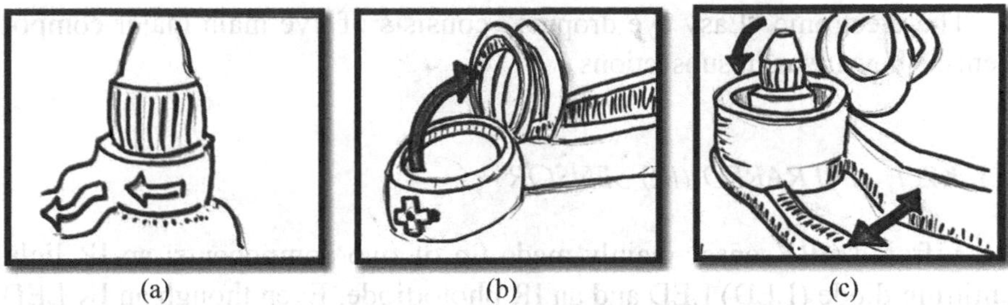

FIGURE 13.1 (a) If the bottle is new remove the protective seal; (b) open eye piece towards arms/leverage of the dropper; (c) place the device on the surface and spread arms open, push the eye drop bottle into the device until base is fully depressed into it tight/lose the screw according to the size/diameter of the eye drop bottle.

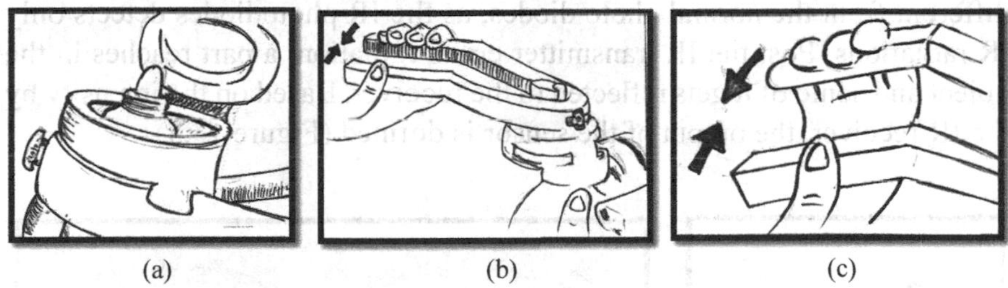

FIGURE 13.2 (a) Open eyepiece and remove bottle cap. Tight the screw or lose the screw according to the size of the bottle; (b) close the eye piece. Rotate the eyepiece towards left or right as per in which eye you are using for the most comfortable administration of the drug; (c) gently press the arms of device to install eye drops one per squeeze.

13.4.2 PHASE-II (EASY EYE DROPPER: ELECTRONIC MODEL)

The steps involved in mechanism of "easy eye dropper" are as follows:

- The device has to be placed on the ocular surface at first, in such a position so that the drop falls on the eye itself;
- Then the user has to Switch on the device;
- The IR sensor which is present beneath the device will detect the ocular surface and it will send the message to the Arduino. Then the Arduino will send this message to the motor to create torque and to start the working of the lever;
- Out of the two levers one will be fixed at a side another will press the drop bottle by creating force onto;
- Then by One press by the lever = one beep sound will be coming out = which will indicate that one drop of medicine from the bottle has come out.

The electronic "Easy eye dropper" consists of five main major components. They are as in subsections.

13.4.2.1 INFRARED (IR) SENSOR

An infrared (IR) sensor mainly made up of two components: an IR light emitting diode (LED) LED and an IR photodiode. Even though an IR LED looks like a regular LED, but the radiation emitted by IR LED is not visible to the human eye. Since the IR sensors can catch hold of the radiation emitted from the IR transmitter, are even called IR receivers. IR receivers come two forms that is called photodiodes and phototransistors. IR photodiodes are different from the normal photo diodes, as the IR photodiodes detects only IR radiations. Post the IR transmitter emits radiation, a part reaches in the object and some of it gets reflected to the receiver. Based on the intensity by the IR receiver, the output of the sensor is defined (Figure 13.3).

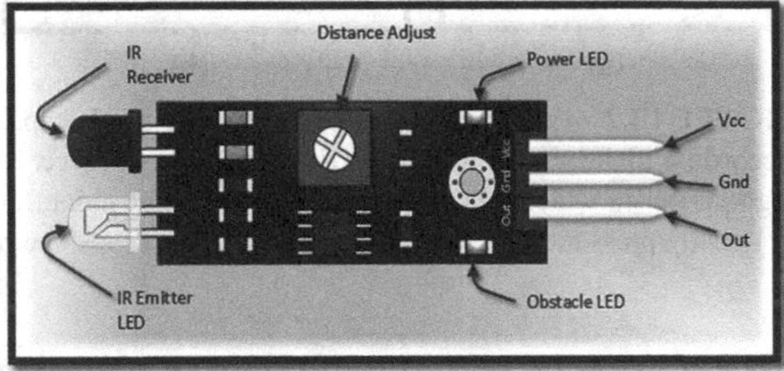

FIGURE 13.3 (a) IR sensor which we have used in our study; (b) diagram of the IR sensor.

Specifications:

- **Operating Voltage Range:** 3.0 V–5.5 V;
- **Device Detection Range:** 2 cm–35 cm;
- **Electricity Consumption:** At 3.0 V: ~23 mA, at 5.0 V: ~43 mA;
- **Output Level:** Output level will reduce /stop on detection of location;
- **Location Detection:** LED indicator.

For the connection of IR obstacle sensor to Arduino Nano, the hardware and software needed are:

- IR obstacle sensor module;
- Arduino nano;
- Arduino IDE (1.0.6 V).

Hardware connections to the sensor:

- Vcc to 5 V;
- Gnd to Gnd;
- Out to digital pin 7.

```
intIRsenseor = 7;
int buzzer = 8;
int lever=4;

void setup()
{
pinMode(IRsenseor, INPUT);
pinMode(lever, OUTPUT);
pinMode(buzzer, OUTPUT);
Serial.begin(9600);
}


void loop()
{
int a =digitalRead(IRsensor);
serial.print(a);
if(a==0)
{
digitalwrite(lever,HIGH);
digitalWrite(buzzer,HIGH);
delay(500);
}
else
{
digitalWrite(lever,LOW);
digitalWrite(buzzer,LOW);
delay(500);
}
}
```

Program for IR Sensor

13.4.2.2 ARDUINO

Arduino nano board is used as the main part of the electronic eye dropper. The Arduino can control the levers by receiving input signals from the IR sensor. The Arduino boards are arranged with sets of digital and analog input/output (I/O) pins that are connected one with the IR sensor, one with the buzzer and one with the lever (Figure 13.4).

FIGURE 13.4 Arduino Nano used in our study.

13.4.2.3 SOUND COMPONENT

1. **Buzzer:** Here we have included a buzzer just because to make the patient aware of that how much drop they have to install on their eye. It helps the patient/the user to get to know that how much amount of drop went off from the drop bottle. A dropper is operated by transducer and it converts electric, oscillating signal in the audible range of 20 Hz to 20 KHz (Figures 13.5 and 13.6).

13.4.2.4 POWER REGULATORS

The input voltage to the Arduino board has been used with 9 volts battery regulated power source (Figure 13.6).

FIGURE 13.5 Buzzer.

FIGURE 13.6 Battery.

13.4.2.5 DC MOTOR

DC motor is also an important part of this project, the main mechanism is mostly dependent on the DC motor, once the IR sensor detects the surface it sends the signal to the Arduino sends the signal to the DC motor, after receiving the signal from the Arduino the DC motors rotates with a speed of 3000 RPM and thus activating the levers. So, from one end the lever will press the drop bottle and from the other end the lever will stay static as a result a force will be created and hence a drop from the bottle will come out. The DC motor will rotate and a small clip will rotate with it which will go and hit a string which is attached with a spring by which the lever will get activate. It will follow the mechanism of cam follower in Creo elements with compression spring with motion (Figure 13.7).

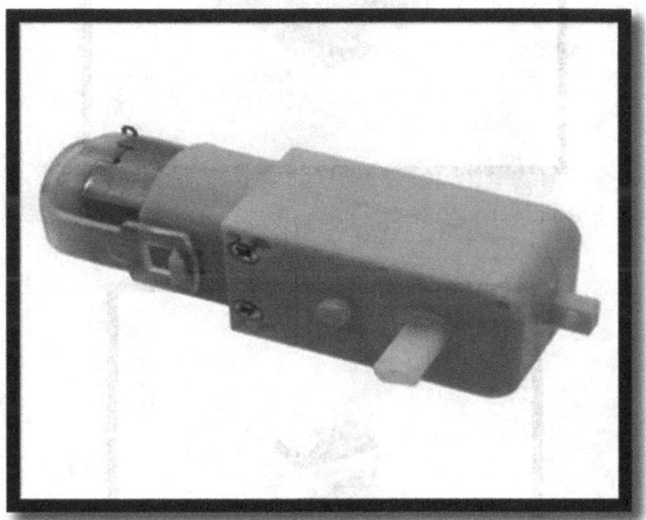

FIGURE 13.7 DC motor used in our device.

13.4.6 PHASE III: FUNCTIONAL APPROACH OF EASY EYE DROPPER

This phase has not been explained in this chapter as this phase will be carried out later. But we have done the planning for this phase. Phase III will be carried out at Pailan Eye Clinic, Kolkata.

We are going to include 30 subjects of various categories (10=glaucoma, 10=Post-Op cataract and 10= Normal Individuals or volunteer). Those subjects will be asked to do an exercise for a period of two months. In the first month they will be asked to use a bottle of approx. 5 ml drug in one eye without the help of any kind of device in the right eye and in left eye to put the drug with the help of non-electronic "Easy Eye dropper" model. Next month they will be given both the non-electronic and electronic model of "Easy Eye dropper" devices and they will be instructed again to put the eye drop with the help of the non-electronic model in right eye and electronic model in left eye. Both the time they have to be use two different bottle for two eyes. After each month's trial or when the drops in one of the bottle finished, the users will be asked to return the used bottle. The amount of the residual drugs which is not been used in the returned bottles will be measured with the help of a measurement syringe.

Also, we are going to ask few general questions to get the feedback on this newly made aid. All procedures and protocols will be standardized. This will be ensured by proper instructional and handy training to the subjects before to hand over the devices. Examiners also will be instructed properly about the assessment procedure for those who will be involved in this study. A Single blinded technique has to be followed during the assessment of feedback round. This part of this study is approved by the institutional review board (IRB) and informed written consent will be taken from all the included participants.

Three parameters which will be mainly observing are: (i) total times of applications without any device and with both non-electronic and electronic devices; (ii) total number of attempts; (iii) wastage of drugs.

13.5 RESULTS AND DISCUSSION

It is worth mentioning that at this point the aim of our study fulfilled as per our objectives. We started our work with three objectives; the first one is to design and implementation of a non-electronic eye dropper aids. We have designed this model with all the possible features of lightweight, handy, easy to use and cheaper cost. These features made this device more unique as compared to all other existing aids which are commercially available in the market. Most of the existing devices are not suitable for patients with severe arthritis or those who have weaker hand grip so that it is very difficult for them to hold those devices properly and to locate exactly over the eye where exactly they need to put the drop. But our design has the advantage of placing it exactly over the eye because of the design of the holder of this device. For the same reason other devices are not equally helpful for the patients with severe visual impairment. Some of the existing devices are fitted with mirror and users have to locate the tip of the bottle on the eye by looking in the fitted mirror, but again this would create some difficulties for the users who have very severe visual impairment. Another disadvantage is the mirror may break easily. In our non-electronic design, we do not have any such kind of mirror arrangement which enhances the durability of this aid.

Few devices designed in such a way that the tip of the bottle needs to touch the ocular surface. This is very harmful in terms of the invasive procedure as well as it can be the source of infections. Another way, the invasive technique may damage the ocular surface also if the user get a little of bit distraction of mind or handshake while installing the drugs. Another disadvantage of the invasive technique is that the user needs to keep the device clean as it may contaminate because of the multiple touches. In our non-electronic device, there is no chance of touching the ocular surfaces because of the design of its holder.

Though most of the devices are commercially available but still very few designs are available in the Indian market. So, most ocular drug user are still unaware of these small technique, hence they are either dependent on others or wasting eye drops while they are trying to put independently.

The phase two design is completely a newer invention. There are no such electronic aids for this task. Our main aim of making this electron design was to remodeling or modifies few features which may add few extra benefits for the users. In non-electronic device the tip of the device has to put manually over the proper location so there is a still chance of missing the proper eye location where exactly one needs to put the drug. The electronic design can detect the exact location with the help of IR sensor which is connected with an Arduino sensor. The sensor will detect the eye position and once detected it will confirm through a sound that will be coming from a buzzer.

Also, the buzzer will create a sound to make the patient aware of that how much drop they have to install on their eye. It helps the patient /the user to get to know that how much amount of drop went off from the drop bottle. These unique features will definitely reduce the number of trials while putting eye drop and equally will reduce the wastage of drugs as compared to any existing non-electronic aids.

We are going to do few additional things in the phase III part of our study. This is completely a new approach that has never been done in most of the previous works. The previous researches show only the constructions and implementations of a non-electronic eye dropper aid with multiple features but most of the studies have not shown the results in real-world situations. So, there will be a bit doubt for the user about the effectiveness of these kinds of aids in a real world situations and how it is operated by the actual needy patients mainly glaucoma, post-operative cases or a severe visual impairment who are the frequent ocular drug users. The results from phase III must be a positive indicator to differentiate among the without device, non-electronic, and electronic one. However, still in this chapter, we have shown only the design and construction phases. Once we will complete the phase III, we will be adding extra information to prove the effectiveness of our invention.

13.6 CONCLUSION

The non-electronic and electronic "Easy Eye dropper" design will definitely add an extra platform for the frequent ocular drug users who are hindering the barriers of putting drugs in day-to-day basis. It leads to good results for all kinds of users including glaucoma, post-operative cases, or nay general drug user. This is reliable, light weight, portable, low power consumption, and low cost. The effectiveness and affordability will make it more acceptable to the users. This new improvisation in this technique will surely add an extra benefit in terms of time saving, eye drop installing effort and cost effective in the process of independent-installation the eye drops.

KEYWORDS

- **easy eye dropper**
- **electronic Arduino**
- **motor**
- **non-electronic**
- **power regulators**
- **sound component**

REFERENCES

Efron, N., (2013). *Putting Vital Stains in Context, 96*(4), 400–421.

https://www.amazon.com/Cress-Dropper-EyeDrops-Professional-Model/dp/B004R8TOGM (accessed on 08 December 2021).

https://www.guldenophthalmics.com/products/index.php/opticare-eye-drop-dispenser.html (accessed on 08 December 2021).

Junqueira, D., Lopes, F., & Souza, F., (2015). *Evaluation of the Efficacy and Safety of a New Device for Eye Drops Instillation in Patients with Glaucoma, 9*, 367–371.

Kahook, M. Y., (2007). *Developments in Dosing Aids and Adherence Devices for Glaucoma Therapy: Current and Future Perspectives, 4*(2), 261–266.

Mukesh, T., Chappidi, K., & Naga, S., (2016). *Innovative Bulls Eye Drop Applicator for Self-Instillation of Eye Drops, 43*(3), 11.

Sharma, R., Singhal, D., Shashni, A., Agarwal, E., Wadhwani, M., & Dada, T., (2016). *Comparison Of Eye Drop Instillation Before and After Use of Drop Application Strips in Glaucoma Patients on Chronic Topical Therapy, 25*(4), e438–440.

www.maddak.com/autodrop-eye-drop-guide-pr-28115.html?page=2 (accessed on 08 December 2021).

https://magic-touch-eye.myshopify.com/ (accessed on 08 December 2021).

https://www.contactlenskit.com/optiaide/ (accessed on 08 December 2021).

CHAPTER 14

AGRO-NUTRITION ALERT (ANA)

RAJNEESH KUMAR,[1] PRIYA ANAND,[2] DEBAPARNA SENGUPTA,[1] and MONALISA DATTA[1]

[1]*Department of Electrical Engineering, Techno International New Town, Kolkata–700156, West Bengal, India,*
E-mail: rajneeshkumar4044@gmail.com (R. Kumar)

[2]*Department of Information Science and Engineering, AMC Engineering College, Bangalore–560083, Karnataka, India*

ABSTRACT

This chapter is in regard of our major project named agro-nutrition alert (ANA). As most of our group members are from a family with agricultural backgrounds, we have seen farmers facing the issue of low fertility of soil due to mismanagement of chemical fertilizers. Everyone knows that evolution of chemical fertilizers was a huge blessing for farmers in India. Farmers were using chemicals blindly, according to their wish and understandings. But they realized that it was a "temporary fertility." Excessive use of these fertilizers turned the crop-fields barren. During floods, these fertilizers polluted the water also. The life in ponds got destroyed following several other consequences. After facing such critical issues, in recent years, government has arranged a chemical laboratory in each district (Note: Each district has minimum of 1,000–1,500 villages under its jurisdiction) for execution of soil test. But it is a time-consuming process with huge economic costs. So, soil needs to be tested twice a year. After this test, a farmer comes to know the actual amount of fertilizers he should use. Realizing that there was not any electronic, cheap, and time-saving method to detect nutrition level of soil and hence, led to the idea for ANA.

Advances in Data Science and Computing Technology: Methodology and Applications. Suman Ghosal, Amitava Choudhury, Vikram Kumar Saxena, Arindam Biswas, & Prasenjit Chatterjee (Eds.)

14.1 INTRODUCTION

14.1.1 DETAILING ON ANA

We experience a day-by-day reality with the end goal that everything can be controlled and worked. Therefore, yet there are relatively few huge parts in our country where computerization has not been gotten or not been put to an undeniable use on account of a few reasons like cost, achievability, and so on. One such field is that of cultivating. Horticulture has been one of the essential occupations of people since early developments and even today manual intercessions in cultivating are unavoidable. Agro nutrition alert (ANA) is an affordable tool for farmers which guides them to use only required amount of chemical fertilizers, adequate irrigation and sometimes luminosity and temperature (in case of greenhouse farming). Being small in size, it is portable and easy to assemble. Whenever used, it simply sends an SMS to the farmer mentioning the amount of water, fertilizers (urea, potassium, and phosphate) required in the crop-field. In case of greenhouse farming, according to temperature and luminosity recorded and required, farmers can decide air exposure and light exposure inside the house.

ANA has already been fed some general crop data (all parameters for Kharif, Rabi, and some specific commercial crops). For some specific crop (Quitain, 1987), data can easily be fed by introducing new variables in program (can be done by anyone with a little programming skills). ANA is using 4G connectivity for communication, so in case of technology availability, farmers can go for GPRS/LTE communication too.

14.1.2 IMPORTANCE OF SOIL MONITORING

Plant monitoring contributes an important part in the field of agriculture sector in our country as they can be used to grow plants under controlled climatic conditions in order to get optimum productivity. Automating a plant monitoring system and controlling of the climatic parameters, directly or indirectly govern the plant growth and hence their productivity. Farmers can even use their *Nokia 1100* handset to read the messages from ANA and that is the beauty as there is not any need of some kind of smartphones.

Soil is one of the most significant common assets. Precise investigation of soil gives data on nature and sorts of soil (Kumar et al., 2014). The pH value of soil is important part of soil health. The purpose of pH study is to determine the degree of acidity or basicity of soil. The nutrients of soil determine the acidity and basicity of a soil sample. If pH is in range of 0 to 7, then soil is acidic. If pH is 7, then soil is neutral. If pH of soil is 7 or above, then soil is basic. Yellow and red color of soil indicates the presence of Iron-Oxide in soil. Dark black or brown color of soil indicates the soil with high organic matter. Minerals present in the soil can also affect the acidity of the soil. Hence, we can determine the nutrient of soil by using color sensing. RGB are the basic colors which are arranged in bands 321 (RGB). It denote the wavelength of electromagnetic radiations in spectrum band. By using this RGB values we can easily determine the nutrient composition of soil.

Samples of soil are collected and after processing, soil pH is determined by using pH meter and NPK (Regalado and DelaCruz, 2016) values are also determined by using chemical analysis in chemical laboratory. Soil samples were then analyzed with ANA and RGB (Chang and Reid, 1996) color sensor (Kimura, Togami, and Hasimoto, 2010), TCS3200, is being used for detection of color composition of soil. The equation for pH index value for each sample can be determined using the formulae:

$$pH\ index = R * (G/B)$$

From above equation value and measured soil pH value (by using pH-meter) are compared. Similarly, to determine the soil nutrients like Nitrogen, potassium, and Phosphorous, we determine RGB composition of soil sample. Then, we can apply different equations to determine the NPK content in soil.

14.2 FEASIBILITY STUDY

As we know, even an excellent idea is worthless, if it is not feasible. Feasibility is the first and foremost study to take in account. So, let us compare ANA's features with manual soil testing on below bases:

14.2.1 COST

In traditional method, farmers have to visit nearby labs (within 100 km area) and pay a sum of almost Rs. 500 for the test. This test needs to be carried twice a year. ANA costs Rs. 2,000 (one time investment) and can help a group of farmers. So, if a group of four farmers have ANA, it costs only the price of one sample/farmer.

14.2.2 TIME CONSUMPTION

While visiting the soil testing lab, minimum two days are required to carry out the process and generating the report. ANA requires a maximum of 15 minutes to setup and getting the digital report (via SMS).

14.2.3 ACCURACY

From district laboratory, accuracy cannot be determined as a small sample represents acres of crop-field. However, after testing at 3–5 points in a certain field, ANA promises 80–90% accuracy.

14.2.4 LIFETIME

We cannot compare the life of a laboratory with that of a tool. ANA has an approx. lifetime of 7–8 years (must be handled with care).

14.3 SYSTEM DESIGN

14.3.1 HARDWARE REQUIRED

- Arduino UNO R3 microcontroller board (Chavan et al., 2018);
- LDR (4.7 kΩ);
- RGB color sensor (TCS3200);
- LM35 temperature sensor (0–500°C);
- Moisture sensor;
- GSM/GPRS module (SIM800A);
- Jumpers and connecting wires.

14.3.2 SOFTWARE REQUIRED

- Arduino IDE;
- Windows 10 (Operating system/64-bit).

14.3.3 BLOCK DIAGRAM (Figure 14.1)

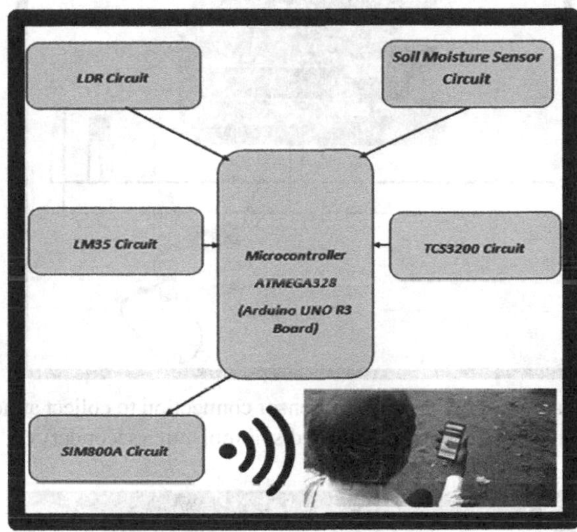

FIGURE 14.1 Diagram showing function of design proposed.

14.3.4 CIRCUIT DIAGRAM (Figures 14.2–14.4)

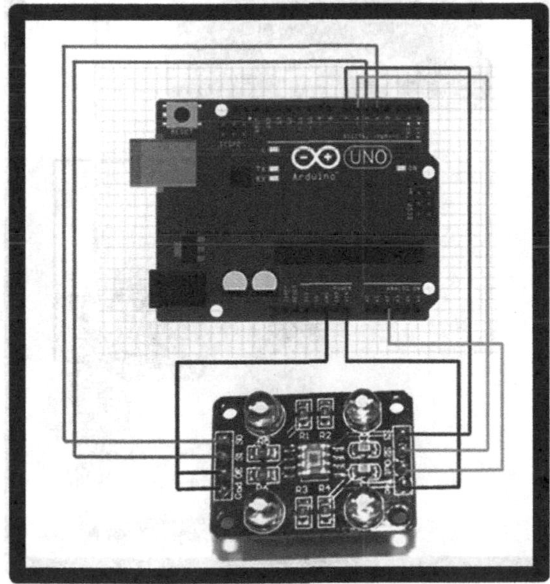

FIGURE 14.2 Diagram showing different sensor connection to collect chemical parameters like nitrogen, phosphorous, and potassium content (primary).

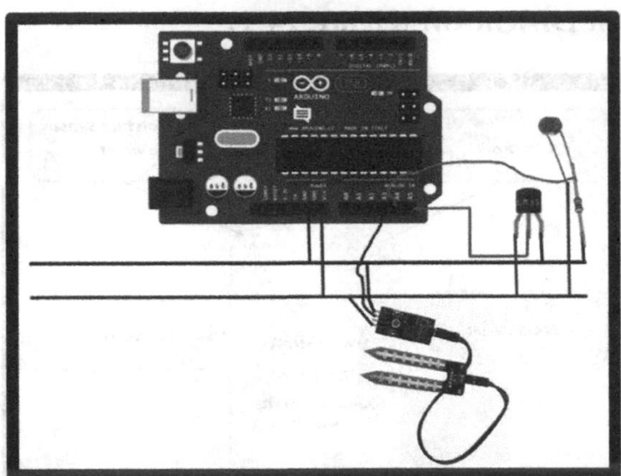

FIGURE 14.3 Diagram showing different sensor connection to collect atmospheric and soil parameters like ambient light, temperature, and soil moisture (secondary).

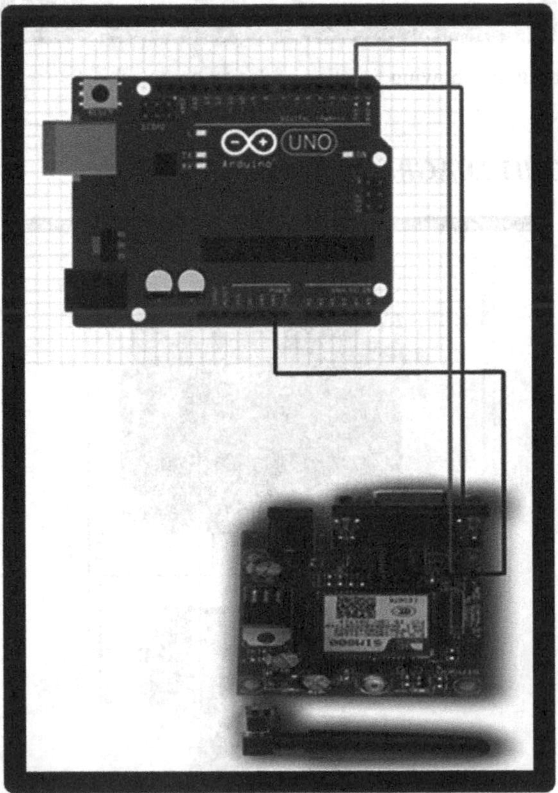

FIGURE 14.4 Diagram showing module connection to transmit collected parameters in form of an SMS (communication).

14.4 RESULTS

14.4.1 *SERIAL MONITOR* (Figure 14.5)

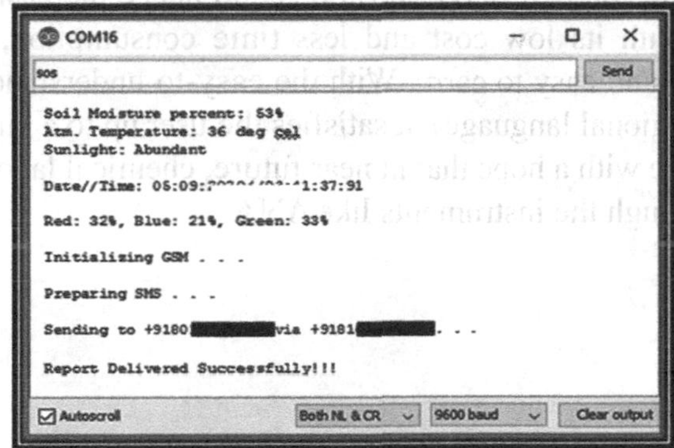

FIGURE 14.5 Diagram showing output in serial monitor of Arduino.

14.4.2 *SMS FORMAT* (Figure 14.6)

FIGURE 14.6 Diagram showing output message screenshot in recipients' mobile phone.

14.5 CONCLUSION

Using the proposed solution, farmers can get their soil report within minutes. They can test the soil at different positions and hence increase the accuracy significantly. With its low cost and less time consumption, it makes the process smooth and easy to carry. With the easy-to-understand SMS format (in English + regional language), it satisfies the user up to a far better extent. We can conclude with a hope that in near future, chemical laboratories could be replaced through the instruments like ANA.

KEYWORDS

- **agro nutrition**
- **agro nutrition alert (ANN)**
- **Arduino**
- **nitrogen**
- **phosphorous**
- **potassium**

REFERENCES

Chang, C. Y., & Reid, J. F., (1996). RGB Calibration for color image analysis in machine vision Image analysis in machine vision. *IEEE Transactions on Image Processing, 5,* 1414–1422.

Chavan, S., Badhe, A., Kharadkar, S., Ware, R., & Kamble, P., (2018). IOT based smart agriculture and soil nutrient detection system. *International Journal on Future Revolution in Computer Science & Communication Engineering, 4*(4).

Kumar, V., Vimal, B. K., Kumar, R., Kumar, R., & Kumar, M., (2014). Determination of soil pH by using digital image processing technique. *Journal of Applied and Natural Science, 6*(1), 14–18.

Quitain, B. M., David, R. M. (1987). Reability of soil test kit (Philippines). Bureau of Soils, Manila (Philippines). *Philippine Journal of Crop Science, 12*(1), 21.

Regalado, R. G., & DelaCruz, J. C., (2016). Soil pH and Nutrient (Nitrogen Phosphorous and potassium) analyzer using colorimetry. *Proceedings of IEEE Region 10 International Conference – TENCON.*

Yoshitsugu Kimura, Kyosuke Yamamoto, Togami, T., Hashimoto, A., Kameoka, T., & Yoshioka, Y. (2010). *Construction of the Prototype System for the Chromatic Image Analysis Using Color Distribution Entropy*, 2438–2442.

PART V

Optimization Technique Using MATLAB Platform

CHAPTER 15

SELECTION OF OPTIMAL MOTHER WAVELET FOR FAULT ANALYSIS IN INDUCTION MOTOR USING STATOR CURRENT WAVEFORM

ARUNAVA KABIRAJ THAKUR,[1] PALASH KUMAR KUNDU,[2] and ARABINDA DAS[2]

[1]*Department of Electrical Engineering, Techno Main Salt Lake, Sector V, Bidhannagar, Kolkata–700091, West Bengal, India, E-mail: arunava.kabiraj007@gmail.com*

[2]*Department of Electrical Engineering, Jadavpur University, Kolkata–700032, West Bengal, India*

ABSTRACT

Current signature analysis is used successfully for induction motor fault detection. FFT based spectrum analysis of current signal is applied for fault detection but it cannot be applied for detection of non-stationary signals. Signal analysis using Wavelet transform provides time and frequency domain information both of signals. It is widely used for signaled noising, signal compression and features extraction. Here, mother wavelet has been selected from different discrete wavelet transform (DWT) families using wavelet coefficients of three phase stator current signals. The data samples of stator current are collected from different faulty (electrical and mechanical faults) induction motors to perform the analysis. Different mother wavelets (db(3–10), sym(3–8), coif(1–5)) are employed for signal decomposition of the distorted stator current signals of different faulty induction motors which

Advances in Data Science and Computing Technology: Methodology and Applications. Suman Ghosal, Amitava Choudhury, Vikram Kumar Saxena, Arindam Biswas, & Prasenjit Chatterjee (Eds.)

are called the wavelet coefficients of the signals. The decomposed signals are assembled back using reconstruction program and square root of mean square error (RMSE) are found for every mother wavelet. The correlation coefficients are also calculated for each mother waveform individually from the original signal and the coefficients of different mother wavelets. The optimal mother wavelet are selected comparing the values of RMSE and correlation coefficients of different mother wavelet. The optimal mother wavelet are found among the 19 mother wavelets which has smallest RMSE value and largest value of correlation coefficient for all faulty conditions and for all phases.

15.1 INTRODUCTION

Fault prediction of induction motor is an important task for modern industry to prevent the industries suffering from downtime cost and lost revenue. The faults have been categorized according to the different components of induction machines-stator, bearing, rotor, and other faults (Singh and Saad, 2003). Various techniques have been applied previously for different faults detection in induction motor (Bhavsar and Patel, 2013; Tavner, 2008; Sangheeta, and Hemamalini, 2017). Among them, signature analysis of stator current is very useful technique to detect faults and to localize different abnormal conditions occurred due to different electrical and mechanical faults in the induction motor (Benbouzid, 2000; Blödt and Rostaing, 2008; Henao, Razik, and Capolino, 2005; Mehala, and Dahiya, 2007; Benbouzid, Vieira, and Theys, 1999; Singh, Kumar, and Kumar, 2014; Kalaskar and Gond, 2014; Gómez and Sobczyk, 2013). MCSA can detect the different motor problems at an early stage to prevents it from secondary damage and complete failure (Henao, Razik, and Capolino, 2005; Amel, Laatra, Sami, and Nourreddine, 2013). MCSA is a less expensive technique than others because electrical signals are cheaper, simpler, and easier to measure and multiple sensors are not required in this technique. Fast Fourier transform (FFT) usually converts time to frequency of signals and vice versa but it does not provide simultaneous time and frequency information. It has limited time duration and frequency bandwidth hence it is not efficient for representing discontinuities. It represents a signal with a few coefficients. Short term Fourier transform (STFT) can be applied for no stationary signal analysis which gives time-frequency informational both but there is a problem with STFT because for all frequencies it provides constant resolution using the same window for entire signal analysis. To overcome the drawback of

FFT and STFT the wavelet transform is used for accuracy in fault analysis and it has gained widespread acceptance in signal processing (Yan, Gao, and Chen, 2014). It has been shown that wavelet analysis can improve the fault diagnosis (Ye, Wu, and Sadeghian, 2003; Poshtan and Zarei, 2007). Discrete wavelet transform (DWT) decomposes the signal into several wavelet coefficients ('approximation' and 'detail') and the decomposed coefficients can be implemented by using a filter (Ponci, et al., 2007; Burrus, Gopinath, and Guo, 1998; Bhola et al., 2013). Monique (1997) compared between non-orthogonal and orthogonal wavelet and he has shown that usually orthogonal wavelet have better de-noising performances than the non-orthogonal. Orthogonal wavelet decomposition is useful for the non-stationary signal, meaning that its variance depends on the window of the data under consideration. Its decomposition provides information on the variability of wave height at the various timescales with time (Daviu et al., 2006). The improvement of de-noising performances with the increase of data length has been shown for all thresholding schemes and transforms. There are different orthogonal wavelets (Haar, Daubechies (db), Symlets, Coiflets (Coif)) which are normally used for machine fault analysis. Despite the satisfactory performance of DWT, the selection of optimal mother wavelet is challenging task for different mechanical and electrical faults in the machines. Researchers are applied different techniques previously to select the optimal mother wavelet for detect of one particular fault in the machines using vibration signals (Rafiee et al., 2009; Wang and McFadden, 1995) but they are not selected optimal mother wavelet for analysis of more than one type of fault using current signals. Optimal mother wavelet is selected in this work among 19 mother wavelet functions for the thee phase current signals of six different faulty induction motors. Haar (db1) wavelet is not used here because the wavelet family is formed by sequence of rescaled 'square-shaped' functions together and it is not continuous and hence it cannot be used for distorted continuous signal analysis. Distorted stator currents are decomposed using MATLAB program applying different mother wavelets functions (db(3–10), sym(3–8), coif(1–5)) for different faulty conditions and the decomposed signals are assembled back using reconstruction program. The square root of mean square error (RMSE) are found from actual and reconstructed signals and the coefficients of correlation are also calculated using the actual signal and denoised signal of different mother wavelets for each phase of each faulty motor. The optimal mother wavelet has been selected based on the lowest RMSE and highest correlation coefficient value of each phase of each faulty I.M.

15.2 METHOD

15.2.1 WAVELET ANALYSIS

Wavelet transform is normally used to overcome the drawback of FFT and STFT. Wavelet transform has become one of the fast-evolving signal processing, emerging, and mathematical tools in the recent years for many distinct merits and the MRA led to the wavelet transform famous for its simplicity and recursive filtering algorithm to compute the wavelet decomposition of the signal from its finest scale approximation. It has gained widespread acceptance for fault analysis (Aktas and Turkmenoglu, 2010) in the machines for signal decomposition into a set of basis functions providing variable window size, the basic functions are called wavelets. Applying translations and scaling (stretch/compress) on the 'mother' wavelet ψ(t) the basis can be constructed.

$$\psi_{t,T}(t) = \frac{1}{\sqrt{S}}\psi\left(\frac{t-T}{S}\right) \tag{1}$$

where; 'S' and 't' are the scaling and shifting parameter, respectively. Wavelet transform is separately computed for different time-domain signal segments at different frequencies. Wavelet coefficients are obtained, at a first level of signal decomposition by applying a mother wavelet. The repetition of this process depends on the scaling and translation of mother wavelet. The mother is named due to formation the basis for various processes of transformation. A mother wavelet can be imagined as a windowed function which moves/shifts along the signal of time-series from the time $t = 0$ to $t = T$. In the window the portion of the signal is multiplied by the mother wavelet and then it is integrated over all times to get the coefficients of wavelet. Wavelet family can represent that signal in as few coefficients as possible is generally considered to be best suited for a particular application. Every application would have one mother wavelet that would be most suited to it. In wavelet theory it is tried to decompose the spectra into high and low frequency content in some peculiar way for making some information extraction process simpler. DWT decomposes the signal into a set of mutually orthogonal wavelet basis functions.

The functions of wavelet are dilated, translated, and scaled versions of a common function Ψ which is known as the mother wavelet. There are different mother wavelet functions (db), Coiflets (Coif), Symlets (Sym)) which are used for different fault detection in the machines. In DWT based decomposition the results in useful data contained in 'approximate' and 'details' parts as shown in Figure 15.1. The low frequency signal components called 'approximate' and the high frequency signal components called 'detail.' Computing the 'n' level decomposition higher detail parts being removed, thereby the overall frequency of the resulting data is reduced. The DWT is implemented in this work to decompose a current signal into scales with resolution of different time and frequency using a multi-resolution signal decomposition algorithm.

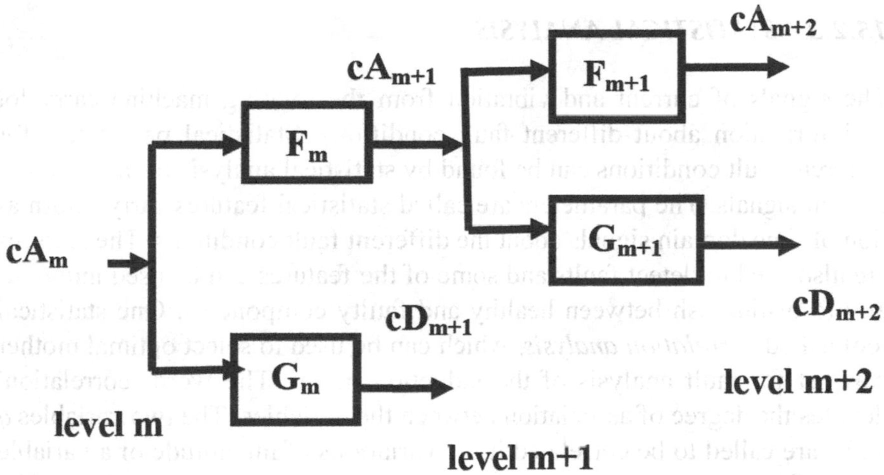

FIGURE 15.1 m-th level wavelet decomposition.

15.2.2 CURRENT SIGNAL RECONSTRUCTION BASED ON DECOMPOSITION OF WAVELET

The decomposition analysis is described in DWT and the other half of the story is how the actual signal comes back into original form by assembling the decomposed components of the signal without losing any information. This name of the process is called synthesis of reconstruction. The mathematical manipulation of the process of reconstruction is called the inverse discrete wavelet transform (IDWT). The process of wavelet reconstruction consists of filtering and upsampling. Upsampling (or interpolation) is done by zero padding between every two coefficients. The filters design for decomposition and reconstruction is based on the well-known technique called 'Quadrature Mirror Filters.' The reconstructed approximations and details are the actual constituents of original signal. In fact, it is found when the approximate coefficient vector cA1 and detail coefficient cD1 are combined because the coefficient vectors are produced by the help of down sampling. Before combining the approximations and details reconstruction is necessary. The process can be extended to the multi-level component analysis; it is found the similarity in relationships hold for all the reconstructed signal constituents. That is, the original signal can be reassembled by several ways.

15.2.3 STATISTICAL ANALYSIS

The signals of current and vibration from the rotating machine carry lot of information about different fault conditions. Statistical parameters for different fault conditions can be found by statistical analysis of the acquired current signals. The parameters are called statistical features carry information of time domain signals about the different fault conditions. The features are also used to detect faults and some of the features can be used individually to distinguish between healthy and faulty components. One statistical tool called *correlation analysis,* which can be used to select optimal mother wavelet for fault analysis of the induction motor. The word 'correlation' denotes the degree of association between the variables. The two variables q and r are called to be correlated if the variations of magnitude of a variable tend to be accompanied by magnitude variations of the other variable. The two variables are called positively correlated if r increases when q trends to increase. The two variables are called negatively correlated if r increases as q tends to decrease. The two variables are called uncorrelated if the values of q are not affected by changes in the values of r. R is the matrix of correlation coefficients calculated from an input matrix P as input whose rows and columns are observations and variable respectively. R matrix is related to the covariance matrix $C_V = \text{cov (P)}$ by:

$$R(q,r) = \frac{C_V(q,r)}{\sqrt{C_V(q,q)C_V(r,r)}} \tag{2}$$

Correlation technique has been used to calculate correlation coefficient from original signal and denoised signal and to detect optimal mother wavelet (Al-Qazzaz et al., 2007). The block diagram represents the steps of correlation coefficients from different mother wavelet family has been shown in Figure 15.2.

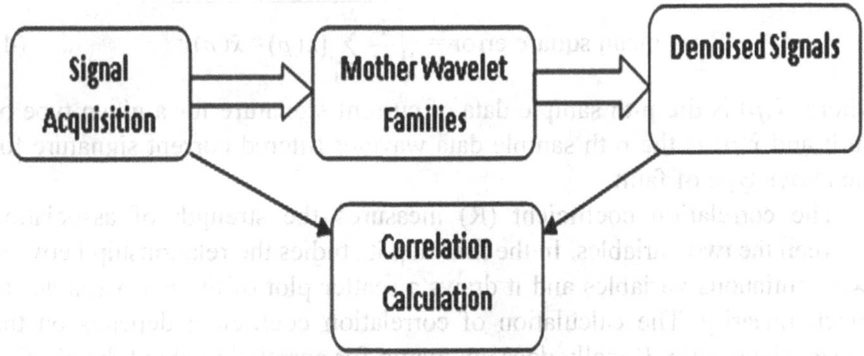

FIGURE 15.2 The correlation block diagram between the noisy and de-noised signals using mother wavelet families.

15.2.4 SELECTION OF OPTIMAL MOTHER WAVELET

Selection of optimal mother wavelet function is an essential task for wavelet analysis for demonstration of its advantages in coefficient reconstruction, denoising, feature extraction and component separation from the time domain and frequency domain signal both. This transform involves correlating between the signal being analyzed and a prototype wavelet function. Thus, the selection of mother wavelet function can influence the fault analysis performance of induction motor. There is a variation of qualities of different mother wavelets criteria viz., (i) symmetry, (ii) speed of convergence to zero, (iii) vanishing moments number, (iv) regularity, (v) existence of scaling function, etc. Optimal mother wavelet selection is also essentials to find the accuracy of fault analysis.

In this work optimal mother wavelet has been selected comparing the values of correlation coefficients and root mean square error (RMSE). The current sample of original and reconstructed signal has been used for analysis and square RMSE has been computed. The mean squared error (MSE) measures the average of the squares of deviations or errors, i.e., the difference between the original and reconstructed signal. The RMSD or RMSE is found taking the square root of MSE. For an unbiased estimator the RMSE has the same units as the quantity being estimated. The RMSE is also known as the standard deviation (SD) as it is the square root of the variance. Mean square error and root mean square error can be defined as:

$$\text{Mean square error} = \frac{1}{M}\sum_{n=1}^{M}[x(p)-\tilde{x}(p)]^2 \quad (3)$$

$$\text{Root mean square error} = \sqrt{\frac{1}{M}\sum_{n=1}^{M}[x(p)-\tilde{x}(p)]^2} \quad (4)$$

where; $\tilde{x}(p)$ is the p-th sample data of current signature for a given type of fault and $\tilde{x}(p)$ is the p-th sample data wavelet filtered current signature for the above type of fault.

The correlation coefficient (R) measures the strength of association between the two variables. In the first step it studies the relationship between two continuous variables and it draws a scatter plot of the two variables to check linearity. The calculation of correlation coefficient depends on the linear relationship. It really does not matter for correlation about the plotting of variables in the axes. However, on the x-axis the independent (explanatory) variable is plotted and, on the y-axis, the dependent (response) variable is plotted conventionally. The higher strength of association between the variables depends on how nearer the scatter points are to a straight line. It does not matter about the units are used for measurement. Computed correlation coefficients and computed RMSE using decomposition by wavelet transform and reconstructed by inverse wavelet transform are compared for selecting optimum mother wavelet. The coefficients of different mother wavelets (Daubechies(db3-db10), Coiflets(Coif1-Coif-5), Symlets(Sym3-Sym8)) are used for correlation analysis. Correlation coefficients are computed using noisy and denoised signal. The mother wavelet is optimal which has minimum RMSE value and largest correlation coefficient. The steps for optimal mother wavelet selection have been described in flowchart as shown in Figure 15.3.

15.3 EXPERIMENTAL SET UP

As shown in Figure 15.4, laboratory test set up is made for experimental work with inverter fed three phase induction motor (2 pole, 60 Hz, 1/2 H.P, 3,450 rpm) acts as a prime mover of a dc generator. Four incandescent bulbs are used as loads the dc generator. The three phase currents of stator of six induction motors with different fault conditions are used for the analysis, viz. Bearing fault, Broken rotor bar, Misaligned rotor, Stator winding fault, Unbalanced rotor, Single phase voltage unbalance. One three phase digital power analyzer is used to capture three phase currents of each faulty induction motor. An essential data acquisition system is used to interface with motor-generator system to PC through USB cable. The captured three phase currents of each induction motors are converted to numerical data samples (amplitude vs time) by the help of computer software and the data samples are stored into pc. The current wave of each phase of each faulty induction motor is converted to 1,002 numerical data sample and stored as CSV file.

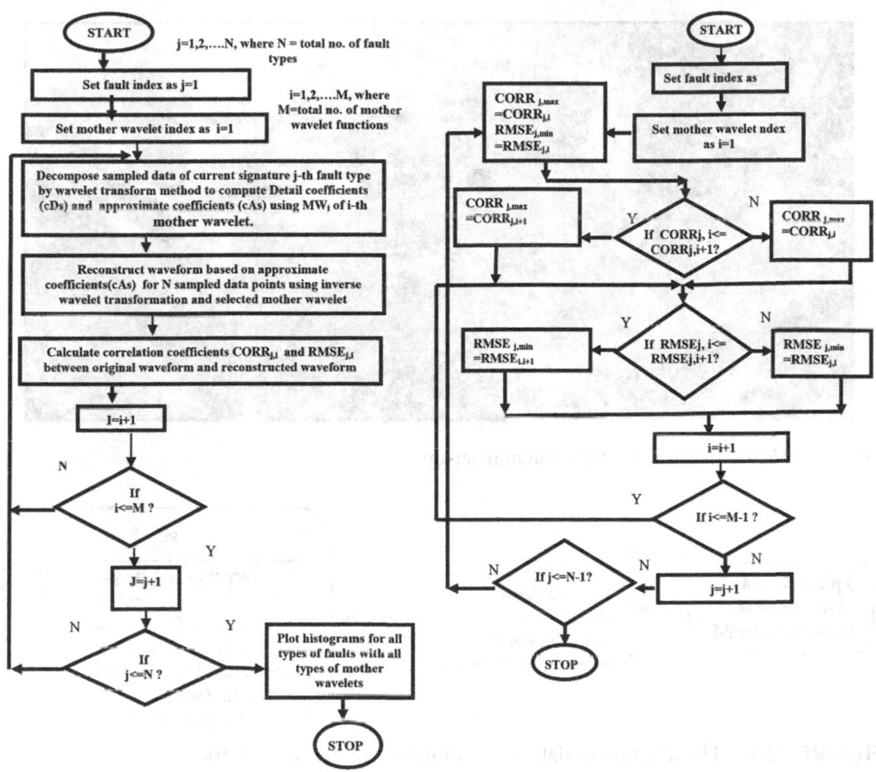

FIGURE 15.3 (i) Computation of correlation coefficients and RMSE using decomposition by wavelet transform and reconstruction by inverse wavelet transform; (ii) selection of optimum mother wavelet function.

15.3.1 DATA ACQUISITION SYSTEM

One digital three phase power analyzer (Yokogawa WT 500) is used as a data acquisition system which captures the three phase stator currents from induction motors with different fault conditions and it interfaces with PC through USB cable as shown in Figure 15.5. Zero-Flux (TM) current sensors are used in this system because it allows the precise measurement of currents providing accurate phase-shift information. The system (WT 500) provides and supports a wide range of voltage from 1.5 V to 1,000 V. The sampling time is set to 5 ms for data acquisition task. The numerical values of current waves are essential for the analysis. Therefore, the WT viewer software is used in the PC to convert the current waveforms to numerical value and the data samples are stored into.'CSV' file format.

FIGURE 15.4 Laboratory experimental set-up.

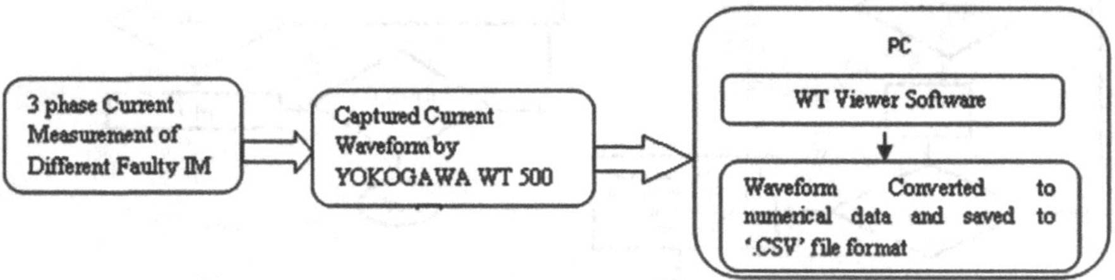

FIGURE 15.5 The diagram of data acquisition by Yokogawa WT 500.

15.4 RESULT AND ANALYSIS

The data samples (amplitude vs. time) of three phase currents are collected from six induction motors with different types of fault (bearing fault, broken rotor bar, misaligned rotor, unbalanced rotor, stator winding fault, single phase voltage unbalance) for the analysis. The wavelet coefficients are computed using different mother wavelet functions (db(3–10), coif(1–5), sym(3–8)) from the data samples of phase currents of each faulty induction motor by MATLAB routine. The signals are reconstructed from different 'approximate' coefficients. Root mean square error (RMSE) has been computed from original and reconstructed signal for different mother wavelets in different faulty conditions. Correlation coefficients are also computed for original signal and denoised signal using MATLAB routine. The histograms (Figures 15.6–15.23) are plotted using RMSE and correlation coefficient values in different faulted conditions for mother wavelet families. From the histograms and the tables (Tables 15.1 and 15.2), it may be concluded that 'Symlet 5' has minimum RMSE value and has maximum correlation coefficients for all phases in all faulty conditions.

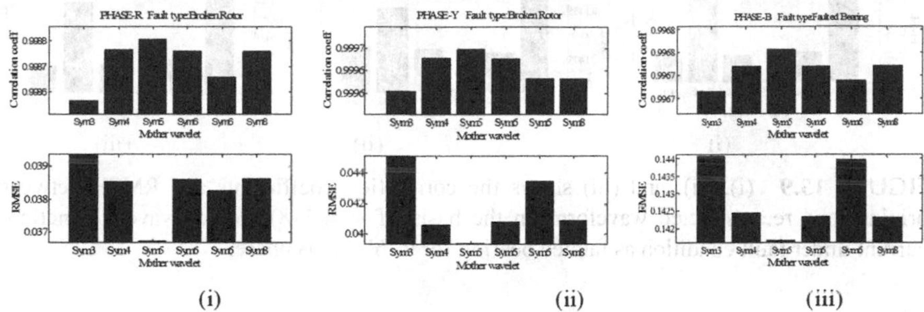

FIGURE 15.6 (i), (ii) and (iii) shows the correlation coefficients and RMSE between original and reconstructed current waveform on the basis of Sym(3–8) mother wavelet functions under fault condition as broken rotor bar for R, Y, and B phase.

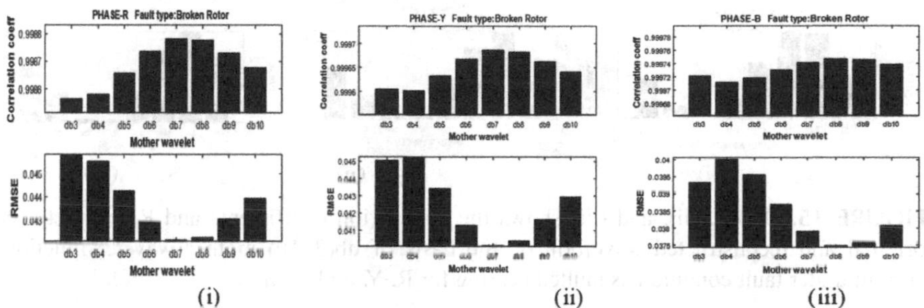

FIGURE 15.7 (i), (ii), and (iii) shows the correlation coefficients and RMSE between original and reconstructed waveform on the basis of db(3–10) mother wavelet functions current under fault condition as broken rotor bar for R, Y, and B phase.

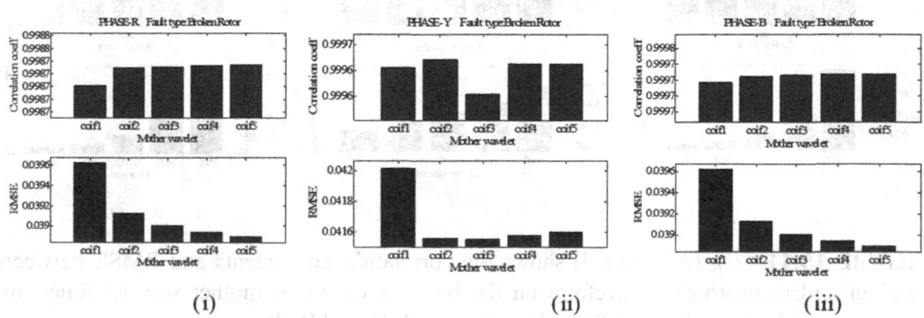

FIGURE 15.8 (i), (ii), and (iii) shows the correlation coefficients and RMSE between original and reconstructed waveform on the basis of coif(1–5) mother wavelet functions current under fault condition as broken rotor bar for R, Y, and B phase.

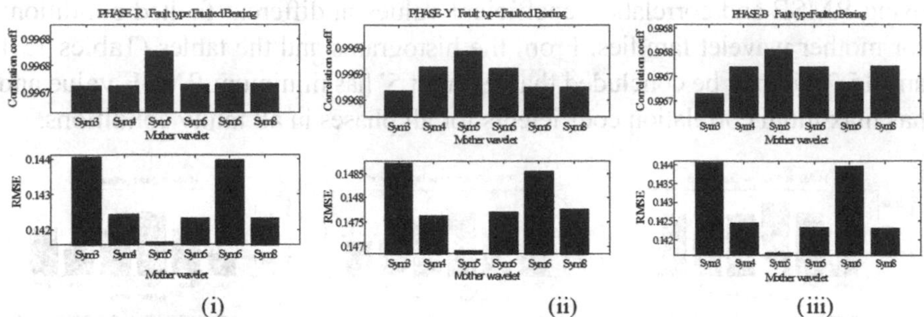

FIGURE 15.9 (i), (ii), and (iii) shows the correlation coefficients and RMSE between original and reconstructed waveform on the basis of sym(3–8) mother wavelet functions current under fault condition as faulted bearing for R, Y, and B phase.

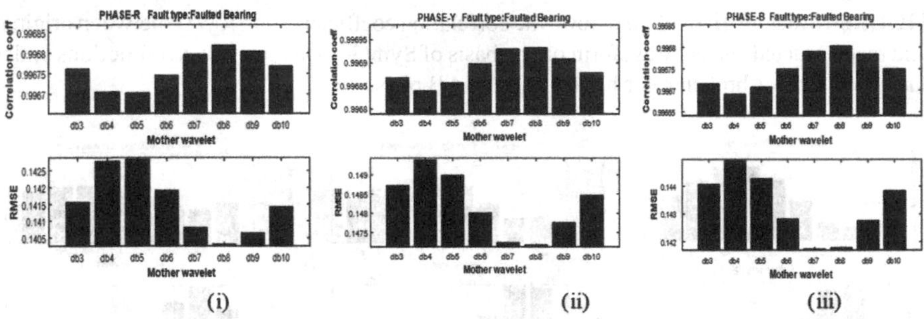

FIGURE 15.10 (i), (ii), and (iii) shows the correlation coefficients and RMSE between original and reconstructed waveform on the basis of db(3–10) mother wavelet functions current under fault condition as faulted bearing for R, Y, and B phase.

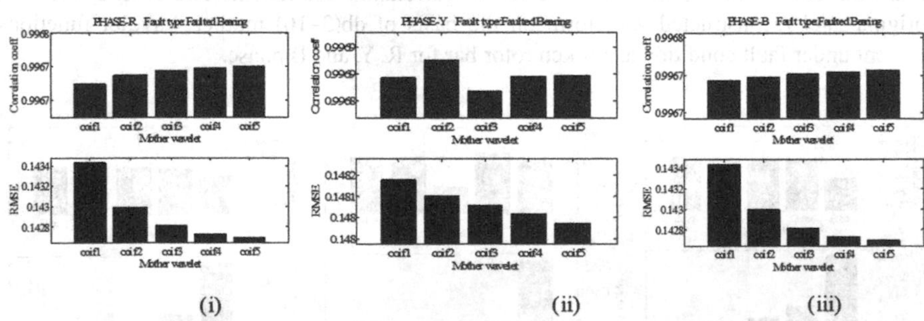

FIGURE 15.11 (i), (ii), and (iii) shows the correlation coefficients and RMSE between original and reconstructed waveform on the basis of coif(1–5) mother wavelet functions current under fault condition as faulted bearing for R, Y, and B phase.

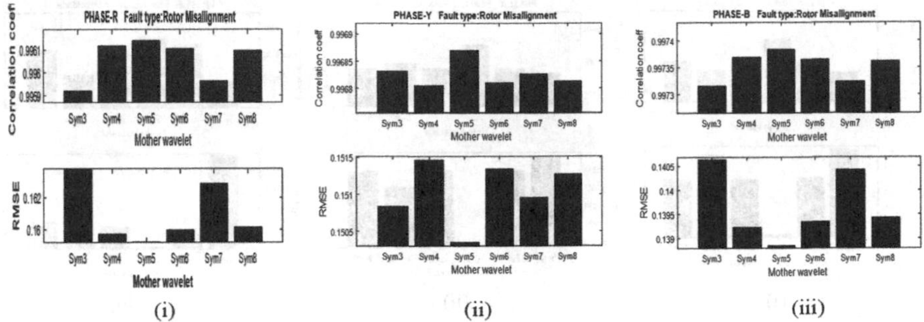

FIGURE 15.12 (i), (ii), and (iii) shows the correlation coefficients and RMSE between original and reconstructed waveform on the basis of sym(3–8) mother wavelet functions current under fault condition as rotor misalignment for R, Y, and B phase.

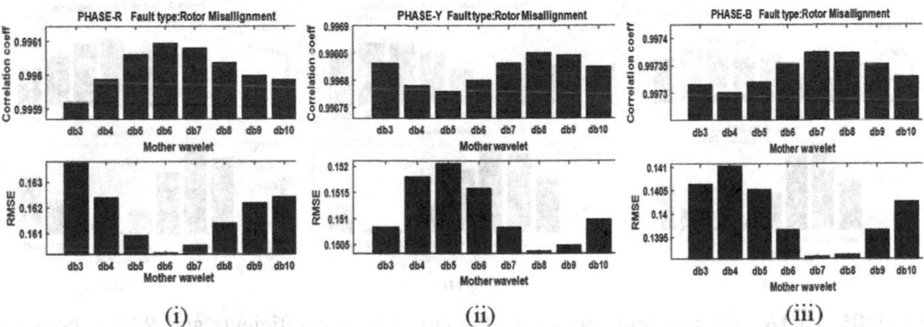

FIGURE 15.13 (i), (ii), and (iii) shows the correlation coefficients and RMSE between original and reconstructed waveform on the basis of db(3–10) mother wavelet functions current under fault condition as rotor misalignment for R, Y, and B phase.

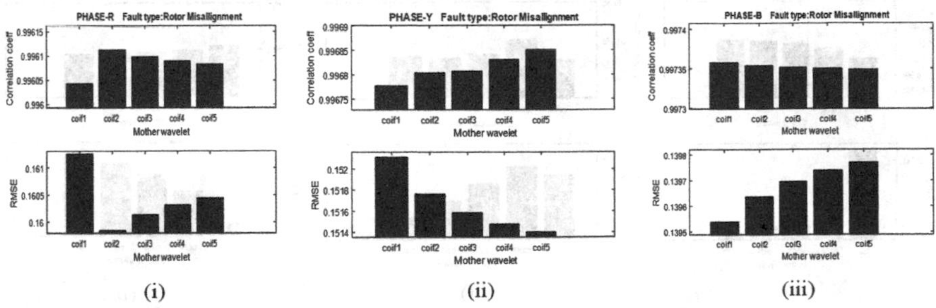

FIGURE 15.14 (i), (ii), and (iii) shows the correlation coefficients and RMSE between original and reconstructed waveform on the basis of coif(1–5) mother wavelet functions current under fault condition as rotor misalignment for R, Y, and B phase.

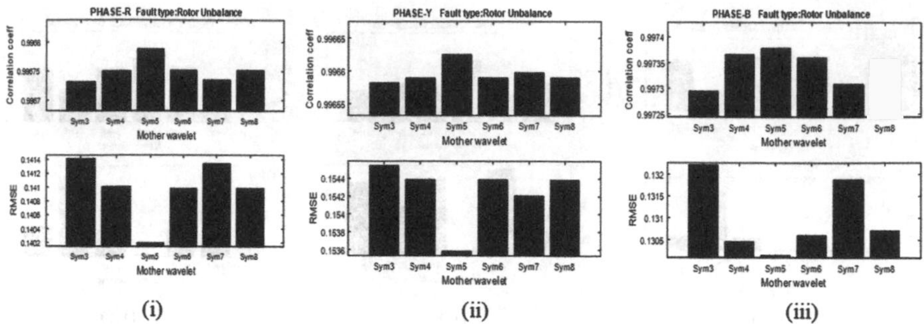

(i) (ii) (iii)

FIGURE 15.15 (i), (ii), and (iii) shows the correlation coefficients and RMSE between original and reconstructed waveform on the basis of sym(3–8) mother wavelet functions current under fault condition as rotor unbalance for R, Y, and B phase.

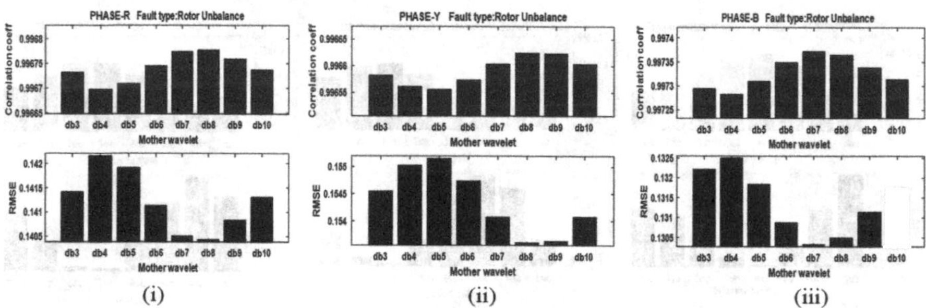

(i) (ii) (iii)

FIGURE 15.16 (i), (ii), and (iii) shows the correlation coefficients and RMSE between original and reconstructed waveform on the basis of db(3–10) mother wavelet functions current under fault condition as rotor unbalance for R, Y, and B phase.

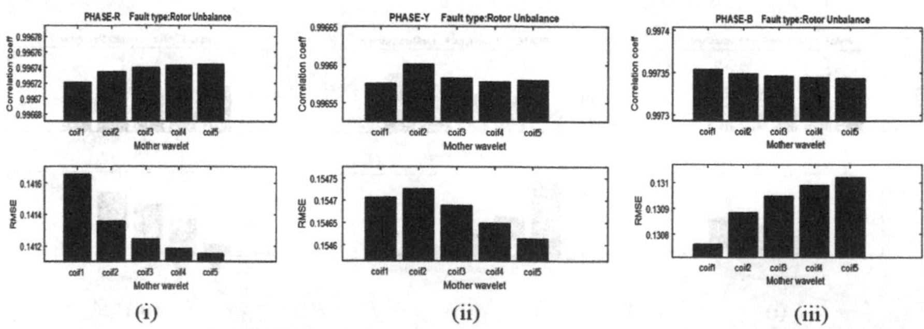

(i) (ii) (iii)

FIGURE 15.17 (i), (ii), and (iii) shows the correlation coefficients and RMSE between original and reconstructed waveform on the basis of coif(1–5) mother wavelet functions current under fault condition as rotor unbalance for R, Y, and B phase.

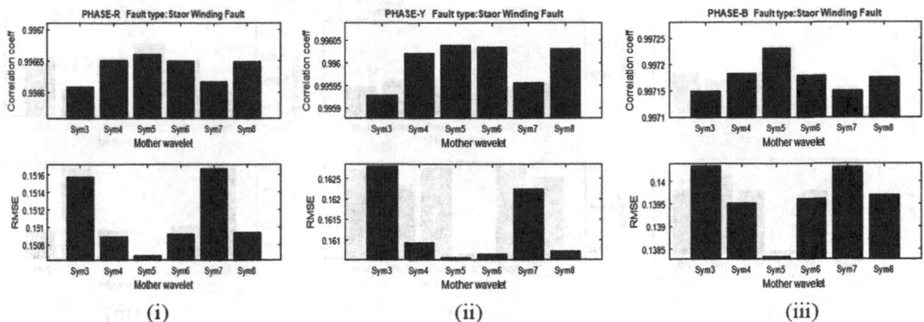

FIGURE 15.18 (i), (ii), and (iii) shows the correlation coefficients and RMSE between original and reconstructed waveform on the basis of sym(3–8) mother wavelet functions current under fault condition as stator winding fault for R, Y, and B-phase.

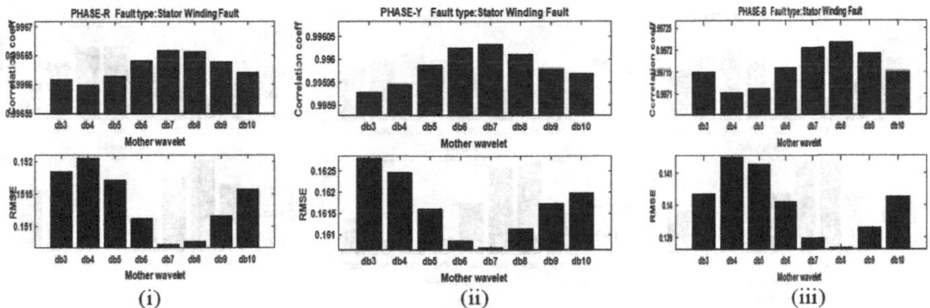

FIGURE 15.19 (i), (ii), and (iii) shows the correlation coefficients and RMSE between original and reconstructed waveform on the basis of db(3–10) mother wavelet functions current under fault condition as stator winding fault for (a) R, Y, and B-phase.

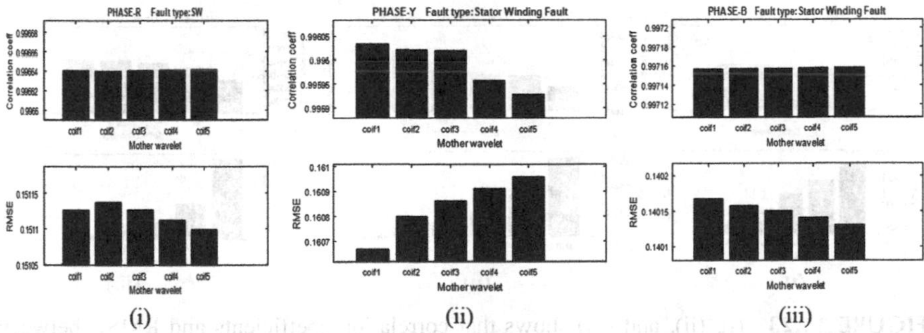

FIGURE 15.20 (i), (ii), and (iii) shows the correlation coefficients and RMSE between original and reconstructed waveform on the basis of coif(1–5) mother wavelet functions current under fault condition as stator winding fault for R, Y, and B phase.

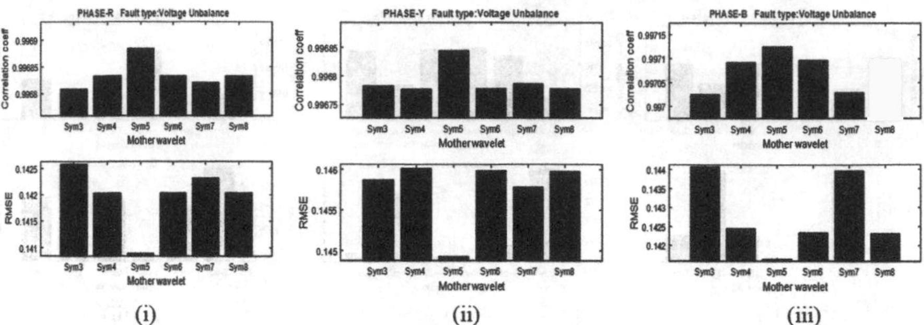

FIGURE 15.21 (i), (ii), and (iii) shows that correlation coefficients and RMSE between original and reconstructed waveform on the basis of sym(3–8) mother wavelet functions current under fault condition as single phase voltage unbalance for R, Y, and B phase.

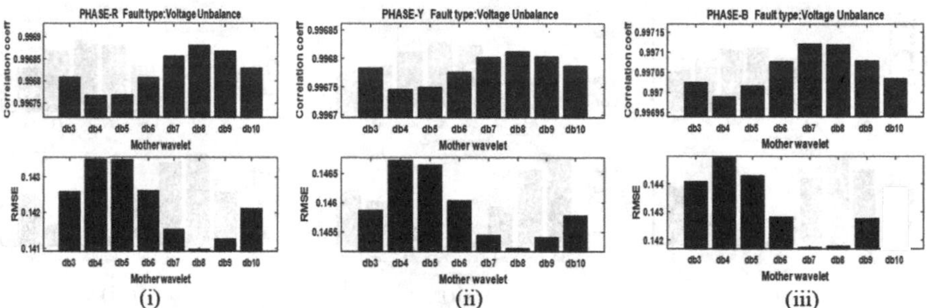

FIGURE 15.22 (i), (ii) and (iii) shows that correlation coefficients and RMSE between original and reconstructed waveform on the basis of db(3–10) mother wavelet functions current under fault condition as single phase voltage unbalance for R, Y, and B phase.

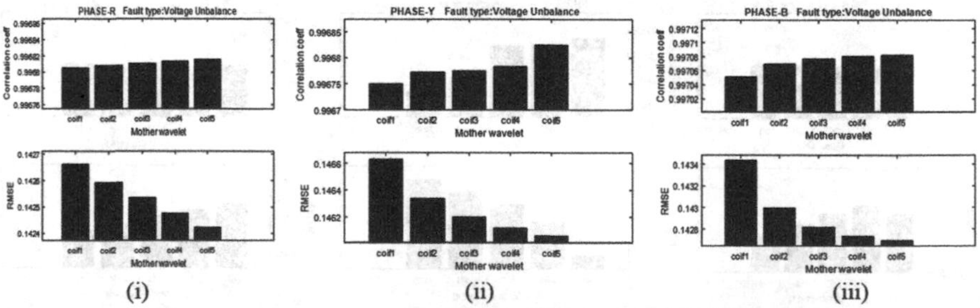

FIGURE 15.23 (i), (ii), and (iii) shows that correlation coefficients and RMSE between original and reconstructed waveform on the basis of coif(1–5) mother wavelet functions current under fault condition as single phase voltage unbalance for R, Y, and B phase.

TABLE 15.1 Correlation Coefficients between Original and Reconstructed Waveform Using Mother Wavelet Sym5 Under Different Faulty Conditions for R-Y and B Phases

Mother Wavelet	BRF	BRB	RML	RTU	SWF	FVU
R	0.996827021	0.998806583	0.996141276	0.996788131	0.996660663	0.996883995
Y	0.996944169	0.999697144	0.996869577	0.996625068	0.996039162	0.996844161
B	0.996806649	0.999756636	0.997383615	0.997378658	0.99723101	0.997124796

TABLE 15.2 Root Mean Square Error between Original and Reconstructed Waveform Using Mother Wavelet Sym5 Under Different Faulty Conditions for R-Y and B Phases

Mother Wavelet	BRF	BRB	RML	RTU	SWF	FVU
R	0.140121385	0.04173592	0.15925463	0.140197881	0.150676488	0.140892016
Y	0.14689829	0.039426956	0.150351419	0.153585049	0.16056938	0.144927614
B	0.14163817	0.03674286	0.13884757	0.130144044	0.13832768	0.141638165

Note: BRB: Broken rotor bar; BRF: Bearing fault; RML: Misaligned rotor; RTU: Unbalanced rotor; SWF: Stator winding fault; FVU: Single phase voltage unbalance.

15.5 DISCUSSION AND CONCLUSION

Optimal mother wavelet selection is an essential task for accuracy in different types of electrical and mechanical fault detection of induction machine using current analysis. It is the critical task to choose the optimal mother wavelet from different mother wavelet for induction motor fault analysis. In this work the optimal mother wavelet has been identified for fault detection in the induction motor analyzing the three phase stator currents. The original distorted current signal from different faulty induction motor and the denoised current signal using different mother wavelet have been compared and root mean square error (RMSE) has been found. Comparing the RMSE and the values of correlation coefficient it has been found that the Symlet5 mother wavelet is the optimal one for detection of electrical and mechanical type faults in the induction motor. As shown in the tables for all phases Sym5 has minimum RMSE values and the maximum values of correlation coefficient than other wavelet and hence it can be said that Sym5 is the optimal one.

KEYWORDS

- **correlation coefficient**
- **discrete wavelet transform (DWT)**
- **mother wavelet selection**
- **root of mean square error (RMSE)**

REFERENCES

Aktas, M., & Turkmenoglu, V., (2010). Wavelet based switching fault detection in direct torque control induction motor drives. *IET Science, Measurement, and Technology, 4*(6), 303–310.

Al-Qazzaz, N. K., Ali, S. H. B., Ahmad, M. S. A., Islam, M. S., & Escudero, J., (2007). Selection of mother wavelet function for multi-channel EEG signal analysis during a working memory task. *Sensors, 15*(11), 497–507.

Amel, B., Laatra, Y., Sami, S., & Nourreddine, D., (2013). Classification and diagnosis of broken rotor bar faults in induction motor using spectral analysis and SVM. In: *8th International Conference and Exhibition on Ecological Vehicles and Renewable Energies (EVER)* (pp. 1–5).

Benbouzid, M. E. H., (2000). A review of induction motors signature analysis as a medium for faults detection. *IEEE Transactions on Industrial Electronics, 47*(5), 983–993.

Benbouzid, M. E. H., Vieira, M., & Theys, C., (1999). Induction motors fault detection and localization using stator current advanced signal processing techniques. *IEEE Transactions on Power Electronics, 14*(1), 14–22.

Bhavsar, R. C., & Patel, R. A., (2013). Various techniques for condition monitoring of three phase induction motor: A review. *International Journal of Engineering Inventions, 3*(4), 22–26.

Bhola, J., Yadav, R. B., Rao, K. R. M., & Yadav, H. L., (2013). Selection of optimal mother wavelet for fault detection using discrete wavelet transform. *International Journal of Advanced Research in Electrical, Electronics and Instrumentation Engineering, 2*(6), 2338–2343.

Blödt, M., & Rostaing, G., (2008). Models for bearing damage detection in induction motors using stator current monitoring. *IEEE Transactions on Industrial Electronics, 55*(4), 1813–1822.

Burrus, C. S., Gopinath, R. A., & Guo, H., (1998). *Introduction to Wavelets and Wavelet Transforms*. A Primer, Englewood Cliffs, NJ: Prentice-Hall.

Daviu, A. J., et al., (2006). Validation of a new method for the diagnosis of rotor bar failures via wavelet transformation in industrial induction machines. *IEEE Transactions on Industry Applications, 42*(4), 990–996.

Fargues, M. P., Barsanti, R. J., & Hippenstiel, R., (1997). Wavelet-based denoising: Comparisons between orthogonal and non-orthogonal decompositions. *Proceedings of 40th Midwest Symposium on Circuits and Systems* (pp. 929–932). Dedicated to the Memory of Professor Mac Van Valkenburg.

Gómez, A. J. F., & Sobczyk, T. J., (2013). Motor current signature analysis apply for external mechanical fault and cage asymmetry in induction motors. In: *9th International Symposium on Diagnostics for Electric Machines, Power Electronics and Drives (SDEMPED)* (pp. 136–141).

Henao, H., Razik, H., & Capolino, G. A., (2005). Analytical approach of the stator current frequency harmonics computation for detection of induction machine rotor faults. *IEEE Transactions on Industry Applications, 41*(3), 801–807.

Kalaskar, C. S., & Gond, V. J., (2014). motor current signature analysis to detect the fault in induction motor. *International Journal of Engineering Research and Applications, 4*(6), 58–61.

Mehala, N., & Dahiya, R., (2007). An approach of condition monitoring of induction motor using MCSA. *International Journal of Systems Applications, Engineering & Development, 1*(1), 13–17.

Ponci, F., et al., (2007). Diagnostic of a faulty induction motor drive via wavelet decomposition. *IEEE Transactions on Instrumentation and Measurement, 56*(6), 2606–2615.

Poshtan, J., & Zarei, J., (2007). Bearing fault detection using wavelet packet transform of induction motor stator current. *Tribology International, 40*(5), 763–769.

Rafiee, J., Tse, P. W., Harif, A., & Sadeghi, M. H., (2009). A novel technique for selecting mother wavelet function using an intelligent fault diagnosis system. *Expert Systems with Applications, 36*(3), 4862–4875.

Sangheeta, P., & Hemamalini, S., (2017). Dyadic wavelet transform-based acoustic signal analysis for torque prediction of a three phase induction motor. *IET Signal Processing, 11*(5), 604–612.

Singh, G. K., & Sa'ad, A. S. A. K., (2003). Induction machine drive condition monitoring and diagnostic research—A survey. *Electrical Power Systems Research, 64*(2), 145–158.

Singh, S., Kumar, A., & Kumar, N., (2014). Motor current signature analysis for bearing fault detection in mechanical system. *Procedia Materials Science, 6*, 171–177.

Tavner, P. J., (2008). Review of condition monitoring of rotating electrical machines. *IET Electric Power Applications, 2*(4), 215–247.

Wang, W. J., & Mc Fadden, P. D., (1995). Application of orthogonal wavelets to early gear damage detection. *Mechanical Systems and Signal Processing, 9*(5), 497–507.

Yan, R., Gao, R. X., & Chen, X., (2014). Wavelets for fault diagnosis of rotary machines: A review with applications. *Signal Processing, 96*(Part A), 1–15.

Ye, Z., Wu, B., & Sadeghian, A., (2003). Current signature analysis of induction motor mechanical faults by wavelet packet decomposition. *IEEE Trans on Industrial Electronics, 50*(6), 1217–1227.

CHAPTER 16

HIGH EFFICIENCY HIGH GAIN SEPIC-BUCK BOOST CONVERTER BASED BLDC MOTOR DRIVE FOR SOLAR PV ARRAY FED WATER PUMPING

K. M. ASHITHA[1] and D. THOMAS[2]

[1]*Department of Electrical and Electronics Engineering, Amal Jyothi College of Engineering, Kanjirappally, Kerala–686518, India, E-mail: ashithakm@ee.ajc.i*

[2]*Amal Jyothi College of Engineering, Kanjirappally, Kottayam, Kerala–686518, India*

ABSTRACT

Renewable energy sources like solar power may be employed for water pumping applications where Solar irradiation availability is plenty and where requirement of pumped water is frequent. BLDC motor drives are preferably employed for pumping, owing to their better power delivery, dynamic response, and efficiency, when compared to Induction motor and other dc drives. In this chapter, studied the performance of a DC-DC converter to reduce overall power consumption in a Solar powered BLDC motor drive-based water pumping system. analyze and design the converter so as to enhance the efficiency of the system. A high efficiency, derived non-isolated SEPIC-based converter is designed for use in between the SPV and voltage switching inverter systems. The converter mentioned in this chapter is derived from SEPIC based buck boost converter. P&O methods have been

Advances in Data Science and Computing Technology: Methodology and Applications. Suman Ghosal, Amitava Choudhury, Vikram Kumar Saxena, Arindam Biswas, & Prasenjit Chatterjee (Eds.)

used to collect solar energy efficiently. Brushless DC motor is controlled by switching pattern of VSI. The necessary switching arrangements for the voltage switching inverter are made by means of electronic commutation methods. Mathematical design of proposed system is also explained here. The proposed system was verified by simulation using MATLAB.

16.1 INTRODUCTION

Due to increasing population and industrialization, global energy crisis as well as crude oil prices are rising day by day. Renewable energy sources, fuel cells and likewise energy storage systems are resources to bridge the gap between energy demand and supply. The devaluation of electronic devices and solar plants may attract researchers to these areas. Solar power generation technology is ceaseless, clean, silent, and do not release any air or water pollution to the environment (Kuthsiyat, Chandru, Dhanapriyan, Kishore, and Vinothraj, 2017; Siva and Sampathkumar, 2015). Among the various applications of PV source, water pumping has been gaining significance. Nowadays, solar powered water pumping is used for industries, agricultural sectors, and household applications.

The conventional DC motors and induction motors are commonly used for water pumping applications. But, low efficiency, higher acoustic noise, complexity in control, etc., have reduced their advantages. In last decades, the BLDC motors were used as an alternative to conventional motors. Later, the use of BLDC motor catered to the large scale industrial and agricultural applications, due to its unique features like better efficiency, lesser noise, light weight, good dynamic response, longer life, etc. (Kumar and Singh, 2017; Singh and Kumar, 2015). As a rule, the BLDC motors are feed by voltage switching inverters (VSI). In this chapter, the motor's speed is controlled by the variations in the inverter's output voltage so no additional controlling circuits are needed here.

There are many options for draining maximum energy from solar plants which are explained by Eltawil and Zhengming Zhao (2013); Bendib, Belmili, and Krim (2015). The important output voltage of the solar power sources, allude to the importance of DC-DC converters. Such converters are the intermediaries of the solar panel and the inverter. Traditional buck-boost converters have some disadvantages like pulsating output and input current, inverting output voltage, etc. In-order to negotiate these annoyances, the high gain converters have come into force (Zhang, Zhang, See, and Zhang, 2018; Banaei and Sani, 2018). But power losses of high gain converters are much more flagrant subject. Some converters are capable to providing higher and lower output voltage levels, but it may be inverting or negative. "Voltage gain of the fly back converter can be adjusted by its turn's ratio" (Hwu and Yau, 2012). But leakage inductance leads to power losses and reduces the efficiency. "Another peculiarity of dc-dc converter is reduced voltage stress across switches. In the absence of input side inductors this may become large" (Miao and Gao, 2019).

Solar powered BLDC motor drive for water pumping system is presented in this chapter. MATLAB implementation of system model with electronic commutation is also established. This chapter is divided in to five sections. Introduction is in Section 16.1. System configuration and system design are discussed in Sections 16.2 and 16.3. Section 16.4 deals with the control parameters of the system. Further results and findings are included in Section 16.5. Finally, we conclude the chapter in Section 16.6.

16.2 SYSTEM OUTLINE

Figure 16.1 shows schematic diagram of the proposed system. High efficiency SEPIC based buck boost converter is an intermediate of SPV panel and VSI.

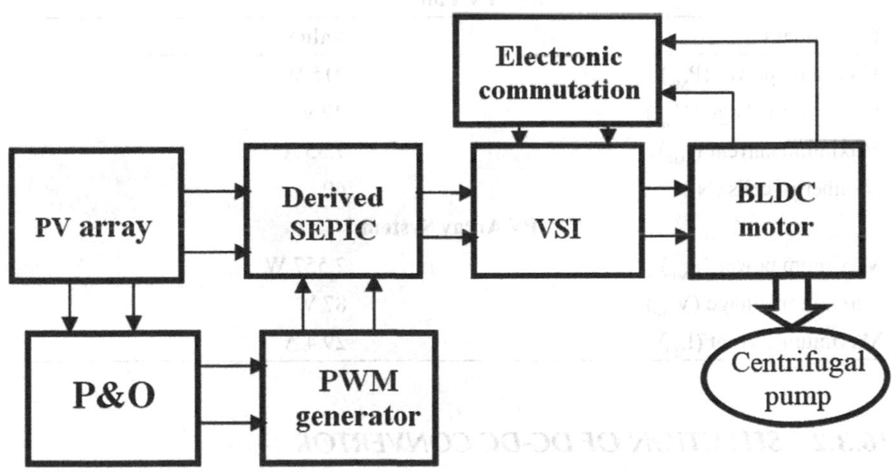

FIGURE 16.1 Schematic diagram of entire system.

Switching pulses of High gain converter is generated by P&O MPPT method, P&O tracks the most power from PV panel. Output of high gain converter is fed into voltage switching inverter and also the BLDC motor is driven by controlling the switching pattern of voltage switching inverter. The switching pulses for VSI is generated by Hall Effect signal position sensors.

16.3 SYSTEM DESIGN

The system consists of PV panel, high efficiency SEPIC based buck boost converter, voltage switching inverter, BLDC motor and centrifugal pump. A 2.2 kW centrifugal pump and 2.5 kW PV array were selected for the design and simulation.

16.3.1 SELECTION OF PV PANEL

A 2.5 kW PV array is selected, to feed sufficient power for the 2.2 kW pump. The output of the PV array is fed into the high gain converter. Details of PV panel per string is given in Table 16.1. Three such panels in series make up a string. Four such strings are connected in parallel to complete the SPV array.

TABLE 16.1 Parameters of PV System

PV Panel	
Parameters	**Value**
Maximum power (P_{MP})	215 W
Maximum voltage (V_{MP})	29 V
Maximum current (I_{MP})	7.35 A
Number of cells (N_s)	60
PV Array System	
Maximum power (P_{MP})	2,557 W
Maximum voltage (V_{MP})	87 V
Maximum current (I_{MP})	29.4 A

16.3.2 SELECTION OF DC-DC CONVERTOR

The high gain, derived SEPIC based converter utilized in between solar array and also the three-phase voltage switching inverter is shown in Figure 16.2. The converter consists of two diodes D1, D2, inductors L1, L2, L3 and capacitors C1, C2, C01, C02. Figures 16.3(a) and (b) show the 2 modes of operation of the converter. Mode 1 is with the switch in turned ON position and Mode 2 is with the switch in turned OFF position.

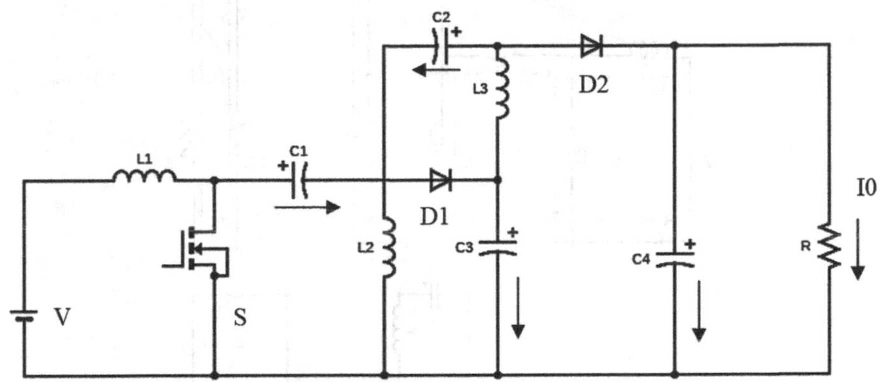

FIGURE 16.2 Derived SEPIC-based buck boost converter.

Mode1: Switch S is turned ON, inductor L_1 charges through input source V. Inductor L2 capacitor C1, and L3 charges through C01. Here capacitor C2 is also charging. All diodes D1, D2 are reverse biased in this mode, therefore, output capacitor C02 feeds the load. Figure 16.3(a) shows mode 1 operation of proposed converter.

The volt-second balance equation applied to the inductors L1, L2 and L3:

$$V_{L1} = V \tag{1}$$

$$V_{L2} = -V_{C1} \tag{2}$$

$$V_{L3} = -V_{C1} - V_{C2} + V_{01} \tag{3}$$

Mode 2: Switch S turned OFF, diodes D1, D2 are forward biased. Inductor L1 discharges through C1. Capacitors C01 and C02 are charging. Related equations are given below:

$$V_{L1} = V - V_{C1} - V_{L2} \tag{4}$$

$$V_{L2} = -V_{C2} + V_{C02} \tag{5}$$

$$V_{L3} = -V_{C2} \tag{6}$$

$$\frac{1}{T_S}\left(\int_0^{DT_S} V dt + \int_{DT_S}^{T_S} \left(V - V_{C1} - V_{L2}\right) dt\right) = 0 \tag{7}$$

$$\frac{1}{T_S}\left(\int_0^{DT_S} -V_{C1} dt + \int_{DT_S}^{T_S} \left(-V_{C2} + V_{C02}\right) dt\right) = 0 \tag{8}$$

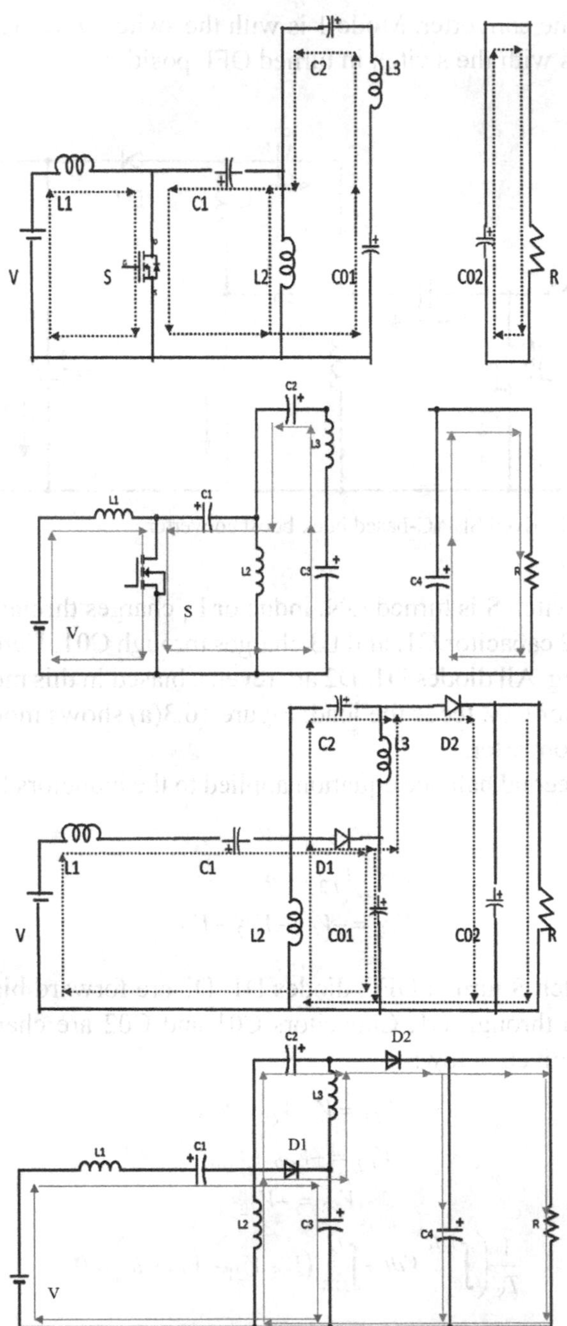

FIGURE 16.3 (a) Mode 1 operation of proposed converter; (b) mode 2 operation of proposed converter.

16.3.2.1 VOLTAGE TRANSFER RATIO OF THE PROPOSED CONVERTER

$$\frac{1}{T_S}\left(\int_0^{DT_S}\left(V_{C1}-V_{C2}+V_{C01}\right)dt+\int_{DT_S}^{T_S}V_{C2}dt\right)=0 \tag{9}$$

Substitute Eqn. (5) in Eqn. (7). We get:

$$\frac{V}{V_{C01}}=\frac{1-D}{D} \tag{10}$$

Substitute Eqn. (10) in Eqn. (9). We have:

$$V_{C2}=\frac{D}{1-D}V \tag{11}$$

Substitute Eqn. (11) in Eqn. (8):

$$V_{C02}=2V_{C01} \tag{12}$$

$$V_{C02}=\frac{2D}{1-D}V \tag{13}$$

Voltage transfer ratio M is obtained:

$$M=\frac{V_0}{V}=\frac{2D}{1-D} \tag{14}$$

16.3.3 CALCULATION OF VOLTAGE AND CURRENT RIPPLES OF CONVERTOR

The amount of current changes during switching cycle is known as current ripple through inductor, which is calculated by applying volt-second balance equation:

$$V=L\frac{di}{dt} \tag{15}$$

Current ripple through the inductor L1, L2, L3

$$\Delta I_{L123}=\frac{VD}{f_S L_{1.2.3}} \tag{16}$$

The switching pulse frequency for the converter is taken to be sufficiently large at about 20,000 Hz. The standardized value for current ripple is taken to be 30% at input and output. For the designed system, the converter receives about 87 V at its input and delivers 240 V at output, giving a Duty ratio of 0.5. Voltage ripple across the capacitor is mainly due to current flowing through the capacitor's equivalent series resistor, and charging-discharging of capacitor. Voltage ripple is assumed to be present at standardized value of 0.3%.

Voltage ripple through capacitor C1:

$$\Delta V_C = \frac{2DI_0}{f_S C_1} \tag{17}$$

Here I_0 is the converter output current.

Voltage ripple across C2, C01, C02:

$$\Delta V_C = \frac{DI_0}{f_S C_{2.01.02}} \tag{18}$$

Table 16.2. shows the value of components in the converter.

TABLE 16.2 Parameters of Converter

Parameters	Value
L1	3.7×10^{-4} H
L2, L3	8.1×10^{-4} H
C1	17×10^{-5} F
C2, C01, C02	8.6×10^{-5} F
fs	20,000 Hz

16.3.4 SELECTION OF DC LINK CAPACITANCE OF VSI

In order to select DC link capacitor, we determine the lowest and highest value of VSI output voltage frequency. Let w_n be the highest frequency and w_m be the lowest frequency (rad/sec):

$$w_n = 2\pi f = 2\pi \frac{N_S \times P}{120} = 2\pi \frac{28000 \times 4}{120} = 586\text{rad/sec} \tag{19}$$

Here f is the frequency of output voltage of VSI, N_S is the maximum rated speed of motor. "The 6th harmonic component of motor voltage appears on DC link" (Kumar and Singh, 2017), whose corresponding ripple voltage is ΔV_d. The ripple amount is assumed to be 8% of VSI input voltage. The DC link capacitance at this rotor speed is given as:

$$C = \frac{I_d}{6w_n \Delta V_d} = 148\mu F \tag{20}$$

The minimum speed of the motor is N_l, then w_m will be:

$$w_m = 2\pi f = 2\pi \frac{N_l \times P}{120} = 230\text{rad/sec} \tag{21}$$

Corresponding value for capacitance is:

$$C = \frac{I_d}{6w_m \Delta V_d} = 377\mu F \tag{22}$$

From the above values of capacitances, we choose the higher value of capacitance, approximately equal to 470 μF.

16.3.5 SELECTION OF WATER PUMP

A 3 hp (approx. 2.2 kW) water pump is chosen as the system load. Its output power is given as:

$$P = K_w N_m^3 \tag{23}$$

"K_w is the water pump constant (Watt/(rad/second)^3) and N_m is the rotor mechanical speed (rad/sec)" (Kumar and Singh, 2017). Mechanical torque of the BLDC motor is 7.5 Nm at the rated speed of 2,800 rpm.

$$K_w = \frac{T_l}{N_m^2} = 8.7 \times 10^{-5} watt(rad/sed)^3 \tag{24}$$

We select a suitable water pump for the system with above data.

16.3.6 CONTROL OF PROPOSED SYSTEM

The switching pattern of derived converter is controlled by P&O MPPT method. Gate pulses of VSI is produced by electronic commutation method.

16.3.7 P&O MPPT ALGORITHM

MPPT method is employed for tracking and acquiring maximum power from the PV panel. Many algorithms to extract maximum power like perturb and observe (P&O), Incremental Conductance, Parasitic capacitance and Constant Voltage Methods are commonly available. In this chapter, P&O MPPT method is chosen for the proposed system. Oscillations around the peak power are avoided and perturbation size is 0.001.

16.4 ELECTRONIC COMMUTATION

BLDC motor requires three hall sensors to detect the rotor position. Hall sensor signal values and VSI switching pattern are given in Table 16.3.

TABLE 16.3 Hall Signals and Switching Position

Hall Signals	Switching Pattern
101	S1; S4
001	S1; S6
011	S3; S6
010	S3; S2
110	S5; S2
100	S5; S4

16.5 RESULT AND DISCUSSION

Figure 16.3 shows the MATLAB model of the proposed system. P&O algorithm generated as a sub system is shown in Figure 16.3(a). In P&O method, instantaneous power of PV array is compared to the previous sample and duty for the converter is generated. Gate signals for VSI is generated as another sub system which is shown in Figure 16.3(b). Electronic commutation method is used to produce these gate signals. We measure the hall signals generated by the motor and determine the rotor positions or rotation by controlling the switching pattern of VSI. Motor with rated speed of 2,800 rpm and output power of 2.2 kW is simulated in this model.

16.5.1 PERFORMANCE OF THE PROPOSED SYSTEM IN STANDARD IRRADIATION LEVEL 1,000 W/M²

Performance of solar PV system, high gain converter, and brushless dc motor shown in Figure 16.3. Explanations of this are covered in further subsections. In standard irradiation the motor reaches its maximum speed.

16.5.2 ANALYSIS OF SOLAR PV ARRAY

In steady state condition, PV array attains its maximum output power. Maximum voltage output at irradiation 1,000 W/m^2 is 87 V and maximum current will be 29.4 A. Also, the out-put power of PV array in standard condition is 2.5 kW which is shown in Figure 16.3(a).

16.5.3 ANALYSIS OF DC-DC CONVERTER

High gain SEPIC based buck boost converter operates in continuous conduction mode. Current through inductor L1 is continuous in nature which is shown in Figure 16.3(b). The output power of the conductor varies with respect to the change in irradiation level. However, converter gives maximum output power of 2.2 kW at 1,000 W/m^2 solar irradiation.

16.5.4 ANALYSIS OF BRUSHLESS DC MOTOR

Figure 16.4(c) shows the steady state performance of BLDC motor at 1,000 W/m^2. Along with the DC link voltage, back EMF, and stature current increases from zero to a steady state value it indicates, the soft starting of motor. Electromagnetic torque produced by the BLDC motor is same as that of load torque in steady state condition which is nearly 7.5 Nm. Also, the rotor speed reaches its rated value in standard conditions which is approximately 2,800 rpm.

16.5.5 PERFORMANCE OF SYSTEM IN DYNAMIC STATE

Dynamic state performance of solar PV system, high gain converter, and brushlesss dc motor are shown in Figure 16.5. Explanations of these are covered in further subsections; the system designed irradiation levels and time duration are listed in Table 16.4.

TABLE 16.4 Time Duration and Irradiation Level

Time in Seconds	Solar Irradiation in *W/m*^{2}
0–1	200
1–1.5	400
1.5–2	600
2–2.5	1,000
2.5–3	800
3–4	600

16.5.6 ANALYSIS OF SOLAR PV ARRAY

PV array is analyzed in different irradiation levels. The output power of PV is nearly 500 W at 200 with an output voltage of 100 V. Output of PV array changes with respect to change in irradiation level which is shown in Figure 16.5(a).

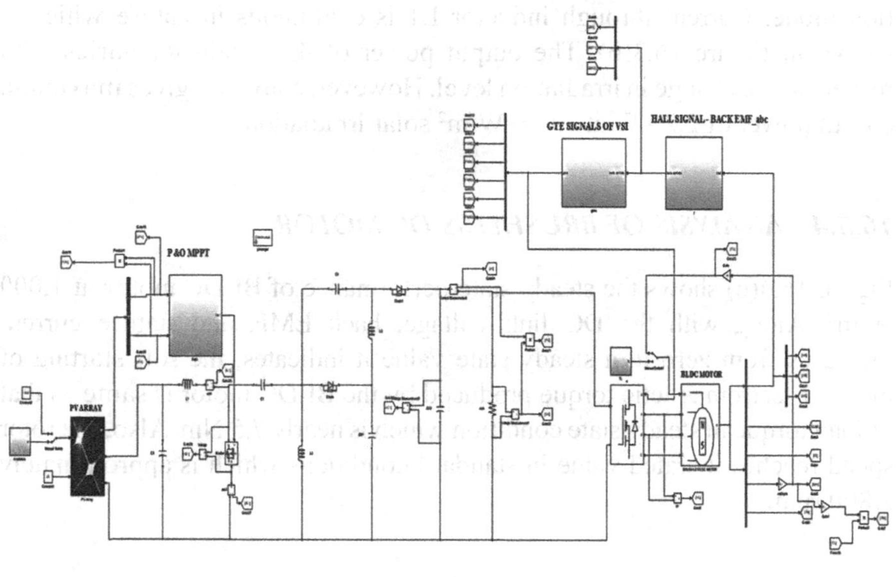

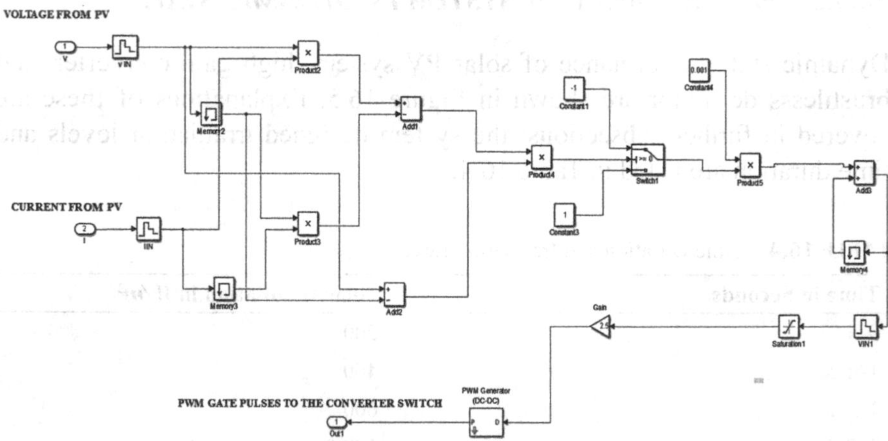

FIGURE 16.4 *(Continued)*

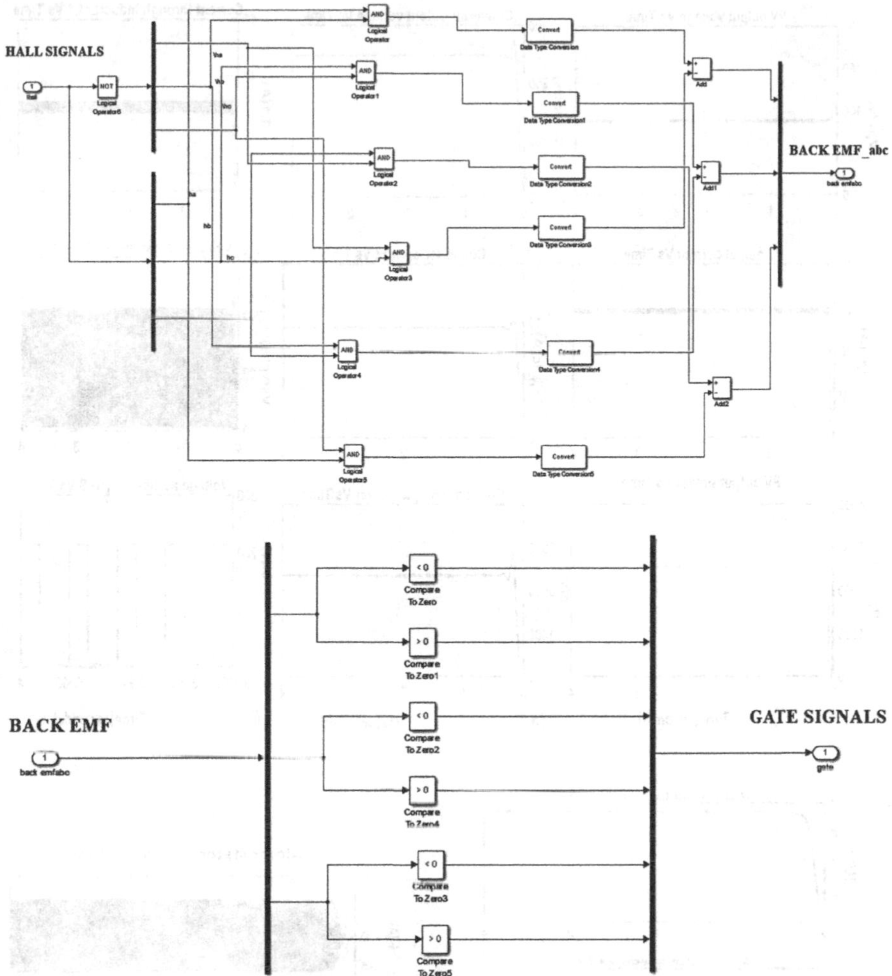

FIGURE 16.4 Simulation diagram: (a) P&O MPPT; (b) electronic commutation method.

16.5.7 ANALYSIS OF DC-DC CONVERTER

High gain SEPIC based buck boost converter operates in continuous conduction mode at all different irradiation level. Current through inductor L_1 is continuous in nature which is shown in Figure 16.5(b). The output power of the conductor varies with respect to the change in irradiation level. The converter gives 120 V output at 200 W/m^2 solar irradiation level. As we can see from the figures the converter performs well dynamically to adjust to the varying irradiation levels.

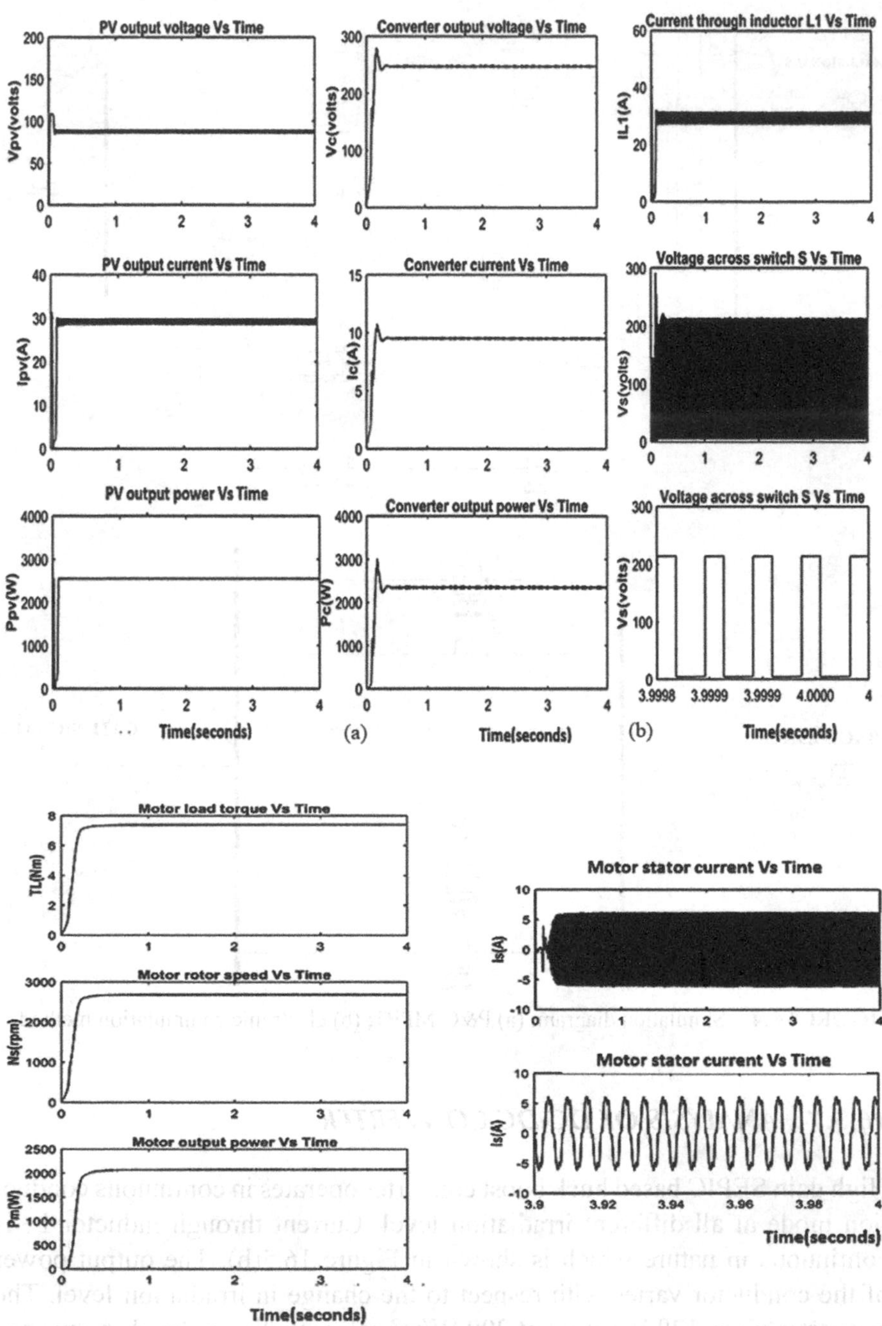

FIGURE 16.5 Steady state analysis at 1,000 W/m^2: (a) PV outputs; (b) converter output; (c) motor output.

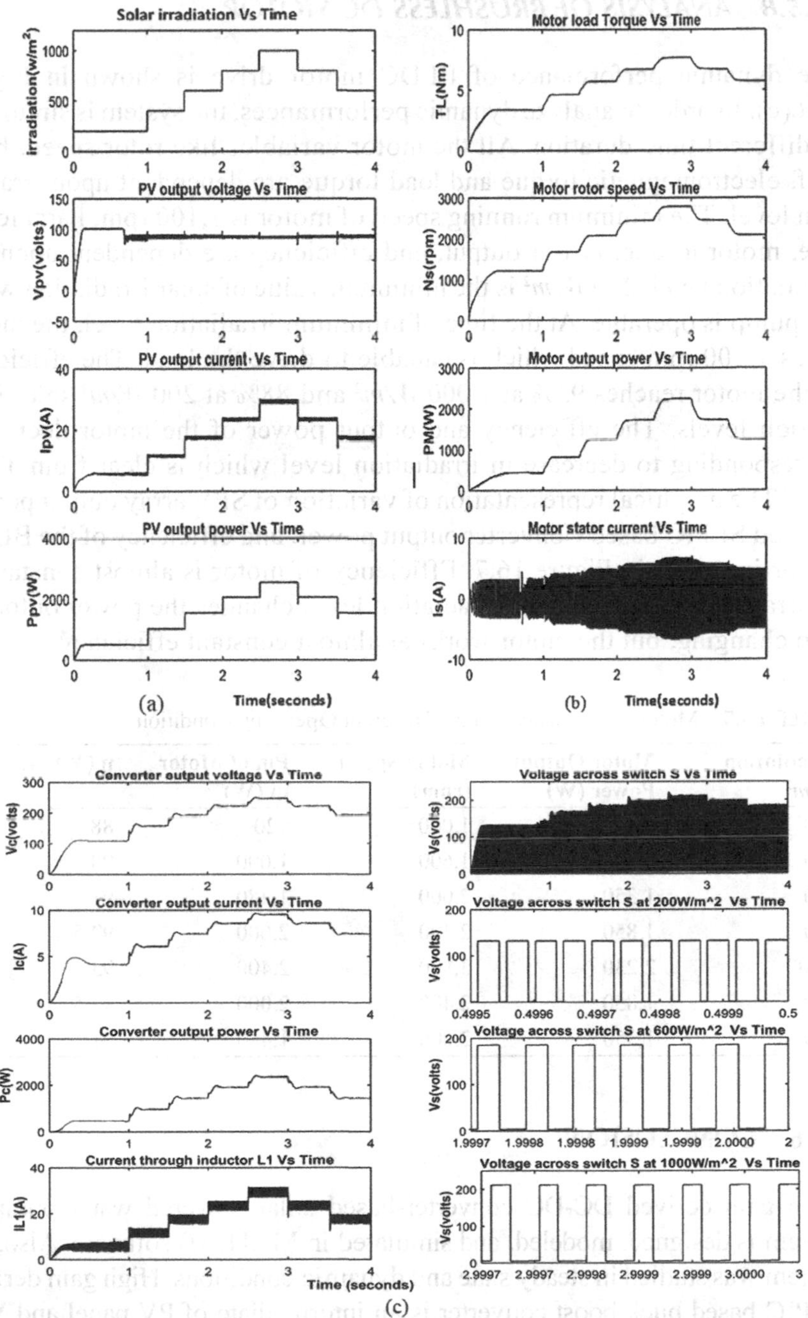

FIGURE 16.6 Dynamic state analysis: (a) PV outputs; (b) motor output; (c) converter output.

16.5.8 ANALYSIS OF BRUSHLESS DC MOTOR

The dynamic performance of BLDC motor drive is shown in Figure 16.6(c). In order to analyze dynamic performances, the system is simulated in different time duration. All the motor variables like rotor speed, back emf, electromagnetic torque and load torque are dependent upon irradiation level. The minimum running speed of motor is 1,100 rpm. Parameters like, motor torque, power output, and efficiency are dependent upon the irradiation level. 200 W/m^2 is the minimum value of solar irradiation when the pump is operable. At the time of minimum irradiation level, the motor gives 1,100 rpm speed which is capable to drive the load. The efficiency of the motor reaches 93% at 1,000 W/m^2 and 88% at 200 W/m^2 solar irradiation levels. The efficiency and output power of the motor decreases corresponding to decrease in irradiation level which is clear from Table 16.5. The graphical representation of variation of SPV array output power, derived SEPIC based Converter output power, and efficiency of the BLDC motor is shown in Figure 16.7. Efficiency of motor is almost constant at all irradiation levels. As the irradiation level changes the power output is also changing, but the motor works at almost constant efficiency.

TABLE 16.5 Motor Performances Under Different Operating Condition

Irradiation W/m^2	**Motor Output Power (W)**	**Motor Speed (rpm)**	**Pin of Motor in (W)**	**ή (%)**
200	460	1,000	520	88
400	900	1,600	1,000	90
600	1,350	2,000	1,480	91
800	1,850	2,400	2,000	92.5
1,000	2,230	3,800	2,400	93
800	1,850	2,400	2,000	92.5
600	1,350	2,000	1,480	91

16.6 CONCLUSION

High gain derived DC-DC converter-based solar powered water pumping system is designed, modeled, and simulated in MATLAB software. Also, the system was studied in steady state and dynamic conditions. High gain derived SEPIC based buck boost converter is an intermediate of PV panel and VSI. Reduced power losses, ease of implementation, high efficiency are the main advantages of this converter. The performance of BLDC motor was simulated at different irradiation levels. From the studies it is clear that the BLDC motor is suitable for all environmental conditions. The efficiency, speed, and torque outputs of the motor is dependent upon the irradiation level.

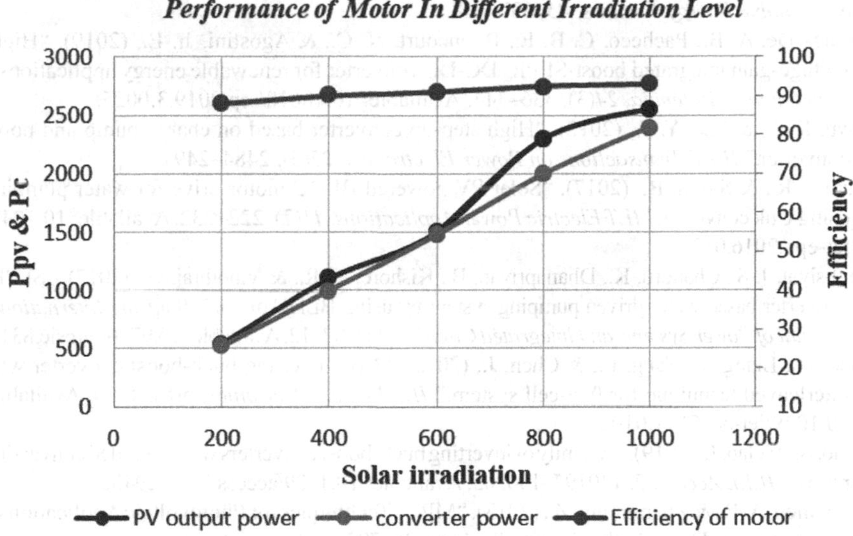

FIGURE 16.7 Graphical analysis of motor performance in different irradiation level.

KEYWORDS

- **BLDC motor**
- **converter**
- **MATLAB**
- **SEPIC**

REFERENCES

Aleena, P. K., Siny, P., & Babu, P., (2017). "Transformer less buck-boost converter with positive output voltage and feedback." *International Journal of Engineering Research, 6*(06). Available: 10.17577/ijertv6is060337.

Banaei, M., & Bonab, H., (2017). "A novel structure for single-switch nonisolated transformer less buck-boost DC–DC Converter." *IEEE Transactions on Industrial Electronics, 64*(1), 198–205. Available: 10.1109/tie.2016.2608321.

Banaei, M., & Sani, S., (2018). "Analysis and Implementation of a new SEPIC-based single-switch buck-boost DC–DC converter with continuous input current." *IEEE Transactions on Power Electronics, 33*(12), 10317–10325. Available: 10.1109/tpel.2018.2799876.

Bendib, B., Belmili, H., & Krim, F., (2015). "A survey of the most used MPPT methods: Conventional and advanced algorithms applied for photovoltaic systems." *Renewable and Sustainable Energy Reviews, 45*, 637–648.

Gomes De, A. B., Pacheco, C. B. E., Bitencourt, N. C., & Agostini, Jr. E., (2019). "High-voltage-gain integrated boost-SEPIC DC-DC converter for renewable energy applications." *Eletrônica de Potência, 24*(3), 336–344. Available: 10.18618/rep.2019.3.0025.

Hwu, K. I., & Yau, Y. T., (2012). "High step-up converter based on charge pump and boost converter." *IEEE Transactions on Power Electronics, 27*(5), 2484–2494.

Kumar, R., & Singh, B., (2017). "Solar PV powered BLDC motor drive for water pumping using Cuk converter." *IET Electric Power Applications, 11*(2), 222–232. Available: 10.1049/iet-epa.2016.0328.

Kuthsiyat, J. S., Chandru, K., Dhanapriyan, B., Kishore, K. R., & Vinothraj, G., (2017). "SEPIC converter based water driven pumping system by using BLDC motor." *Bonfring International Journal of Power Systems and Integrated Circuits, 7*(1), 07–12. Available: 10.9756/bijpsic.8317.

Liao, H., Liang, T., Yang, L., & Chen, J., (2012). "Non-inverting buck-boost converter with interleaved technique for fuel-cell system." *IET Power Electronics, 5*(8), 1379. Available: 10.1049/iet-pel.2011.0102.

Miao, S., & Gao, J., (2019). "A family of inverting buck-boost converters with extended conversion ratios." *IEEE Access, 7*, 130197–130205. Available: 10.1109/access.2019.2940235.

Mohamed, A. E., & Zhengming, Z., (2013). "MPPT Techniques for Photovoltaic Applications." *Renewable and Sustainable Energy Reviews, 25*, 793–813.

Roberto, F. C., Walbermark, M. D. S., & Denizar, C. M., (2012). "Influence of power converters on PV maximum power point tracking efficiency." In: *10th IEEE/IAS Inter. Conf. on Industry Applications (INDUSCON)* (pp. 1–8).

Sarikhani, A., Allahverdinejad, B., Hamzeh, M., & Afjei, E., (2019). "A continuous input and output current quadratic buck-boost converter with positive output voltage for photovoltaic applications." *Solar Energy, 188*, 19–27. Available: 10.1016/j.solener.2019.05.025.

Selva, K. R., Vignesh, C. J., Gayathri, D. V. P., & Naveena, P., (2016). "Design and comparison of quadratic boost converter with boost converter." *International Journal of Engineering Research, 5*(01). Available: 10.17577/ijertv5is010650.

Singh, B., & Kumar, R., (2016). "Solar photovoltaic array fed water pump driven by brushless DC motor using Landsman converter." *IET Renewable Power Generation, 10*(4), 474–484. Available: 10.1049/iet-rpg.2015.0295.

Siva, T., & Sampathkumar, A., (2015). "Sensor less control of brushless DC motor using modified back-EMF detection." *Indian Journal of Science and Technology, 8*(32). Available: 10.17485/ijst/2015/v8i32/87320.

Thrishna Jayaraj, & Justin Sunil Dhas, G. (2020). [Online]. Available: https://www.researchgate.net/publication/333352436_KY_Based_C_DC_Converter_for_Standalone_Photovoltaic_Water_Pumping_System_Employing_Four_Switch_BLDC_Drive (accessed on 08 December 2021).

Zhang, N., Zhang, G., See, K., & Zhang, B., (2018). "A single-switch quadratic buck-boost converter with continuous input port current and continuous output port current." *IEEE Transactions on Power Electronics, 33*(5), 4157–4166. Available: 10.1109/tpel.2017.2717462.

CHAPTER 17

A CLASSY FUZZY MPPT CONTROLLER FOR STANDALONE PV SYSTEM

RAHUL KUMAR SINGH,[1] RAHUL KUMAR PRASAD,[1] RAHUL SINGH,[1] AAYUSH ASHISH,[1] DEBAPARNA SENGUPTA,[1] RISHIRAJ SARKER,[2] and MONALISA DATTA[1]

[1]*Techno International New Town, Kolkata–700156, West Bengal, India*

[2]*Jadavpur University, Kolkata–700032, West Bengal, India,*
E-mail: sarker.rishiraj88@gmail.com

ABSTRACT

Maximum power point tracking (MPPT) is a powerful technique to extract the maximum amount of power from the PV panel under different ambient conditions. Several types of MPPT techniques have been invented and applied by researchers in course of time to improve their performance indices. This chapter presents a classy fuzzy logic based MPPT controller, which is highly efficient in tracking the maximum power point within a stipulated time period at a high accuracy. The proposed intelligent controller is smart enough to respond very quickly to the abrupt environmental changes and precisely tracks the maximum power point.

17.1 INTRODUCTION

Conventional energy crisis is a burning issue in today's world. Day-by-day, due to the upliftment of standard of living, energy demands are increasing rapidly (Seddik et al., 2010), but the conventional energy resources are

Advances in Data Science and Computing Technology: Methodology and Applications. Suman Ghosal, Amitava Choudhury, Vikram Kumar Saxena, Arindam Biswas, & Prasenjit Chatterjee (Eds.)

diminishing in the reverse order. This has created an unresolved problem. The demand is greater than supply. To eliminate this imbalance in supply and demand, renewable energy resources are playing a vital role as an alternative of fossil fuel.

In between the different types of renewable energy sources, the most commonly used resources are solar and wind energies. Renewable energy resources are easily available in nature in ample amount. The additional advantages associated with the newly developed renewable energy stations are they are eco-friendly. Thus, the various socio-economic problems like global warming, greenhouse gas emission, pollution, health issues, etc., are being resolved with the promotion of renewable energy-based power stations. Solar energy (photo voltaic energy) is the kind of nonconventional energy, which is plentiful in nature. It has the advantage that it is being directly converted into DC type electrical energy by the PV panel and thus doesn't require additional mechanism to convert the energy into DC form, before storing it into a battery bank for future use. The main concern related to PV based energy station is the variation in environmental condition such as irradiation and temperature, which directly effects the power supplied by solar PV module. If any measure for maximum Power Point tracking (MPPT) is not initiated within the PV integrated system, the load power delivered will not always be maximum. It will vary with the irradiance and ambient temperature, whenever there will be deviations from the standard test conditions (STC) and finally maximum power extraction will be disrupted. To eliminate this problem, different types of MPPT controllers are used till now (Salas et al., 2006; Esram and Chapman, 2007; Bhatnagar and Nema, 2013; Zegaoui et al., 2011; Mirbagheri, Mekhilef, and Mirhassani, 2013; Rekioua, Achour, and Rekioua, 2013; Oulad-Abbou et al., 2018). But these controllers have complicated configurations or mechanisms. Moreover, finely tuned tracking mechanism cannot be materialized with these types of controllers. Fuzzy logic controllers (FLCs) are robust, simple, and most importantly, it does not require any detailed knowledge about the huge complicated system model to be controlled before designing this type of controller. They are capable of precisely controlling the control variable subjected to variation in input parameters. Henceforth, Fuzzy MPPT controller is an ideal choice for tracking the maximum power point. In 2014 (Bendib et al., 2014), fuzzy MPPT controller integrated PV system is presented with DC-DC buck converter. In this proposed work, a Fuzzy MPPT controller is used for tracking the Maximum Power Point of PV module, where the system has a DC-DC boost converter. In Villalva, Gazoli, and Ruppert (2009); Salameh,

Casacca, and Lynch (1992); Nafeh (2011); Coelho, Concer, and Martins (20100; Hua and Shen (1997), the different components of the system like PV array, Battery characteristics, DC-DC converters are discussed, study on which are necessary to acquire knowledge about the system for execution of the proposed work.

17.2 SYSTEM CONFIGURATION

The entire system configuration with the Fuzzy MPPT controller is presented by Figure 17.1. The PV module is cascaded with a DC-DC boost converter for charging an energy storage unit (ESU) to be used for charging electric vehicles (EVs). After the boost converter, comes the DC link, which relates to the ESU for charging it. The fuzzy MPPT controller is the sub system designed for controlling the action of the boost converter by controlling its duty cycle. Error and derivative of error are the two input signals fed to the fuzzy MPPT controller. They are generated utilizing the voltage and current signals coming out from the PV panel. The voltage and current outputs are constantly varying with corresponding variations in irradiance and temperature. Accordingly, the inputs to the fuzzy MPPT controller are also varying. The controller has been designed in such a way that it can generate appropriate control signal according to the variations in the inputs. The control signal produced by the controller is fed into the gate of the IGBT used inside the DC-DC boost converter. Thus, converter duty cycle is being controlled very smoothly and precisely for getting desired voltage and current signals at the boost converter output, which corresponds to maximum power delivered to the ESU. The objective of this project is to charge the ESU in the most efficient manner so that it can be used efficiently for EV charging purpose. ESU will be installed in electric vehicle charging stations (EVCSs). This is the preliminary work on a part of a huge system to be designed further. The future scope of this work will be discussed in the future scope section.

17.3 MATHEMATICAL MODEL OF PHOTOVOLTAIC MODULE

Figure 17.2(a) presents the equivalent circuit for solar (PV) cell. It consists of a current source connected in parallel with a diode. R_S and R_{Sh} are the series and parallel resistances connected with the circuit. R_{Sh} is infinite and R_S is zero for an ideal PV cell.

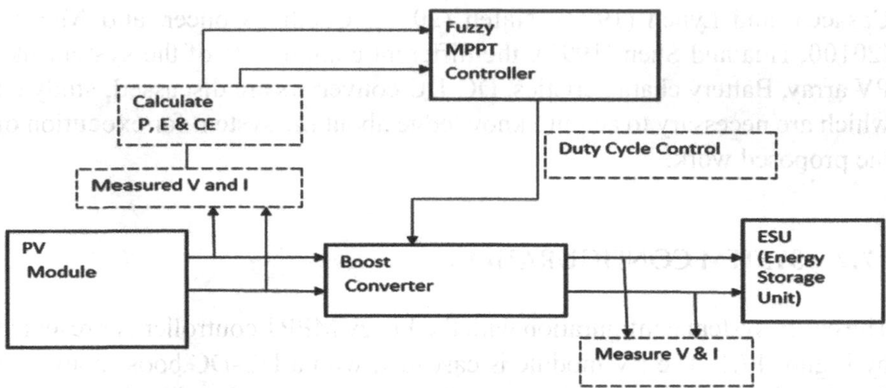

FIGURE 17.1 Block diagram of the stand-alone PV system.

In actual practice, many PV cells are assembled together in series fashion to form larger unit called PV module. A cluster of such modules are coupled together in series/parallel to form PV arrays. These PV arrays are the backbone of the PV generation plant. Figure 17.2(b) presents the equivalent circuit of PV array.

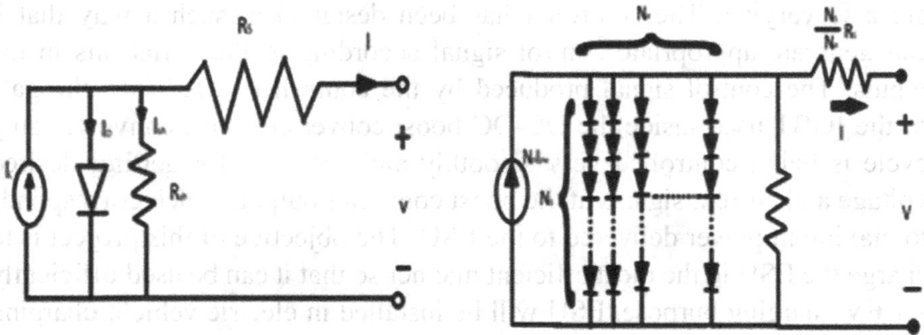

FIGURE 17.2 (a) Equivalent circuit of a PV cell; and (b) equivalent circuit of a PV array.

The equation relating voltage and current output of a PV cell is depicted by:

$$I = N_p I_{ph} - N_p I_o \left[\exp\left(\frac{\left(V + I(\frac{N_S}{N_p})R_s \right)}{N_S.\alpha.V_T} \right) - 1 \right] - \frac{(V + I\left(\frac{N_S}{N_p}\right)R_s}{\left(\frac{N_S}{N_p}\right)R_{sh}} \quad (1)$$

where; I_{ph} is the photo current; I_o is the diode's reverse saturation current; α is the diode ideality factor; V is the PV array output voltage of, I is the PV array output current; R_s is the PV cell's series resistance; R_{sh} is the PV cell's shunt resistance; N_P is the parallel path counts of the PV array; N_S is the PV cell numbers connected in series in each parallel path; V_T is the thermal voltage of the PV cell.

$$V_T = \frac{K.T}{q} \tag{2}$$

where; K is the Boltzmann constant (1.3805×10^{-23} J/K); T is the operational temperature (in Kelvin); q is the electron charge (1.6×10^{-19} C).

Photo current is given by:

$$I_{ph} = \left(I_{ph-STC} + K_i\left(T - T_{STC}\right)\right)\frac{G}{G_{STC}} \tag{3}$$

where; $I_{ph\text{-}STC}$ is the short circuit photo current generated at STC, T_{STC}, and G_{STC} imply the temperature and irradiance at STC. K_i is the short circuit current coefficient of PV cell; T is the operating temperature in Kelvin; G is the irradiance (W/m²); K_i is the short circuit current coefficient of the cell usually provided by manufacturer.

The saturation current of a cell varies with cell temperature, which is presented by:

$$I_o = \frac{\left(I_{ph-STC} + K_i\left(T - T_{STC}\right)\right)}{\left[\exp\left(\frac{\left(V_{OC\_STC} + K_V\left(T - T_{STC}\right)\right)}{\acute{a}.V_T}\right) - 1\right]} \tag{4}$$

where; V_{OC-STC} is the cell open circuit voltage at STC; K_v is the open circuit voltage coefficient.

In this proposed work, the standard test condition parameter values, T_{STC}, and G_{STC}, are taken as 25°C and 1,000 W/m² respectively. A PV array of 47 parallel strings with each string consisting of 15 series connected PV modules are taken. Number of cells per module is chosen as 60. PV module is used for study using Simulink MATLAB. The P-V and I-V characteristics of the PV panel at STC are detected after setting the required parameters as per the values mentioned in Table 17.1.

From the P-V characteristics, it is clearly seen that there is a specific voltage of the PV array, where the maximum power is achieved (Figure 17.3). That voltage is the V_{mp}. From the I-V curve, the corresponding current value

I_{mp} can be noted down. As the PV array output voltage and current depends on the irradiance and temperature directly, henceforth, with variations in irradiance or temperature, the maximum power point varies. Figures 17.4(a) and (b) present the I-V, P-V characteristics of the PV array for different irradiances at T_{STC} (25°C) and different temperatures at G_{STC} (1,000 W/m^2). From the graphs of Figures 17.4(a) and (b), it can be observed that the current and power changes abruptly keeping the output voltage V almost constant with variations in irradiance. On the contrary, current value remains constant with a great variation in voltage in case of temperature change keeping the irradiance fixed.

TABLE 17.1 Electrical Characteristics Data of PV Module

Electrical Characteristics Values
Short-circuit current (I_{sc}) 7.84 A
Current value maximum power point (I_{mp}) 7.35 A
Voltage value at maximum power point (V_{mp}) 29 V
Open-circuit voltage (V_{oc}) 36.3 V
Shunt Resistance (R_{sh}) 313.3991 ohms
Series Resistance (R_s) 0.39383 ohm
STC condition's Maximum power (P_{max}) 213.15 W
I_{sc}'s temperature coefficient 0.102%/°C
V_{oc}'s temperature coefficient – 0.36099%/°C
Photo current (I_{ph}) 7.8649 A
Saturation current of diode (I_0) 2.9259*exp^{-10}
ideality factor of diode 0.98117

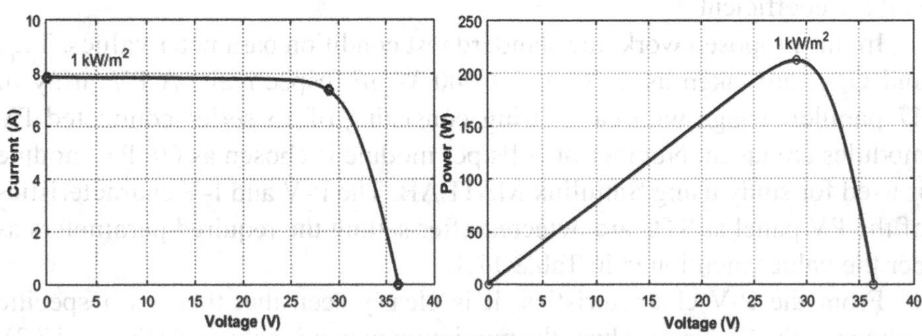

FIGURE 17.3 I–V and P–V characteristics of a PV module under STC.

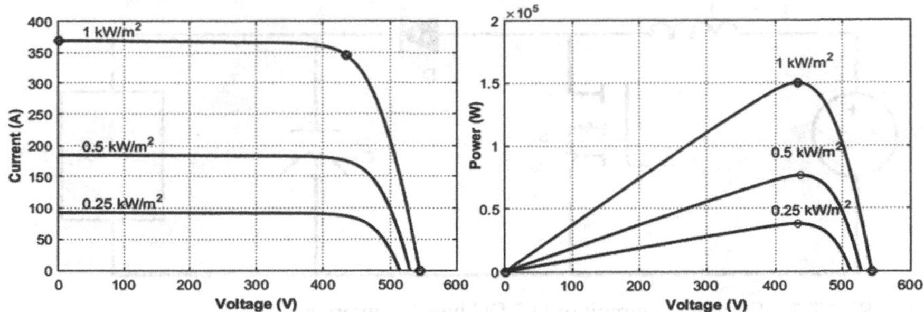

FIGURE 17.4(a) I–V and P–V characteristics of a PV module for different irradiances at T_{STC} (25°C).

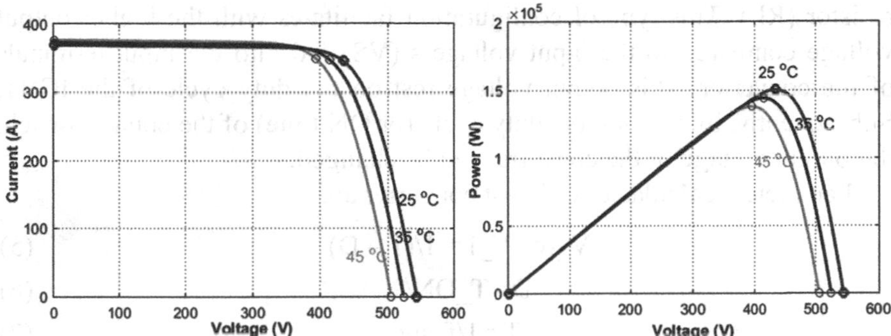

FIGURE 17.4(b) I–V and P–V characteristics of a PV module for different temperatures at G_{STC} (1,000 W/m²).

17.4 DC-DC BOOST CONVERTER

DC-DC converters are generally used for boosting up the voltage level for high voltage applications. It is mainly used in PV integrated system for proper regulation of the PV array output voltage. DC-DC boost converters are used to step up the output voltage of the PV array, if the load voltage requirement is higher than the voltage produced by the PV panel. It is situated in between the PV array and the load (here the energy storage unit (ESU)). The main purpose for installing the boost converter is to facilitate the provision of MPPT by varying the duty cycle of the boost converter. DC-DC boost converter's circuit diagram is presented in Figure 17.5.

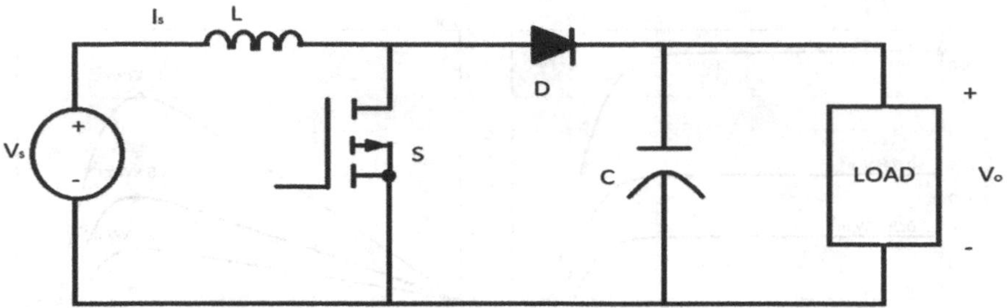

FIGURE 17.5 Equivalent circuit of DC-DC boost converter.

The main components used in a DC-DC boost converter are: an inductor (L), a capacitor (C), a control switch, i.e., IGBT (S), a diode (D) and a load resistor (RL), This type of configuration facilitates with the higher output voltage compared to the input voltage s (VS) fed into the input terminals of the converter. This output voltage rest on the duty cycle of the IGBT. Subsequently, by varying the duty cycle (or ON time) of the control switch, the output voltage of the converter can be changed.

Parameters calculated for boost converter are:

$$V\_(o)/V\_i = 1/(1 - D) \tag{5}$$

$$D= T\_ON/T \tag{6}$$

$$T= 1/f\_sw \tag{7}$$

where; D denotes the duty cycle. Vi(or VS) and V0 are the input and output voltages of the converter. TON is the on time and T is the total time period of the IGBT. fsw is the switching frequency (Table 17.2).

TABLE 17.2 Operating Values of Boost Converter

Electrical Characteristics Values
Time period ($1/30\times10^3$) sec
Inductance $1.559e^{-4}$ H
Capacitance $2.767e^{-4}$
IGBT internal resistance $1e^{-3}$ ohms
IGBT snubber resistance 1e5 ohms
IGBT snubber capacitance infinite
Diode resistance 0.001 ohms
Diode forward voltage 0.8 V

17.5 FUZZY BASED MPPT

Among the various techniques articulated for tracking the maximum power point of the PV integrated system, the modest and prevalent technique is the perturb and observe (P&O) technique. This method facilitates good results under steady state condition where there are no such considerable variations in irradiation and temperature. But if there are huge variations in these parameters at a frequent and faster rate, this method fails to track the maximum power point precisely and quickly.

Fuzzy control is a newly developed heuristic method that facilitates the construction of nonlinear controllers from vague information collected by the expert. Moreover, the controller design does not require to know the system model to be controlled accurately. The knowledge and expertise of the designer play vital role in the successful implementation of these types of controllers. FLC is more robust than conventional controller (Figure 17.6). FLC can be classified three steps:

- fuzzification;
- inference engine with proper rule base; and
- defuzzification.

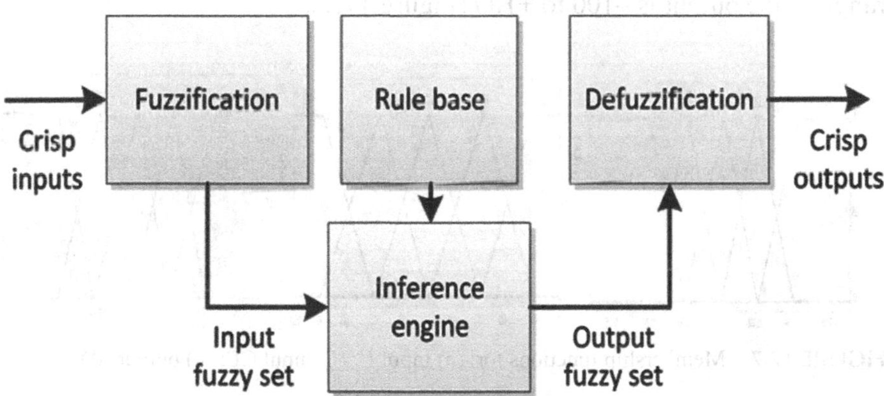

FIGURE 17.6 Block diagram of fuzzy MPPT controller.

17.5.1 FUZZIFICATION

"Fuzzification" is the technique of converting crisp inputs into fuzzy or linguistic inputs. For this proposed Fuzzy MPPT controller, two inputs are taken namely error (E) and change in error (CE). The membership functions are assigned to both inputs with the help of linguistic variables using five fuzzy subsets called negative big (NB), negative small (NS), zero (ZE), positive small (PS) and positive big (PB). The voltage (V) and current (I) outputs from the PV array is constantly monitored by the Fuzzy MPPT controller and instantaneous power is calculated. The governing equations of the inputs E and CE to the controller are:

$$E(k)=\{[P(k)-P(k-1)]/[I(k)-I(k-1)]\}=\Delta P/\Delta I \tag{8}$$

$$CE(k)=E(k)-E(k-1)=\Delta E \tag{9}$$

where; E denotes the slope of P-I characteristics curve and CE presents the change of this error; k is the sampling instant; E(k) indicates whether the operating point at any instant k is situated on the left or right side of MPP on the P-I characteristics; CE(k) gives the direction of displacement of this point. The control output from the fuzzy MPPT controller is the change in duty ratio (ΔD) of the DC-DC boost converter. Henceforth, the control action by the controller is materialized by controlling the change in duty ratio in an efficient, robust, and smooth manner. The membership functions assigned to the output variable (ΔD) are NB, NS, zero (ZE), PS and PB. The variation ranges of the inputs are taken from –0.07 to +0.07; whereas the variation range of the output is –100 to +100 (Figure 17.7).

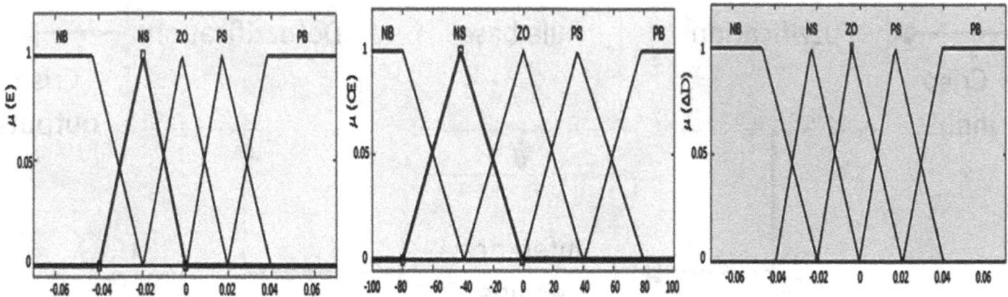

FIGURE 17.7 Membership functions for: (a) input E; (b) input CE; (c) output ΔD.

17.5.2 INFERENCE ENGINE

Fuzzy inference engine is the tool used for relating the various fuzzified inputs and outputs with the help of fuzzy rules. This mapping provides a set of rules by which decisions are taken about the output values of the controller based on the controller input values. The procedure of fuzzy inference encompasses all the linguistic variables used for the inputs and output mentioned in the fuzzification section along with logical operations in between the different inputs and if-then rules. Table 17.3 presents the fuzzy rules used for this proposed work. There are 25 rules altogether involving E, CE, and ΔD.

TABLE 17.3 Fuzzy Rule Table

	CE →				
E ↓	**NB**	**NS**	**ZO**	**PS**	**PB**
NB	ZO	ZO	NB	NB	NB
NS	ZO	ZO	NS	NS	NS
ZO	NS	ZO	ZO	ZO	PS
PS	PS	PS	PS	ZO	ZO
PB	PB	PB	PB	ZO	ZO

17.5.3 DE-FUZZIFICATION

The procedure of converting fuzzy values into crisp values is called de-fuzzification. Defuzzification is comprehended by a decision-making criterion that chooses the best crisp value based on a fuzzy set described in terms of membership in fuzzy sets. There are plentiful defuzzification methods including mean of maximum (MOM), center of gravity (COG) and center average methods. The MOM method can be viewed as the method which returns the point of balance on a curve, whereas the COG method yields the value of the center of area under the curve. Amongst those several methods, centroid method is the most predominant one. In the proposed work, the COG method is used for defuzzification. Centroid defuzzification returns the COG of the fuzzy set for the output variable along the x-axis. The centroid is computed using Eqn. (9), where $\mu(\Delta D_i)$ is the membership value for point ΔD_i in the universe of discourse. After the defuzzification, the ΔD is scaled by $S_{\Delta D}$ and finally converted into actual duty ratio by the formula mentioned in Eqn. (10).

$$\Delta D = \frac{\sum_{i=1}^{n} \mu(\Delta D_i).\Delta D_i}{\sum_{i=1}^{n} \mu(\Delta D_i)} \tag{10}$$

$$D(k) = D(k-1) + S_{\Delta D}.\Delta D(k) \tag{11}$$

17.6 SIMULATION RESULTS AND DISCUSSIONS

The simulation results are tabulated in table IV for different irradiance values keeping temperature fixed at 25°C. Table 17.4 presents the voltage, current, and maximum power values from the Fuzzy MPPT controller integrated system for different temperatures, keeping the irradiance fixed at 1,000 W/ m^2 (Figures 17.8 and 17.9).

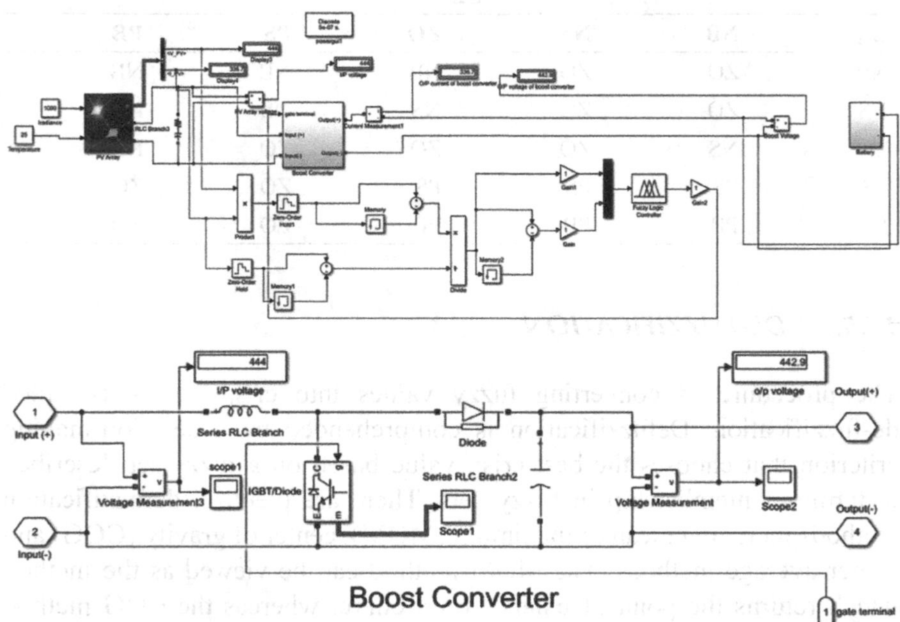

FIGURE 17.8 Simulink model of (a) the entire system; (b) the DC-DC boost converter.

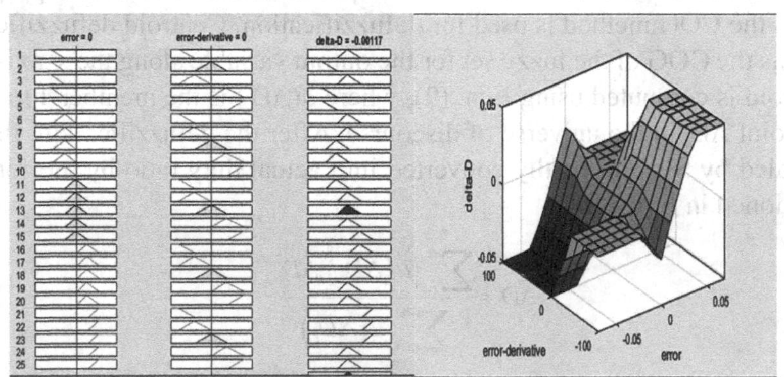

FIGURE 17.9 (a) fuzzy rule viewer; (b) fuzzy surface.

TABLE 17.4 The Voltage, Current, and Power Outputs from the Boost Converter for Different Irradiances Under Standard Temperature 25°C

Temperature (°C)	Irradiance (W/m²)	Boost Converter Output Voltage (V)	Boost Converter Output Current (A)	Power Delivered by the Boost Converter to the ESU (kW)
25	1,000	442.9	336.7	149.124
35	1,000	441.6	310	136.896
45	1,000	439.7	267.8	117.751

Comparing the simulation results tabulated in Tables 17.4 and 17.5 with the maximum power point values V_{mp}, I_{mp}, and P_{mp} obtained from the PV module characteristics for different input parameter conditions (irradiance and temperature) mentioned in Figures 17.4(a) and 17.4(b), it is observed that the fuzzy MPPT controller is capable of tracking the maximum power point very closely. The results obtained in this proposed work, is compared with the results of Hua and Shen (1997) and found to be far better in terms of transient response of the boost converter's output voltage and current. The steady state for voltage is achieved in 0.005 seconds and the currents settle down in 0.009 seconds. In Oulad-Abbou et al., (2018), P&O method is used. The voltage settlement is achieved in 0.05 seconds. So, the proposed fuzzy MPPT controller can track the maximum power point accurately as well as very rapidly with a sudden change in parameters (Figure 17.10).

TABLE 17.5 The Voltage, Current, and Power Outputs from the Boost Converter for Different Temperatures Under Standard Irradiance 100 W/m²

Irradiance (W/m²)	Temperature (°C)	Boost Converter Output Voltage (V)	Boost Converter Output Current (A)	Power Delivered by the Boost Converter to the ESU (kW)
1,000	25	442.9	336.7	149.124
500	25	435.3	174.4	75.916
250	25	431.3	87.11	37.57

17.7 CONCLUSION

The proposed fuzzy based MPPT controller is facilitating precise control of the duty cycle of the DC-DC boost converter, thus dragging the system output very close to the maximum power point for the changed values of the irradiance or temperature.

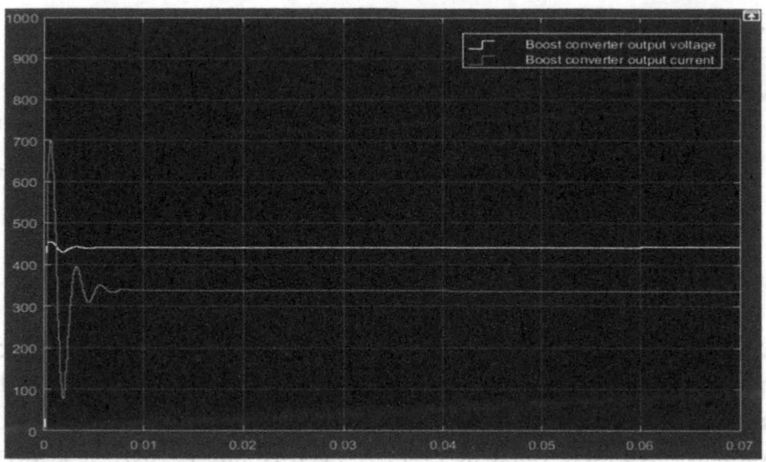

FIGURE 17.10 Transient response of boost converter output voltage and current.

Thus, it can be concluded that, this controller is an efficient one to put a fine control in the domain of MPPT. Moreover, due to the non-requirement of exact system mathematical model to design and implement this controller along with its robust nature, it can be designed and implemented very fast and can be used for a very large and complicated system.

17.8 FUTURE SCOPE

The proposed work can be extended by expanding the system. Integration of EVs charging ports with the existing system for charging the EVs directly from the output of the DC-DC boost converter can be done. The ESU unit and the EV charging ports will be connected in parallel. The charging strategy of EVs along with charging of ESU unit can be designed for proper energy management of the entire system.

KEYWORDS

- **DC-DC boost converter**
- **duty cycle**
- **fuzzy logic controller (FLC)**
- **maximum power point tracking (MPPT)**

REFERENCES

Bendib, B., Krim, F., Belmili, H., Almi, M. F., & Boulouma, S., (2014). Advanced Fuzzy MPPT Controller for a stand-alone PV system. *Energy Procedia, 50*(2014), 383–392.

Bhatnagar, P., & Nema, R. K., (2013). Maximum power point tracking control techniques: State-of-the-art in photovoltaic applications. *Renewable and Sustainable Energy Reviews, 23*, 224–241.

Coelho, R. F., Concer, F. M., & Martins, D. C., (2010). A simplified analysis of DC-DC converters applied as maximum power point tracker in photovoltaic systems. In: *The 2nd International Symposium on Power Electronics for Distributed Generation Systems* (pp. 29–34). IEEE.

Esram, T., & Chapman, P. L., (2007). Comparison of photovoltaic array maximum power point tracking techniques. *IEEE Transactions on Energy Conversion, 22*(2), 439–449.

Hua, C., & Shen, C., (1997). Control of DC/DC converters for solar energy system with maximum power tracking. In: *Proceedings of the IECON'97 23rd International Conference on Industrial Electronics, Control, and Instrumentation (Cat. No. 97CH36066)* (Vol. 2, pp. 827–832). IEEE.

Mirbagheri, S. Z., Mekhilef, S., & Mirhassani, S. M., (2013). MPPT with Inc. Cond method using conventional interleaved boost converter. *Energy Procedia, 42*, 24–32.

Nafeh, A. E. S. A., (2011). An effective and safe charging algorithm for lead-acid batteries in PV systems. *International Journal of Energy Research, 35*(8), 733–740.

Oulad-Abbou, D., Doubabi, S., Rachid, A., García-Triviño, P., Fernández-Ramírez, L. M., Fernández-Ramírez, C. A., & Sarrias-Mena, R., (2018). Combined control of MPPT, output voltage regulation and capacitors voltage balance for three-level DC/DC boost converter in PV-EV charging stations. In: *2018 International Symposium on Power Electronics, Electrical Drives, Automation and Motion (SPEEDAM)* (pp. 372–376). IEEE.

Rekioua, D., Achour, A. Y., & Rekioua, T., (2013). Tracking power photovoltaic system with sliding mode control strategy. *Energy Procedia, 36*, 219–230.

Salameh, Z. M., Casacca, M. A., & Lynch, W. A., (1992). A mathematical model for lead-acid batteries. *IEEE Transactions on Energy Conversion, 7*(1), 93–98.

Salas, V., Olias, E., Barrado, A., & Lazaro, A., (2006). Review of the maximum power point tracking algorithms for stand-alone photovoltaic systems. *Solar Energy Materials and Solar Cells, 90*(11), 1555–1578.

Seddik, M., Zouggar, S., Ouchbel, T., Oukili, M., Rabhi, A., Aziz, A., & Elhafyani, M. L., (2010). A stand-alone system energy hybrid combining wind and photovoltaic with voltage control (feedback loop voltage). *International Renewable Energy Congress* (pp. 227–232). Tunisia.

Villalva, M. G., Gazoli, J. R., & Ruppert, F. E., (2009). Comprehensive approach to modeling and simulation of photovoltaic arrays. *IEEE Transactions on Power Electronics, 24*(5), 1198–1208.

Zegaoui, A., Aillerie, M., Petit, P., Sawicki, J. P., Jaafar, A., Salame, C., & Charles, J. P., (2011). Comparison of two common maximum power point trackers by simulating of PV generators. *Energy Procedia, 6*, 678–687.

CHAPTER 18

ACTIVE AND REACTIVE POWER CONTROL IN VARIABLE WIND TURBINE USING SIMULTANEOUS WORKING OF STATCOM AND PITCH CONTROL

ARRIK KHANNA,[1] PUSHPANJALI SINGH BISHT,[2] and VIKRAM KUMAR SAXENA[3]

[1]*Department of Electrical Engineering, Chitkara University, Punjab–140401, India, E-mail: arrik1433@gmail.com*

[2]*Department of Electrical Engineering, DPGITM Gurugram, Haryana–122001, India*

[3]*Department of Electrical Electronics and Engineering, Pailan College of Management and Technology, West Bengal–700102, India*

ABSTRACT

The energy need is ever attention requiring issue for the under developed and developing countries. These countries are in the figurative of the shortfall of hydrocarbons or fossils energy. Wind farms connected to existing power system poses credible challenges to economic operation of power system. Wind power is undergoing through a very expeditious development. The assimilation of wind energy in the existing power system is a concern to revamp the vitalization of the ability and to maintain the increasing rate of wind generation installation capacity, this is utmost important in achieving

Advances in Data Science and Computing Technology: Methodology and Applications. Suman Ghosal, Amitava Choudhury, Vikram Kumar Saxena, Arindam Biswas, & Prasenjit Chatterjee (Eds.)

the set parameters of guarantee and tenability of uninterrupted supply. Recently, variable-speed wind generators added in wind farm has expanded and more manufacturers of wind generators are assembling "variable-speed" wind generators. Adoption of micro grids has added advantage of including wind generators working with low voltage ride through (LVRT) to stay inter connected when the voltage and frequency drops. Tripping of wind turbines generators in fault conditions has a drastic effect on the stability of power system. This chapter lays light on working of variable-speed wind generators with STATCOM and pitch control. Analysis obtained from simulation shows that wind generators can be used at its optimal energy while using simultaneous Pitch Control and STATCOM for a wide range of wind speeds.

18.1 INTRODUCTION

The evolution in wind energy technology has been exceptional. The cost of electricity generation from wind energy has decayed to about one seventh from the early 1980s. In United States, a substantial growth has been seen in wind power generation capacity, with growth averaging nearly 24% every year during the past 5 years. In India, as on 31st December 2018, the total installed wind power capacity was 34,046 MW with generation of 52,666 GWh. But as on 31st December 2019 in India the installed capacity of wind power has raised to 36,625 MW with generation reaching 62,036 GWh, a subsequent rise of 7.57% in 12 months. In order to expedite the application of new wind power projects, numerous computer algorithms using MATLAB and Homer Pro have been developed for designing and optimization of solar, wind power, microgrids, and other renewable energy sources. But the major three challenges to adoption to wind power's future remains' the problems of grid reliability, voltage stability and wind intermittency (Eisa, 2019). Electrical power generated from wind power has increased exponentially in global market wind turbines are being equipped with a variable pitch technology from speed control and static compensator (STATCOM). This technology is adopted to extract the maximal energy potential and STATCOM provides stable generation at the grid frequency when generators in wind turbines operate on the basis of variable rotation speed (Awad et al., 2014; Fu et al., 2017). As world electricity consumption is on rise steadily, there is an everlasting onus on society to flourish and adopt the renewable source of energy for decreasing world carbon footprint and have positive impact on global environment. In direction of expanding energy intake, wind power along with solar/ PV generation is the one the swiftly growing sector. As per data released by Global Wind Energy Council (GWEC) for year 2019 approximately 651 GW of total wind energy has been installed worldwide with 10% increase from year 2018. China and United States continues to be largest on shore wind energy market and accounting almost 60% of new capacity added in year 2019. The potential impact of COVID-19 globally has nosedived economy and demand for energy for year 2020 was expected to be a record year for wind generation and as per GWEC with a forecast of 76 GW to be added as new capacity.

18.2 WIND POWER IN INDIA

The installed capacity worldwide for wind power has reached 651 GW by the end of 2019. World's top five markets for new wind installations were China, United States, United Kingdom, India, and Spain (Awad et al., 2014). The expansion of wind power in India started in last decade of 1990s and has unquestionably expanded in the last half a decade. In spite of being a relative novice to the wind business correlated with United Kingdom or Spain, India lies fourth biggest equipped country with wind power size in the world. As on 31st December 2019 wind power stands at 3,665 MW in India, mainly spanning crosswise Tamil Nadu (9231.77 MW), Gujarat (7203.77 MW), Maharashtra (4794.13 MW), Karnataka (4753.40 MW), Rajasthan (4299.73 MW), Madhya Pradesh (2519.89 MW), Andhra Pradesh (4077.37 MW), Kerala (62.50 MW), Telangana (128 MW) and other states (30 MW). It is predicted that 60 GW of further wind power capacity will be added in India by 2022. Figure 18.1 show wind map of India for the year 2019.

In order to obtain better flexibility and generate more energy to be injected in system, wind generators of order of MW range are normally interconnected simultaneously to form a wind farm. Since past 2 decades the size of wind turbines has increased now, we have turbines ranging from 8–10 MW with blade span of 160 m (Mahvash, Taher, and Rahimi, 2017). However, integration of such large variable speed wind generators with existing power grid, will increase the energy seepage to an equilibrium above which a compelling brunt effect will be seen on grids operation (Rashad, Kamel, and Jurado, 2018). Previously many tasks are being laid down by regulatory task force for system stability (Rashad, Kamel, and Jurado, 2018) and consistently expanding of this renewable power, transmission system handling operators (TSO) are compelled to lay down new stipulations on the wind farms for the well-operation of the power grid network.

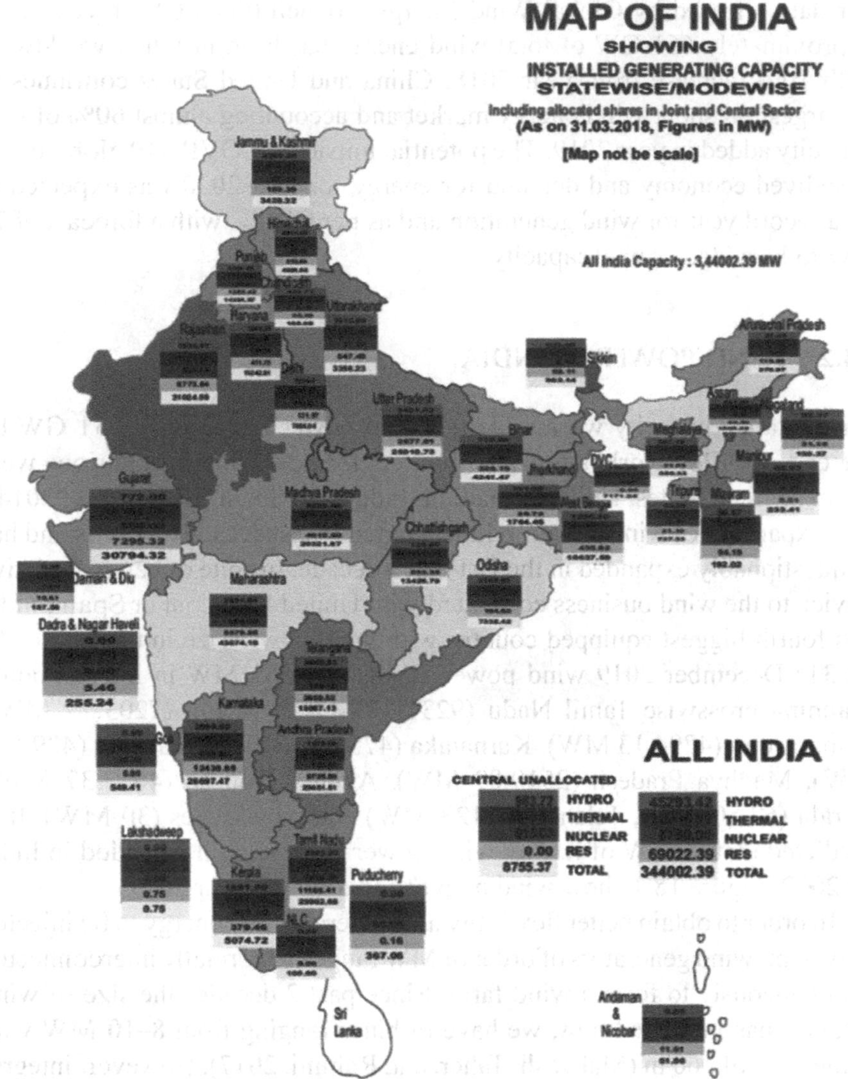

FIGURE 18.1 Energy MAPS of India: Installed generation capacity (Central Electricity Authority, 2019).

18.2.1 INTEGRATION OF WIND POWER WITH EXISTING GRID NETWORK IN INDIA

Integration of wind turbine with existing electric grid network has an effect on the power quality of power being delivered to the grid or end users. The power generated from a wind turbine is full of problems such as active power, reactive power, variation of voltage, flicker, harmonics, voltage swell, voltage sag, etc. (Al-Jowder, 2007). In variable speed and fixed-speed wind generators operation, variation in wind acceleration is imparted as change in the automated torque (Habibi, Rahimi, Howard, 2017). Throughout the traditional working, wind generator outturns an endless variable output power. This power change is mainly result of turbulence, wind shear. Thus, the network connecting wind turbine with existing grid needs to manage for such fluctuations. In order to eliminate some effects of fluctuations generated in wind turbines variable speed turbines are being used these days in conduction with an induction generator (Eisa, 2019). Some of the problems encountered by using variable speed on grid network are (Habibi, Rahimi, and Howard, 2017):

- Change in voltage degree;
- Transient voltages and currents;
- Harmonic changes in the waveforms for AC power.

A STATCOM – Pitch control added supervision technology is recommended in this chapter for developing quality power being managed technically by both effect of variable frequency, voltage, and to reduce mechanical stresses developed by variable wind on wind turbine blades (Lin et al., 2019), so as to increase power quality and reduce stresses on power system, which will help in integrating wind farms in existing power systems.

18.2.2 PROPORTIONAL CONTROL BASED PITCH ACTUATOR SYSTEM (CSA)

A MATLAB based Simulink model (Oh et al., 2014) is proposed for the CSA system is shown in Figure 18.2. Wr1 is taken as input which is run through a congestion filter. The purpose of the congestion filter is to cap the pitch input in the range of 0 to 90° the normal range of valid pitch angle. Second model shown below is the gain block which is used to define the proportional gain of the system, denoted by K-Block 0-pitch_max is also a congestion filter which is used to define the values of pitching speed. The range of maximum and minimum permissible values of pitching speed in this system are set at 3.5° per unit time and –3.5° per unit time.

The feedback of this model was realized and noted at three changes values variable speed matrix. In entire cases pitch angle remained constant at 75° irrespective of speed and the same is shown in Figure 18.3.

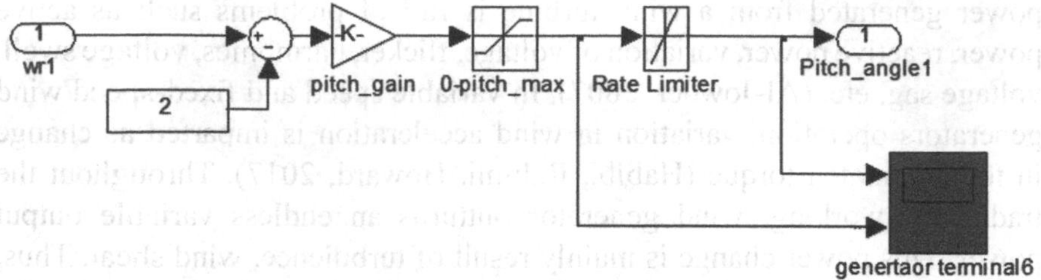

FIGURE 18.2 Block diagram of proportional control based pitch actuator system.

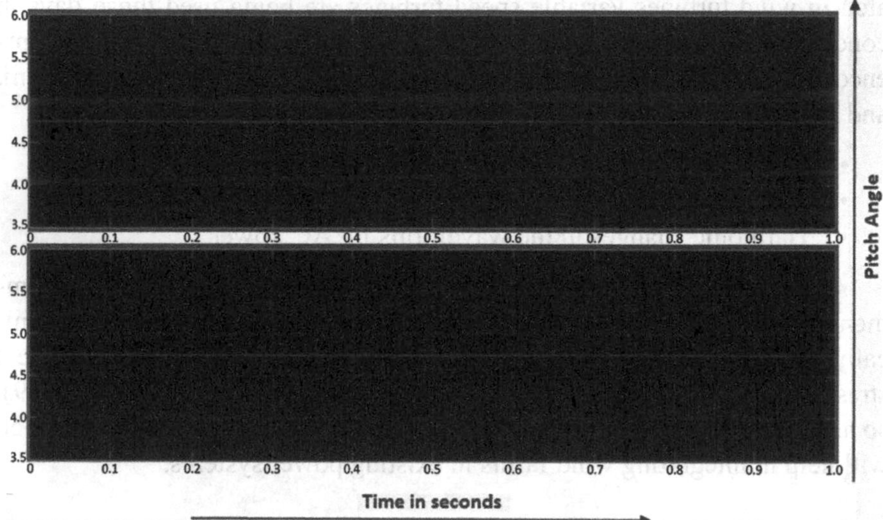

FIGURE 18.3 Pitch angle response of proportional control-based pitch actuator system.

The effect of Pitch angle control on rotor speed, Electrical torque T_e and Mechanical Torque T_m as shown in Figures 18.4 and 18.5.

18.3 DRIVE TRAIN MODEL

The dynamic equations of the two-mass drive-train written on the generator side are (Habibi, Rahimi, and Howard, 2017):

$$T'_{wt} = J'_{wt}\frac{d\omega'_{wt}}{dt} + D(\omega'_{wt} - \omega_{gen}) + K(\theta'_{wt} - \theta_{gen}) \tag{1}$$

$$-T_{gen} = J_{gen}\frac{d\omega_{gen}}{dt} + D(\omega_{gen} - \omega'_{wt}) + K(\theta_{gen} - \theta'_{wt}) \tag{2}$$

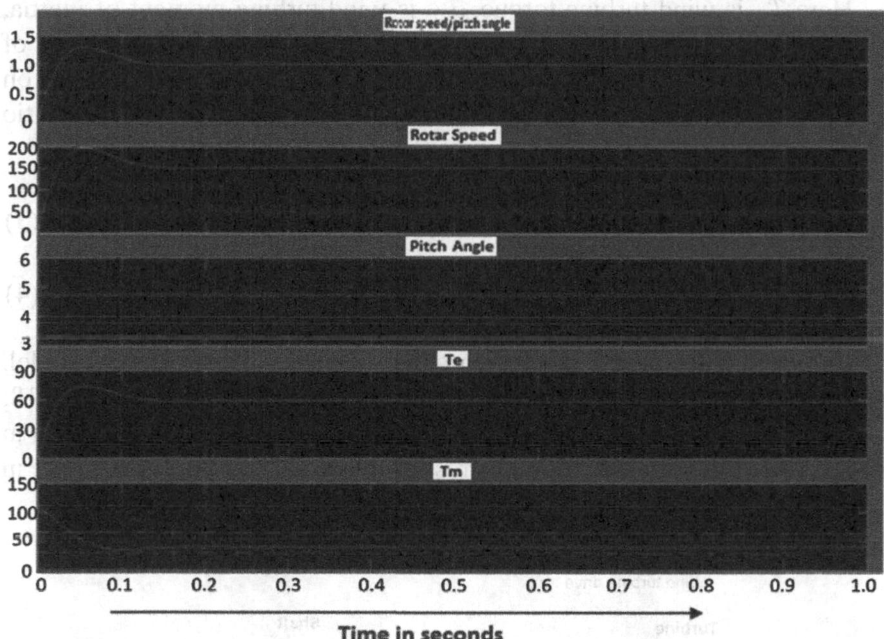

FIGURE 18.4 Depicting effect of using constant pitch control on T_e and T_m.

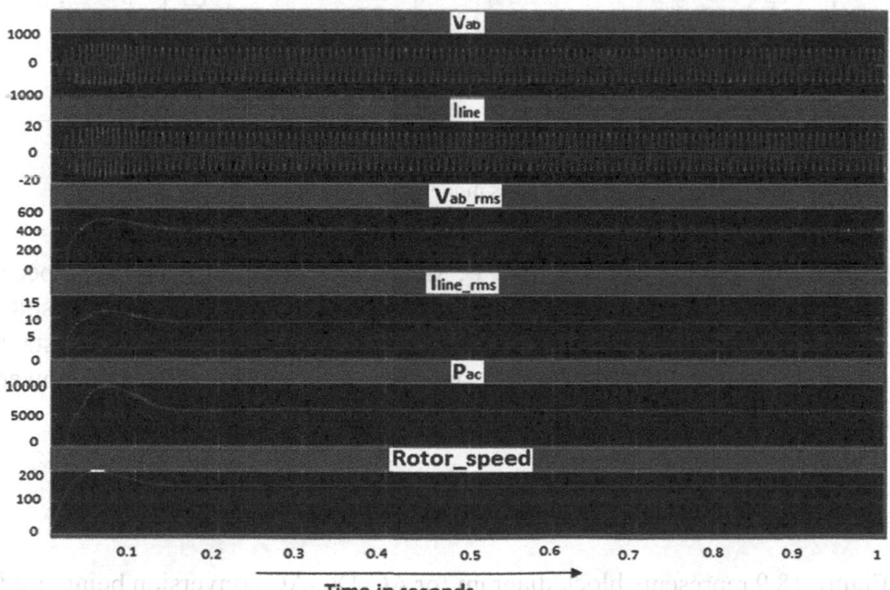

FIGURE 18.5 Pitch angle response of pitch control on V_{ab}, I_{line}, V_{ab_rms}, I_{line_rms}, P_{ac}, and rotor speed W_m w.r.t time on x axis.

Here T_{wt} is wind turbine torque, J_{wtr} is wind turbine moment of inertia, φ_{wtr} is spring constant T_{gen} is generator torque, J_{gen} is generator moment of inertia, and φ_{gen} is spring constant indicating the stiffness of the shaft on generator part. The torque and is transmitted via the gearbox with a gear ratio of *Ng* and angular speed is given as:

$$Rotor\ speed, \omega'_{wt} = \frac{d\theta'_{wt}}{dt} \tag{3}$$

$$Generator\ speed, \omega_{gen} = \frac{d\theta_{gen}}{dt} \tag{4}$$

Figure 18.6 depicts the block diagram of turbine and shaft 2 mass model, in which the mechanical stress on turbine blades are reduced (Li et al., 2018; Vidal et al., 2012) with change in speed and the result for which is shown in Figure 18.7. But with reduction in stress on blades the output shaft torque in pu is kept almost constant as shown in Figure 18.8.

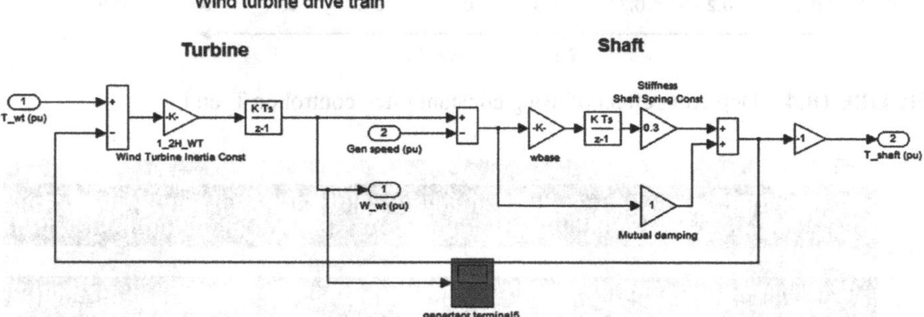

FIGURE 18.6 Turbine and shaft 2 mass model.

The effect of Electrical Torque T_e, Mechanical Torque T_m and rotor speed W_m using constant pitch control actuator is shown in Figure 18.8. The results show that after using pitch control technique T_e, T_m, W_m remains constant irrespective of the variable speed and hence providing a Smooth power generation at variable speed.

18.4 STATCOM (STATIC COMPENSATOR)

Figure 18.9 represents block diagram for AC-DC-AC conversion being used in wind power generation.

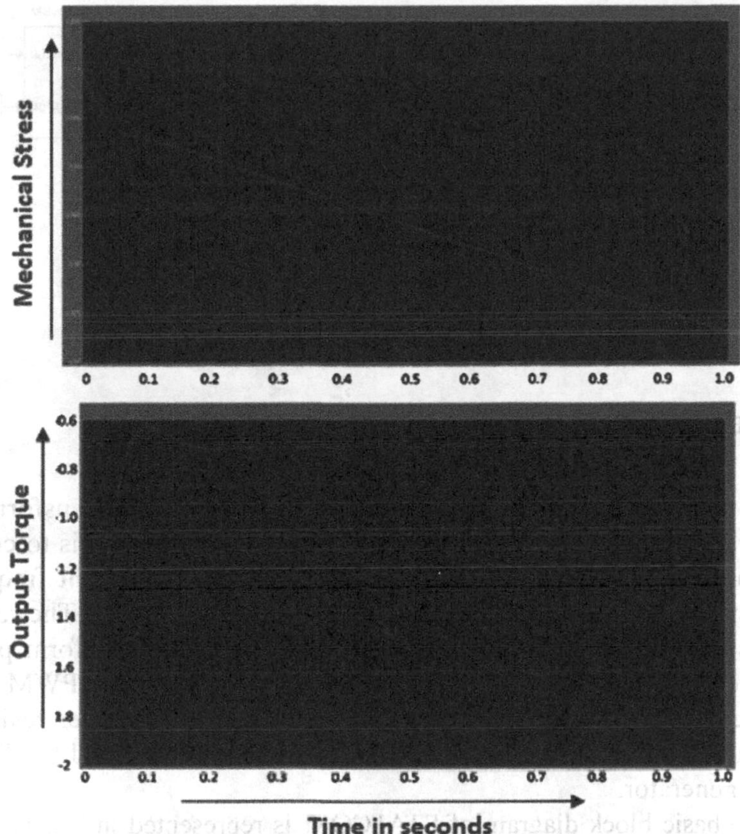

FIGURE 18.7 (a) Reduction of mechanical stresses on turbine blades w.r.t time in sec on x axis; and (b) constant output shaft torque w.r.t time in sec on x axis.

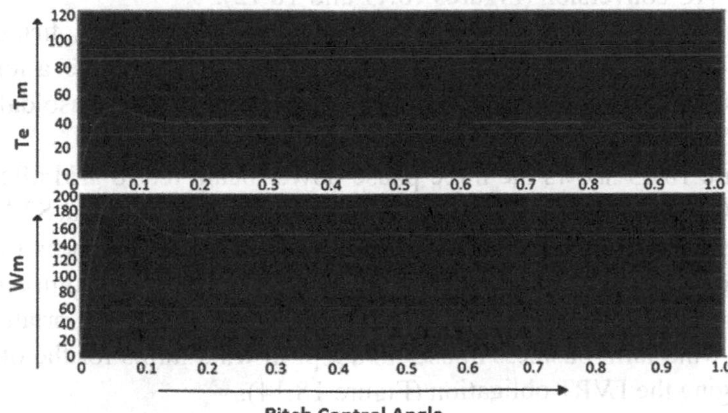

FIGURE 18.8 Effect of pitch control (x axis) on $T_{e,}$ T_m, W_m. on Y axis.

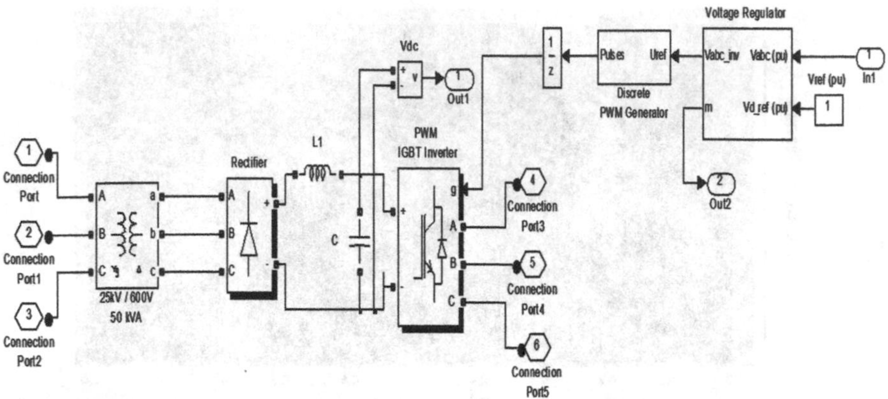

FIGURE 18.9 Simulink block diagram of AC-DC-AC conversion.

In this block diagram a three phase (two winding) transformer is followed by a rectifier unit. The basic principle of working is to convert generated AC-DC-AC with the output AC having a constant frequency and active reactive power control (Liu, Zhao, and Li, 2014). The rectifier unit is followed by a LC filter used to eliminate AC waveform present after rectification; the LC filter network gives its input to PWM IGBT inverter, which is used to convert DC-AC having a snubber resistance of 5,000 ohms. The firing angle control of IGBT is provided by discrete PWM generator.

The basic block diagram of STATCOM is represented in Figure 18.10. This shows the actual working of STATCOM circuit. A three-phase parallel RLC link of 50 KW is provided which acts as load for inverted voltage after AC-DC-AC conversion (Figures 18.11 and 18.12).

Figure 18.13 shows the behavior converted DC voltage, firing angle control, inverted AC voltage and inverted AC load voltage after using STATCOM. The figure clearly depicts that one gets near sinusoidal waveform of load using STATCOM.

Figure 18.13 shows the three phase power being fed to grid after using STATCOM. Results obtained through simulation have shown that STATCOM can drastically raise the restoration of the wind generators because of its brisk recovery of voltage and hence increased stability when compared with traditional voltage control techniques. Therefore, it is an appropriate operation for wind farm business houses to adopt in wind farms for the objective of engaging the LVRT obligation (Figure 18.14).

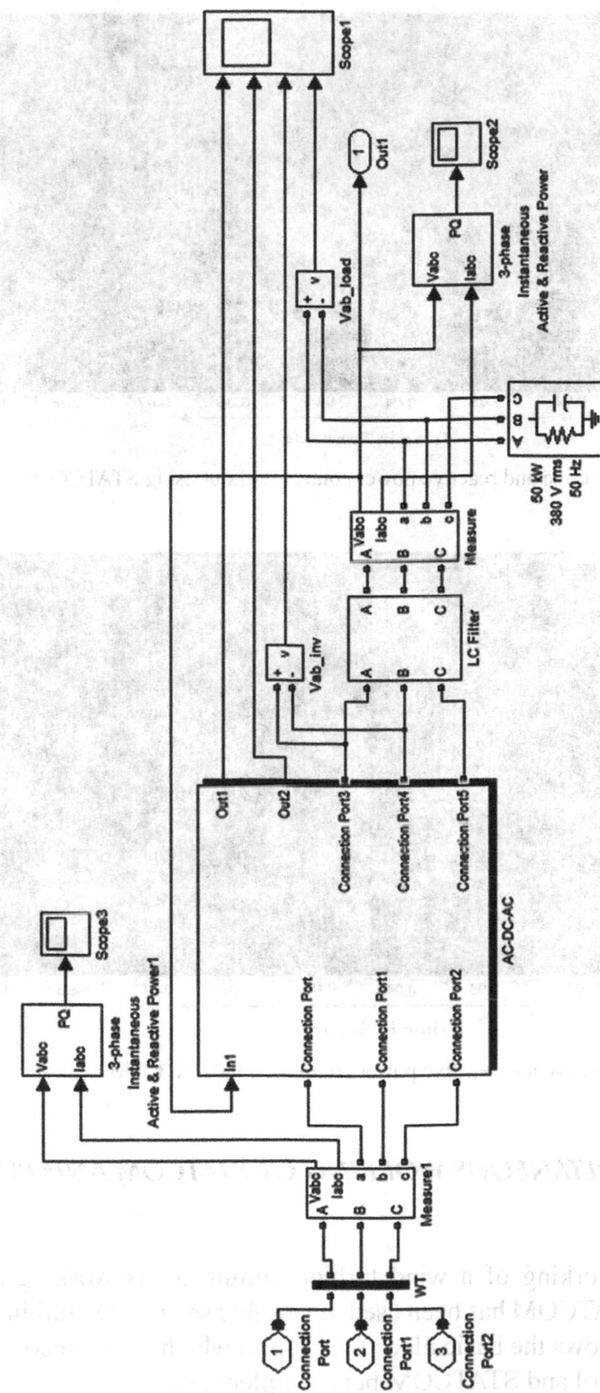

FIGURE 18.10 Basic Simulink block diagram of STATCOM.

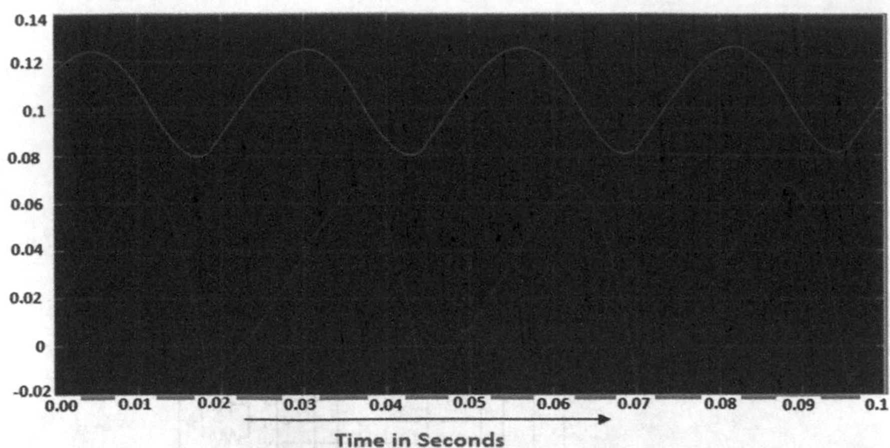

FIGURE 18.11 Active and reactive power control without using STATCOM.

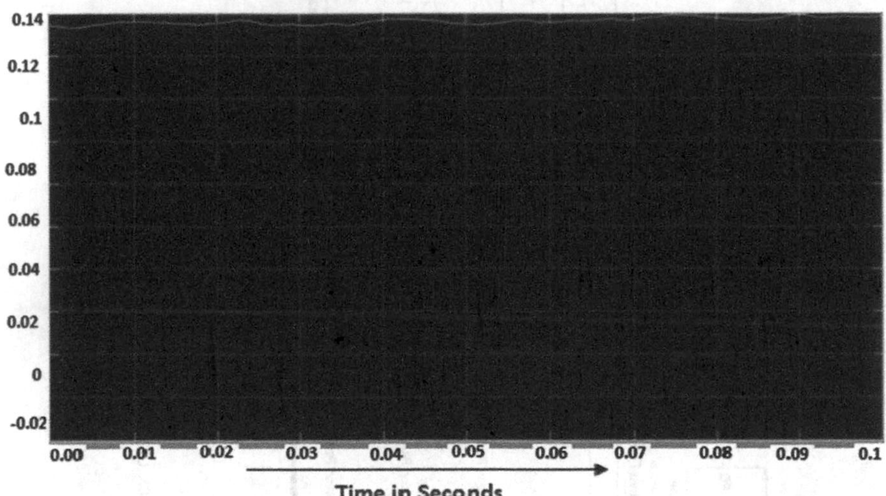

FIGURE 18.12 Active and reactive power control using STATCOM.

18.4.1 SIMULTANEOUS WORKING OF STATCOM AND PITCH CONTROL

For efficient working of a wind turbine simultaneous working a PITCH control and STATCOM has been used in variable speed wind turbine. Here in Figure 18.15 shows the basic block diagram in which simultaneous working of PITCH control and STATCOM being implemented.

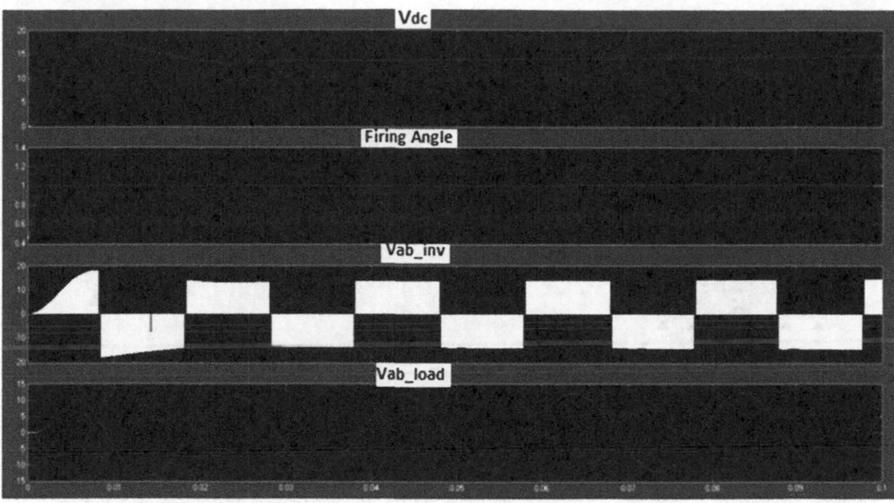

FIGURE 18.13 Vdc, firing angle, Vab_inv, and Vab_load waveforms.

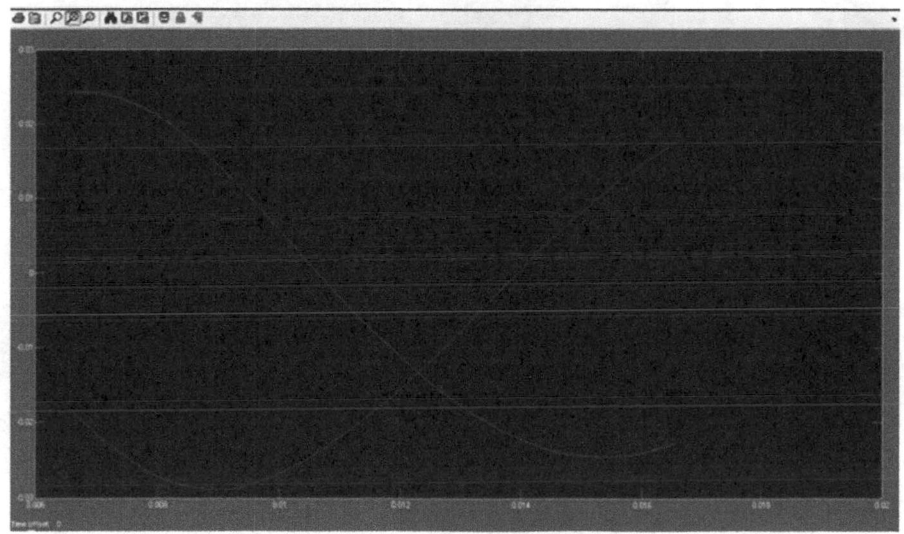

FIGURE 18.14 Three phase power being fed to grid after using STATCOM.

The problem with using only Pitch control is that as the stress on turbine blades is reduced considerably along with smooth operation of Turbine Generator set. But Pitch control has no effect on power quality issues. Similarly, by using only STATCOM we do obtain a near sinusoidal waveform which can be fed to grid at varying speed having constant frequency

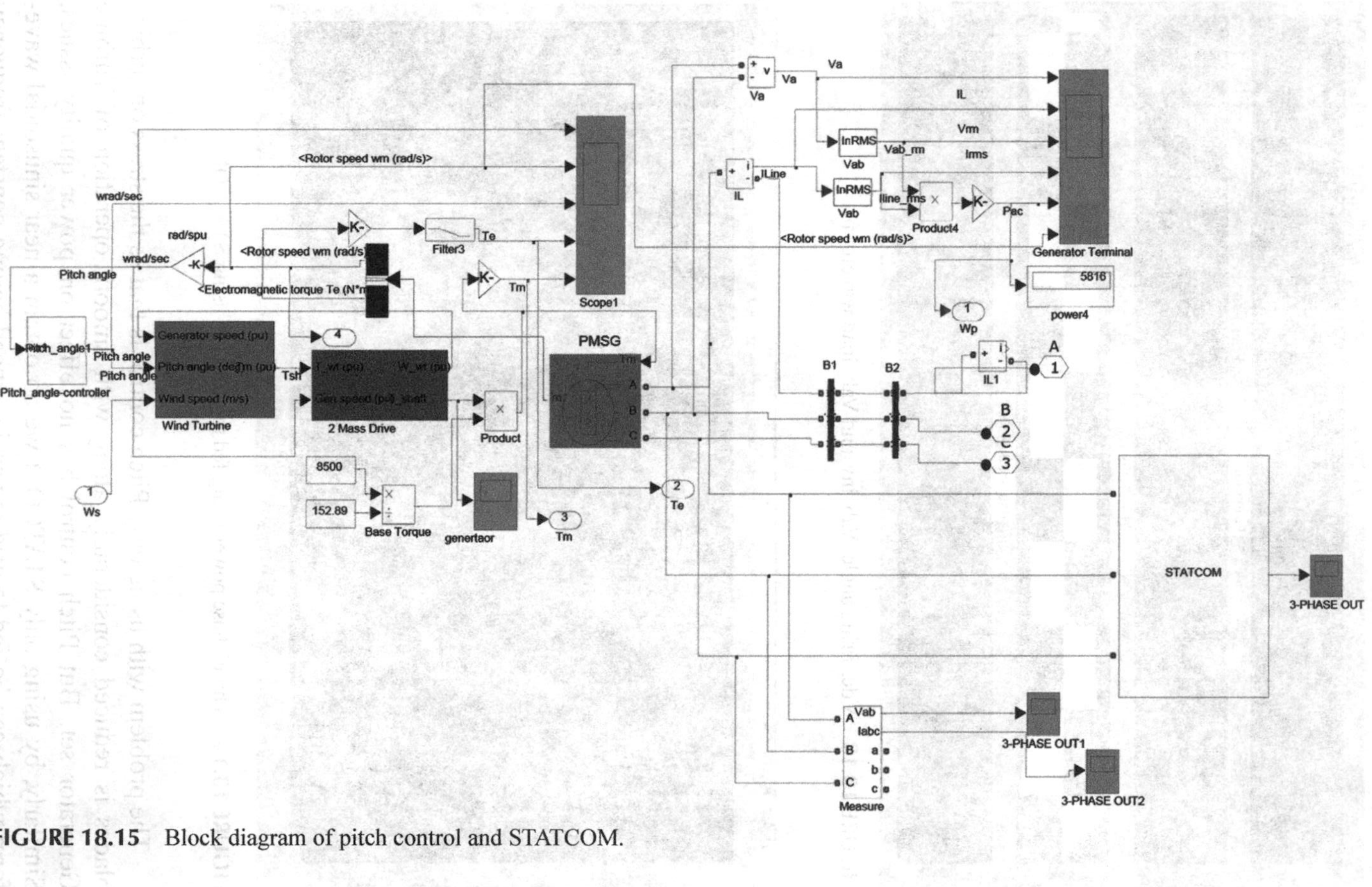

FIGURE 18.15 Block diagram of pitch control and STATCOM.

of operation and providing a quality power (Al-Jowder, 2007) with fewer burdens on transmission lines and reduced losses (Vidal et al., 2012). As Figure 18.16 shows the simultaneous working of both Pitch Control and STATCOM we get a turbine generator set which works on power quality and less stresses on turbine blades at variable wind speed.

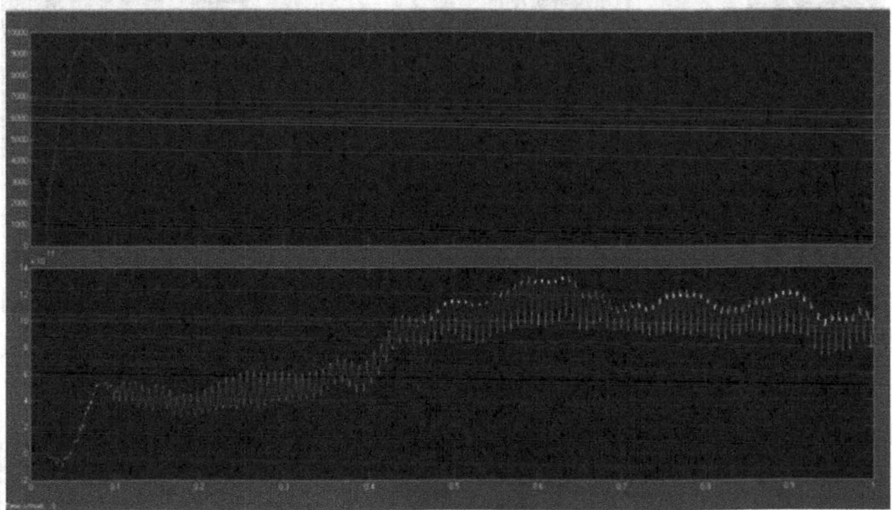

FIGURE 18.16 Effect of active reactive power of turbine using pitch control and STATCOM.

In Figure 18.17 we get the result which depicts the three phase power before being fed with Pitch Control without load and similarly Figure 18.21 shows the power being fed to grid after using both Pitch Control and STATCOM under 50 KW parallel RLC mask link load (Figures 18.17–18.22).

The power generated without using STATCOM and PITCH CONTROL is 5,816 KW and the power generated using PITCH CONTROL and STATCOM is 5340.41 KW.

$$\text{Now } \acute{\eta} \text{ (efficiency)} = \text{Output/Input}$$

Hence ή in comparison with STATCOM and Pitch control are not being used

$$= 5340.41 / 5816 = 91.81\%$$

Which clearly shows that by simultaneous usage of STATCOM and PITCH CONTROL has very less impact over power loss and has a high efficiency of 91.81%.

FIGURE 18.17 Three phase power with pitch control without load.

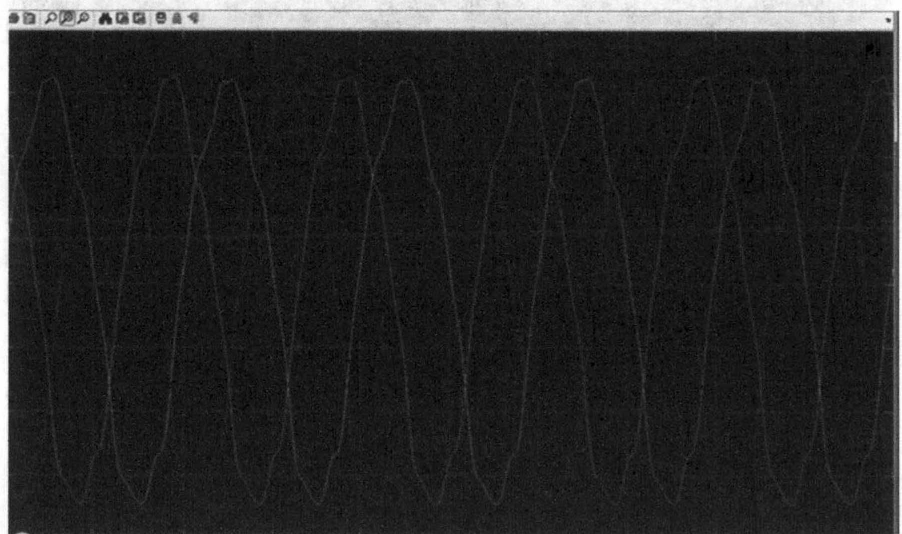

FIGURE 18.18 Three phase power after using STATCOM and pitch control on load.

18.5 CONCLUSION

As the issues that arise due to voltage disturbances, THD, and transients does not allow power generated by variable speed wind turbine to be directly fed to grid, hence an intermediate has to be applied so as to eliminate the effect of THD, transients, voltage swell. This is attained by using PITCH CONTROL and STATCOM.

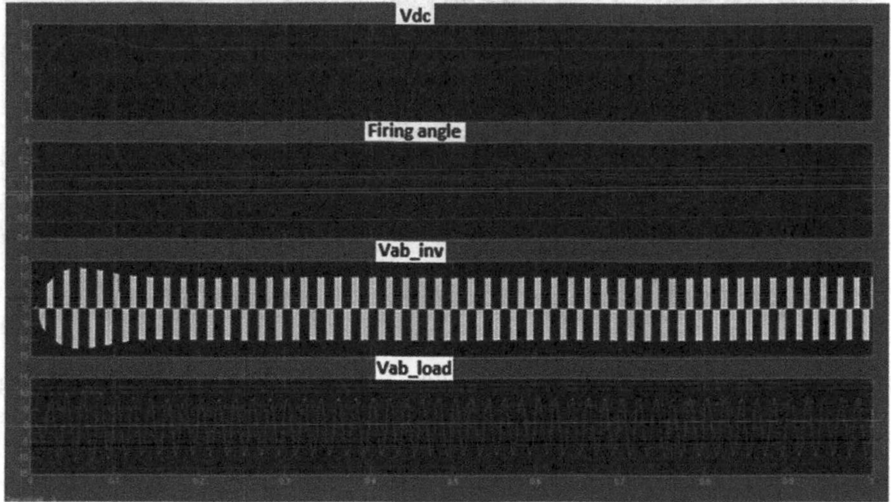

FIGURE 18.19 Vdc, firing angle, Vab_inv, and Vab_load waveforms after using both STATCOM and pitch control technique.

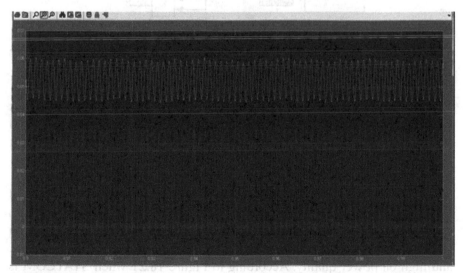

FIGURE 18.20 Active and reactive power being fed to STATCOM using PITCH control.

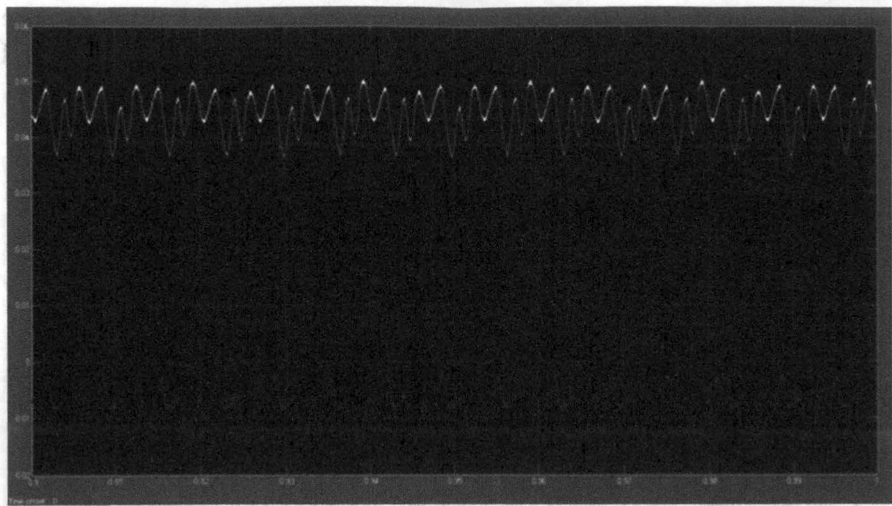

FIGURE 18.21 Compares the waveform of three phase active and reactive power using both pitch control and STATCOM.

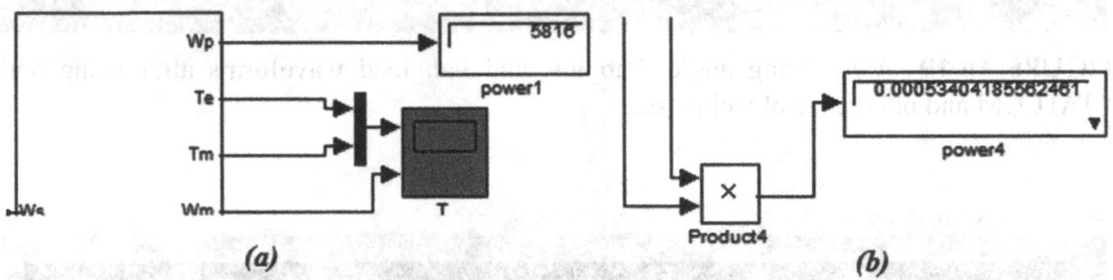

FIGURE 18.22 (a) Power generated without STATCOM; (b) power generated using STATCOM and PITCH control.

On comparing Figure 18.15 and 22.22, we see that on simultaneous working of STATCOM and Pitch Control we get a better controlled and smooth waveform. Similarly, on comparing phase voltage V_{ab}, Line current I_l, RMS voltage V_{ab_rms} and P_{ac} from Figures 18.15 and 18.22 we get a waveform with better frequency control and free from THD and transients.

On studying Figure 18.20 we see that we get a near pure sinusoidal waveform on no load on using PITCH control, but our main goal was to use both PITCH Control and STATCOM and then obtain a waveform on load having reactive power control, which being free from THD, transients, and high on power quality. According to Figure 18.21 when STATCOM is introduced to PITCH CONTROL technique and 50 KW RLC load is applied, we obtain a waveform which is free from ripple/sag/voltage swell effect. Free from transients and THD. When the voltage or power as shown in Figure 18.21 is fed to grid then the problem of power quality is minimized. Similarly, electric devices will work properly without the problem of abrupt/ abnormal overheating, insulation breakdown. This will result in increased life of electric devices, transmission lines and stress on existing Grid network will be reduced.

Hence by using STATCOM and PITCH CONTROL simultaneously we get a better-quality power which can be fed directly to grid. Therefore, instead of feeding 5,816 KW power, we will feed 5340.41 KW power rich in quality power. Moreover, the efficiency of simultaneous working of STATCOM and PITCH CONTROL is very high in order of 91.81% when compared with model without using STATCOM and pitch control simultaneously. Also, reduction on power is easily compensated with increase in quality of power and reduction in THD and transients.

KEYWORDS

- **pitch control**
- **STATCOM**
- **variable speed**

REFERENCES

Al-Jowder, F., (2007). "Improvement of synchronizing power and damping power by means of SSSC and STATCOM." *Electric Power Systems Research, 77*, 1112–1117.

Awad, A. S. A., Shatshat, R. E., Salama, M. M. A., & El-Fouly, T. H. M., (2014). "Low voltage ride through capability enhancement of wind farms' generators: DVR versus STATCOM." *Presented at the 2014 IEEE PES General Meeting | Conference & Exposition.*

Central Electricity Authority, (2019). Available: http://cea.nic.in/ (accessed on 08 December 2021).

Eisa, S. A., (2019). "Modeling dynamics and control of type-3 DFIG wind turbines: Stability, Q Droop function, control limits and extreme scenarios simulation." *Electric Power Systems Research, 166*, 29–42.

Fu, Y., Zhang, X., Hei, Y., & Wang, H., (2017). "Active participation of variable speed wind turbine in inertial and primary frequency regulations." *Electric Power Systems Research, 147*, 174–184.

Habibi, H., Rahimi, N. H., & Howard, I., (2017). "Power maximization of variable-speed variable-pitch wind turbines using passive adaptive neural fault tolerant control." *Frontiers of Mechanical Engineering, 12*, 377–388.

Li, C., Xiao, Y., Xu, Y. L., Peng, Y. X., Hu, G., & Zhu, S., (2018). "Optimization of blade pitch in H-rotor vertical axis wind turbines through computational fluid dynamics simulations." *Applied Energy, 212*, 1107–1125.

Lin, Z., Chen, Z., Liu, J., & Wu, Q., (2019). "Coordinated mechanical loads and power optimization of wind energy conversion systems with variable-weight model predictive control strategy." *Applied Energy, 236*, 307–317.

Liu, Z. J., Zhao, Z. Q., & Li, B., (2014). "Study on application of STATCOM in dynamic reactive power compensation of wind farm integration." *Applied Mechanics and Materials, 672–674*, 255–261.

Mahvash, H., Taher, S. A., & Rahimi, M., (2017). "A new approach for power quality improvement of DFIG based wind farms connected to weak utility grid." *Ain Shams Engineering Journal, 8*, 415–430.

Oh, K. Y., Lee, J. K., Bang, H. J., Park, J. Y., Lee, J. S., & Epureanu, B. I., (2014). "Development of a 20-kW wind turbine simulator with similarities to a 3 MW wind turbine." *Renewable Energy, 62*, 379–387.

Rashad, A., Kamel, S., & Jurado, F., (2018). "Stability improvement of power systems connected with developed wind farms using SSSC controller." *Ain Shams Engineering Journal, 9*, 2767–2779.

Vidal, Y., Acho, L., Luo, N., Zapateiro, M., & Pozo, F., (2012). "Power Control Design for Variable-Speed Wind Turbines." *Energies, 5*, 3033–3050.

PART VI

Technique for Improving Information and Network Security

CHAPTER 19

INFORMATION SECURITY USING KEY MANAGEMENT

SANJIB HALDER,[1] BIJOY KUMAR MANDAL,[2] and ARINDAM BISWAS[3]

[1]*The Bhawanipur Education Society College, Kolkata, West Bengal, India*

[2]*NSHM Knowledge Campus, Durgapur, West Bengal, India, E-mail: writetobijoy@gmail.com*

[3]*Kazi Nazrul University, Asansol, West Bengal, India*

ABSTRACT

To exchange secret messages between two users, while using the processes of encryption and decryption on the message with the help of a secret key. We will see briefly how the key is being generated. Key will be generated by the process of Diffie-Hellman key exchange algorithm, which does not follow the rules of sharing secret key instead two parties will develop the secret key together while exchanging partial information for key generation. After the generation of secret key, we will use DES (data encryption standard) algorithm for sending the message, while using the public media, i.e., Internet as communication channel and in receiver's side show that message is well received without any third party alteration and can be decrypted perfectly into originally sent plain text.

Advances in Data Science and Computing Technology: Methodology and Applications. Suman Ghosal, Amitava Choudhury, Vikram Kumar Saxena, Arindam Biswas, & Prasenjit Chatterjee (Eds.)

19.1 INTRODUCTION

We human beings always tend to keep privacy among us, whether it is about precious treasure or day to day life chores, or our intensions (Hellman and Diffie, 1976; Lin and Wang, 2013). To keep it secret we tend to use means of secrecy, some secret signs that will prevent unwanted public unaware of what's going on. Here comes the use of cryptography (Eric and Chachati, 2014). Around 1900 B.C. Egyptians used non-standard hieroglyphs to inscribe some documents that is the first documented use of cryptography. Cryptography is nothing but an art of hiding secret in plain-site, so that even if it's visible deciphering the meaning behind is impossible (Pervaiz, Cardei, and Wu, 2010). Scientifically saying it conveys the message from sender to receiver, while restricting threat data theft or unauthorized access. To keep the information safe cryptography involves generating codes, it may be in written form or generated by machine. Data is converted into a format that is unreadable for unauthorized user, only allowing it to be transmitted without compromising the data, means restricting decoding it back into a readable format by unauthorized entities (Ballardin, 2011; Carmen, 2012). Modern cryptology exists at a junction of subjects that include the use of various types of rules and discipline like physics, mathematics, communication science, computer science, electrical science. In addition, its application is spread across several fields like military communications, business communications, digital currencies/ transactions, electronic commerce, chip-based payment cards, etc. Cryptography is used on several levels for Information security to fulfill distinct tasks (Chen, 2013; He and Lee, 2008).

There are five primary functions of cryptography nowadays:

1. **Confidentiality:** It ensures that except for intended and authorized user no other person can access the data or read the message.
2. **Key Exchange:** The process, which involves the exchange of keys between sender and receiver.
3. **Integrity:** It assures the receiver that no kind of alteration or changes have occurred in the message, and the message has reached in its original form.
4. **Authentication:** It is the process of proving one's identity, i.e., the message is sent by an intended sender also intended receiver receives it. No third party has accessed the message.
5. **Non-Repudiation:** It is a mechanism to prove that the sender's involvement. When a message is sent and reaches the receiver, during cross checking sender may deny the report of having sent the

message, here non-repudiation comes to work. It is like a signature, which cannot be denied.

Cryptography consists of few simple steps. We start with the unencrypted data, also known as the plain text. Plain text is readable format of the message, which is easily understandable by anyone (Liaw et al., 2005). Now using the secret key, we change the unencrypted date to encrypted form called cipher text. The cipher text is sent from sender to receiver through some channels of communication. Receiver will decrypt the coded message into simple plain text. Cryptography has always been a great means for intelligence gathering (Saeed and Shahriar, 2011). Lastly, cryptography is most closely related with the creation and development of mathematical algorithms, which are used in the process of encryption and decryption of messages. It is the science of analyzing and breaking encryption scheme, also use the deciphered information for greater good of society (Fan, Wen, and Zhang, 2010).

Key Exchange is the process of sharing cryptographic keys between two parties, sender, and receiver prior to exchange of messages, using cryptographic algorithm (Xu, Zhu, and Feng, 2011). As when sender and receiver wish to exchange secret messages, they must be equipped with keys in order to encrypt the message to be sent and to decrypt the message the received. In-band and out-of-band are the two ways in which key exchange can takes place. FTPS, HTTPS, and SFTP, etc., several secure file transfer protocols have to encrypt the data through symmetric encryption to preserve the confidentiality during data transmission (Ali, Sivaraman, and Ostry, 2013; He, Niedermeier, and De Meer, 2013). This type of communication requires both the parties to have the same encryption key. In the world two communicating parties may be globally situated at end two ends of the earth, hence physically meet up for every time they need to do a transaction is not at all convenient. To make life easier we use public communication channels like Internet to do the talk, but data theft can happen anytime, as it is a public server. Here comes the need of secret means of communication (Xu, Zhu, and Feng, 2011). In addition, eventually the need of key exchange that will ensure later send messages do not fall in unauthorized hands. FTPS and HTTPS file transfer protocols which are SSL/TLS protected uses the key exchange occurs by the process of handshake. It works in request basis, where usually a web browser or a client requests a connection to the intended server by sending message, known as Client Hello. Client Hello consists of some random data like session id, compression algorithms and mainly a cipher suite. When the server allows the permission, a connection is established. Each cipher suite will have separate algorithm for key exchange, encryption, and message authentication. As the server receives, the Client Hello it searches for supported cipher suite in its own set of lists, after comparison choose the best cipher suite, and eventually choose the desired key exchange algorithm. In addition, immediately set to exchange key using the chosen key exchange algorithm, and proceed to share messages.

The two most popular key exchange algorithms are RSA and DIFFIE-HELLMAN. It probably would not be too much of a stretch to say that the advancement of these two key exchange protocols accelerated the growth of the Internet. That is because these two protocols allowed clients and server, as well as server and server, to exchange cryptographic keys over insecure medium (Internet) and in turn enable them to transact electronically in a secure manner.

19.2 PREVIOUS WORK

The first invention got the root of key exchange was in the hand of Ralph Merkle who developed the 'public key agreement' technique in 1974. Later, it became popular as 'Merkle's Puzzles' and got published in 1978. Before then, in 1977, Ron Rwest, AdiShamia, and Leonardo Adleman from MIT, discovered a Cock's Scheme autonomously. The well-known protocol was proposed by Bailey Whitfield Diffie, an American cryptographer, with collaboration of Martin Hellman and Ralph Merkle, who became a spearhead in development of public-key cryptography. The concept of Diffie-Hellman was published in the journal 'New Directions in Cryptography' in 1976 which had set a new field in distribution of cryptographic keys over an in secured communication channel. The problem for this key distribution was solved using this public key cryptography algorithm and communication became more secured then before. This also initiated the root of another advanced encryption algorithm, which is asymmetric key algorithm.

Since, Merkle the key exchanging phenomenon, Hellman suggested that the algorithm should be called Diffie-Hellman-Merkle's algorithm in 2002 set the first footmark. The AKE protocol is also called STS protocol developed by Diffie in 1992. This is an advancement of raw Diffie-Hellman Key exchange protocol because it authenticates the two parties in communication. The mechanism follows as, the two parties can calculate a secret key using Diffie-Hellman algorithm and then they get authenticated by the exchange of digital signatures between them. The STS protocols prevents the man-in-middle attack due to authentication. It provides the forward secrecy with the help of Diffie-Hellman algorithm to compute the session key that is short termed. It also provides the facility of unaffected session of previous communications using the long term keys. It shortens the process of communication, as the central authority need not be interfered due to authentication and prevents redundant exchange of elements.

19.3 PROPOSED ALGORITHM

The Diffie-Hellman algorithm is being used to establish a shared secret that can be used for secret communications while exchanging data over a public network. We need to remember the point that we are not sharing secret, instead creating path that will be used later as communicating channel.

Few points to remember before writing the algorithm are as follows:

- ➢ For the simplification and practical and easy implementation of the algorithm, we will use only 4 variables one prime number P, M (a primitive root of P) generator and two private values of two parties' c and d.
- ➢ M and P are publicly available numbers. Users (say Alice and Bob) pick private values c and d and they generate a key and exchange it publicly. The other party receives the key and from that generates a secret key after which both parties will possess same secret key for encryption and decryption.

- Both Alice and Bob set an agreement about the prime number and the generator P and M where (0<M<P).

  e.g., –23 and M=9
- Alice selects a random secret number that is her private key as c and Bob selects the same as d.

  e.g., c=4 and d=3
- With the help of the private key and the other two parameters Alice and Bob compute their public values

  A. Alice→X = (M^ c)mod P
     X = (9^4) mod 23

  B. Bob→Y = (M^ d)mod P
     Y = (9^3) mod 23
- After that, Alice, and Bob exchange their public numbers between each other, i.e., X is sent to Bob and Y is sent to Alice.

- Alice receives public key of Bob as Y, and Bob receives public key of Alice as **X**
- Alice and Bob compute symmetric keys to trick the third party.
  - o Alice→ka = (Y^ c) mod Pka=(((9^3)mod23)^4) mod 23
  - o Bob→kb = (X^ d)mod P kb=(((9^4)mod 23)^3) mod 23
- If K = ka = kb, then K is the shared key generated. K=9
- The shared secret key K is transformed to 64 bits and with the plain text (64 bits) it is given as input to the DES algorithm.

Initial Permutation: This step produces two equal halves of permuted block, i.e., LPT, and RPT with each having 32 bits and 16 rounds of encryption process is being carried out on each LPT and RPT.

1. **Transformation of Key**
   i. For the removal of 1 bit from each byte of block for parity checking, 56 bits are actually available which is the actual key length to be used for encryption. Here we get this key from previously calculated Diffie-Hellman key exchange algorithm.
   ii. For each round of permutation, a different set of key of 48 bits is derived from the above length.
   iii. 56-bit key is divided into two equal halves, which has 28 bits of length in each.
   iv. Depending on the rounds, the bits are circularly shifted to the left by one or two positions.
   v. After appropriate shift, 48 bits from 56 bits are selected. This is called Compression Permutation as it compresses the number of bits.
2. **Expansion Permutation**
   i. The right hand side block RPT is expanded from 32 bits to 48 bits and also, they are permutated.
   ii. RPT having 32 bits is divided into 8 blocks of bits each.
   iii. The above 4 bits of each block is expanded to a corresponding 6-bit block, i.e., 2 more bits are added which is the repetition of 1st and 4th bit (48 bit, 8 blocks each of 6 bits),
   iv. The compressed key of 48 bit from key transformation part is XORed with 4 RPT of 48 bits from the previous step and the output is fed to the S-Box subset.
3. **S-Box Substitution**
   i. Substitution that is performed is in 8 substitution boxes.
   ii. Each 8 S-boxes have input of 6 bits and output of 4 bits.

   iii. The 48 bit input block is divided into 8 sub blocks and each block is given an S-box.
   iv. The output produced from all the S-Boxes are combined to form 32-bit block which is given to the P-Box permutation as input.
4. **P-Box Permutation**
   i. Simple permutation takes place. Each of the bits are swapped with each other.
5. **XOR and Swap**
   i. LPT is XORed with output of P-Box.
   ii. Result produced in the above step is the new right half.
   iii. The old right half now forms the new left half.
      a. Final LPT and RPT are recombined to form a single block and final permutation is performed on the rejoined block where there is a simple transposition of each bit. This is performed at the end of all the 16 rounds.
      b. Four results that are produced is a 64 bits encrypted cipher text.
      c. To reverse or decrypt from the cipher text to the plain text or the original text, the reverse of all the above steps are performed along with the key that is generated in Diffie-Hellmen key exchanging algorithm.

19.4 WORKFLOW

The above workflow gives a clear view of the combination of the Diffie-Hellmen key Exchanging Algorithm with DES algorithm. Alice and Bob agree to a prime number and a generator that is much smaller than the prime number and has to be co-prime with it. They both has a private key with which they can calculate their public keys and exchange them with each other. After the exchange, they calculate their shared key from the parameters present with them. If both the key matches then only they can have a communication with each other, otherwise, not. If the keys are matched, then it is given as an input along with the plaintext to the DES encryption algorithm where it generates the cipher text after going through several phases of transformations. The cipher is decrypted using the same key by DES decryption algorithm (Figure 19.1).

A proper analysis of the above proposed design is being done. The time taken by both combination algorithm is much larger and hence it prevents a third party from interrupting.

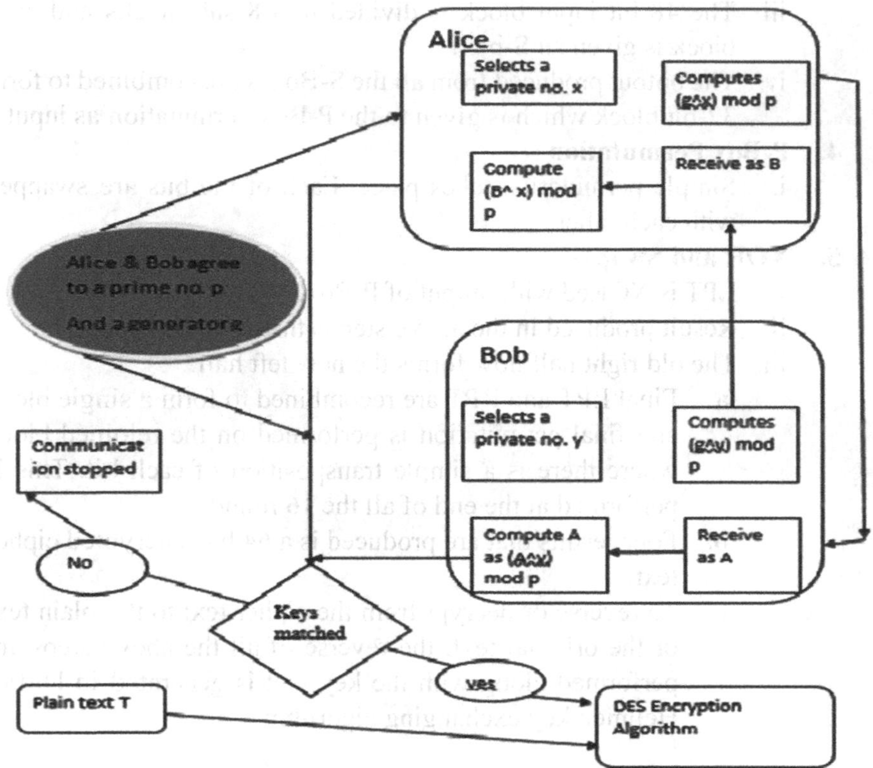

FIGURE 19.1 Workflow.

19.5 ANALYSIS AND DISCUSSION

A key is nothing but a stream of bits generated using some cryptographic algorithms. It is the 'heart' of cryptosystems. A cryptographic key can be compared with a real life lock and key. Suppose, we have locked a baggage using a key and parceled it to another person. The person can only be able to open the lock if he/she has the same key or a key that is able to open the particular lock. The cryptographic keys work in the same manner. Two person communicating with each other has a key with them, which enables them to encrypt or decrypt a message. For that, either the key is same for the sender and receiver or there is a paired key, which is different for both encryption and decryption.

The usage of keys are more than encryption and decryption. They are also used for digital signatures because each signature has a separate identity that is identified by a key. Message Authentication is another utilization of keys (Figures 19.2 and 19.3).

```
Output - CRYPTO (run)
run:
Enter moduolo(p)
23
Enter primitive root of23
9
```

FIGURE 19.2(a) Key generation.

```
run:
Enter moduolo(p)
23
Enter primitive root of23
9
Choose 1st secret no(Alice)
4
Enter the 2nd secret no(Bob)
3
```

FIGURE 19.2(b) Key generation of with clients.

```
run:
Enter plain text : Hello India
Enter secret key: 9
```

FIGURE 19.3(a) Plain text.

```
run:
Enter plain text : Hello India
Enter secret key: 9
Cipher Text: Lx+8Aznolbzjm Y7geOJ27A==
Enter cipher text :
```

FIGURE 19.3(b) Generation of cipher text.

```
run:
Enter plain text : Hello India
Enter secret key: 9
Cipher Text: Lx+8AznolbzjmY7geOJ27A==
Enter cipher text : Lx+8AznolbzjmY7geOJ27A==
Enter secret key: 9
Decrypted text: Hello India
BUILD SUCCESSFUL (total time: 2 minutes 14 seconds)
```

FIGURE 19.3(c) Decrypted message.

The Diffie-Hellman key exchanging algorithm is a key generation and exchanging algorithm where two parties communicating with each other share a common secret key between them before the actual communication process starts. As soon as the communication link is set up, a key is generated and exchanged by any one of them. The Diffie-Hellman design solves the purpose. Before the communication the keys are not known to them. After the interchange of the key, the same is used to encrypt the message. The cipher text generated is then received by the receiver and he/she uses the same key to decrypt the message. Hence, the receiver receives the original message.

The Diffie-Hellmen Key exchanging algorithm follows the mechanism in the given manner. There are two users namely Alice and Bob. They would communicate with each other but before that, they need to exchange a key. Actually, the exchange of key is an illusion. Keys are derived in joint collaboration, which gives an illusion of exchange. They set an agreement with two prime numbers one is the P and other is the generator G. The P has to be quite large in in comparison to the G. The limit should be 0<G<P. Alice has a private key x and Bob has a private key y. They keep this with them. Alice calculates (G^x) mod P and Bob calculates (G^y) mod P. Now these results are exchanged between them, which called public keys. Now Alice computes (G^y)^x mod P and Bob computes (G^x)^y mod P. For both cases, the result becomes (G^xy) mod P. This is the shared key. Now if am intruder knows the values P, G,(G^x) mod P,(G^y) mod P he can figure out x and y which are there private keys. It can only be possible if P value is small. This is the reason the value of P is takes a large prime number which leads to a discrete logarithmic problem. To figure out the private numbers it would be time taking and by the whole process would be completed.

Now both the parties using DES algorithm use for encryption as well as for decryption this shared key. The DES has a drawback of having a very small key length that is 56 bits. But when the key from Diffie-Hellmen is taken it takes quite a large amount of time and a new key is generated for every other session. Hence, the key size does not affect the process much and the communication is done in more secured process.

The same key that is generated is farther given as an input to the DES algorithm. Since, we are giving an external input to the keys; the algorithm need not compute the keys using the round function. This reduces the task of DES algorithm and hence enhance its efficiency. We provide a plaintext and key of 64 bits where only 56 bits are used in actual and the rest are used for checking the parity. Therefore, the actual length of the key becomes 56 bits. Each byte of the key is sent with an odd parity. In each byte.one of the bit is used for checking the error while generating and distributing the key. Generally, the 8th, 16th, …, 64th are used for checking the parity. Several permutations are performed in the plain text with the key to produce the cipher text. To convert the cipher text to the plain text again the reverse order is performed.

Hence, this makes the algorithm efficient in comparison to the individual Diffie-Hellmen and DES algorithms.

19.6 FUTURE SCOPE

Diffie-Hellman is not a symmetric algorithm; it is an asymmetric algorithm used to establish a shared secret for a symmetric key algorithm. It works on secret communication while exchanging data over public network using Elliptical Curve to generate points and get the secret kcy using the parameters. A padlock icon is can be seen next to the URL of your devicc every time you go to a web browser; it can be seen because that server has used D-H:

1. **TSL/SSL:** These two layers use both D-H and symmetric encryption algorithm. D-H was used initially for the creation of the session key (shared secret) that is used afterwards in message communication for that particular session.
2. **Perfect Forward Secrecy:** The algorithm provides a perfect forward secrecy. After the shared secret key is known, it can be used to encrypt a traffic of data. Later, even if the data traffic is known it would be impossible to know the key. The exchange may also be visible but the key would be irrecoverable as it was never transmitted and saved.
3. **Trapdoor Function:** This works as a one-way function. This means you can calculate in one direction but opposite is not possible and can only be done by the help of some special information and this works like a "trap." The Diffie-Hellman design has this disadvantage for the interceptors.
4. **Authentication:** Diffie-Hellman design is well enough to work on symmetric algorithms. The man-in-the-middle attack problem is solved to some extent when it is combined with the DES algorithm as it takes a large amount of time ion exchanging the key, encrypting, and decrypting the messages. However, there is still a question about the authentication of the two parties. They can be masked if somehow or the other the prime number and the generators are very small and

can be cracked by the intruder because of easy discrete logarithmic calculations. For the same reason, this is used with digital signatures.

5. **Time Out:** This is another problem faced by the algorithm. The crackers could not crack the keys until now but the time would finish before the communication is completed. This problem compels NSA to move to the ECC for farther security.

   Small Key Length: Due to the smaller key length, this would not be secured in the future. Moreover, it has to be applied with some other algorithms to enhance the key size to increase the security properties farther.

6. **ECDH:** There is a high chance of the Elliptic Curve Discrete Logarithm Problem to gain popularity in the near future since it operates on a smaller key size but provides a larger security then normal Diffie-Hellman problem. None of the parameters are open to anybody except for the public keys.

19.7 CONCLUSION

When the DES algorithm is used for encryption, the key used in it is provided by some inbuilt function or algorithm within it. This may raise a question to the security because it can be attacked by an outsider while the communication is going on. This is the reason DES algorithm is not used presently in practical life. But when we have combined the Diffie-Hellman Key Exchanging algorithm with the DES, the key generated is instant. This means the key was unknown to both the sender and receiver before the communication link was established. When it is prior unknown to both of them, it does not permit the intruders to know the keys and identify any of the two parties, which would allow them to attack the information. Hence, using the Diffie-Hellman we can create an instant session key and exchange the keys between the sender and receiver. Once the keys are exchanged, the messages are sent by the sender, which is encrypted by the DES algorithm using the key from Diffie-Hellman key exchanging algorithm, i.e., exchanged between the two communicators. The receiver receives the message and decrypts the same with the help of the key that is already present with him/her. Hence, the communication ends here. It is noted that for every time a link is created between them again, a new key is interchanged which means even if the interloper tries to know the key it would be useless since for the next time, he/she has to compute another key for knowing the information. The session key exchange also signifies that the time taken would be more than the normal encryption algorithms. Hence, it would take quite a long time to break the key because of larger computation. The time the intruder would break the key the communication would be completed between the two people. Hence, the combination of the two algorithms would provide a better security then the other symmetric key encryption algorithms.

KEYWORDS

- **data encryption standard (DES)**
- **SFTP**
- **SSL**
- **TLS**

REFERENCES

Ali, S. T., Sivaraman, V., & Ostry, D., (2013). Eliminating reconciliation cost in secret key generation for body-worn health monitoring devices. *IEEE Transactions on Mobile Computing, 13*(12), 2763–2776.

Ballardin, F., (2011). A calculus for the analysis of wireless network security protocols. *Formal Aspects of Security and Trust* (pp. 206–222). Springer: Berlin Heidelberg.

Carmen, R., (2012). *Wireless Network Security*. Ovidius University; Annals Economic Science Series.

Chen, L., (2013). Applications, technologies, and standards in secure wireless networks and communications. *Wireless Network Security* (pp. 1–8). Springer: Berlin Heidelberg.

Eric, G., & Chachati, M., (2014). Analyzing routing protocol performance with NCTUns for vehicular networks. *Indian Journal of Science and Technology, 7*(9), 3191–1402.

Fan, Y. J., Wen, Q., & Zhang, H., (2010). Smart card-based authenticated key exchange protocol with CAPTCHA for wireless mobile network. In: *2nd International Conference on Future Computer and Communication (ICFCC)* (pp. 119–123).

He, X., Niedermeier, M., & De Meer, H., (2013). Dynamic key management in wireless sensor networks: A survey. *Journal of Network and Computer Applications, 36*(2), 611–622.

He, Y. H., & Lee, M. C., (2008). Towards a secure mutual authentication and key exchange protocol for mobile communications; In: *6th International Symposium on Modeling and Optimization in Mobile, Ad Hoc, and Wireless Networks and Workshops* (pp. 225–231). Berlin.

Hellman, M. E., & Diffie, W., (1976). New directions in cryptography. *IEEE Transactions on Information Theory, 22*(6), 644–654.

Liaw, S. H., Su, P. C., Chang, H. C., Lu, E. H., & Pon, S. F., (2005). Secured key exchange protocol in wireless mobile ad hoc networks. In: *39th Annual 2005 International Carnahan Conference on Security Technology* (pp. 171–173).

Lin, Q., & Wang, Y., (2013). Novel three-party password-based authenticated key exchange protocol for wireless sensor networks. *Advances in Wireless Sensor Networks, 334*, 263–270.

Pervaiz, M. O., Cardei, M., & Wu, J., (2010). Routing security in ad hoc wireless networks. *Network Security* (pp. 117–142). Springer, US.

Saeed, M., & Shahriar, S. H., (2011). security analysis and improvement of smart card-based authenticated key exchange protocol with CAPTCHAs for wireless mobile network. *IEEE Symposium on Computers and Communications* (ISCC) (pp. 652–657). Kerkyra.

Xu, J., Zhu, W. T., & Feng, D. G., (2011). An efficient mutual authentication and key agreement protocol preserving user anonymity in mobile networks. *Computer Communications, 34*(3), 319–325.

CHAPTER 20

STRENGTH-BASED NOVEL TECHNIQUE FOR MALICIOUS NODES ISOLATION

PRACHI CHAUHAN[1] and ALOK NEGI[2]

[1]*G.B. Pant University of Agriculture and Technology, Pantnagar, Uttarakhand, India, E-mail:cprachi664@gmail.com*

[2]*National Institute of Technology, Uttarakhand, India*

ABSTRACT

The ad-hoc vehicle network is a distributed and self-configuring network that does not have a central controller. Due to the high mobility of vehicle nodes, route establishment through source to destination is really the biggest challenge on the network. Different techniques have recently been studied for the identification of malicious nodes in any network. The proposed work is based on monitoring mode and signal strength. The simulation was conducted on NS2 and experimental results show that the proposed technique showed descriptive results in terms of different parameters.

20.1 INTRODUCTION

MANET is indeed a mobile ad-hoc network made up of a versatile infrastructure the whole network is often a comprehensive wireless network consisting of specified mobile packet nodes within it. After that, whenever an arbitrary configuration is developed, the connection of nodes occurs to each other on the random basis. The nodes mostly on network may also

Advances in Data Science and Computing Technology: Methodology and Applications. Suman Ghosal, Amitava Choudhury, Vikram Kumar Saxena, Arindam Biswas, & Prasenjit Chatterjee (Eds.)

be known as routers or hosts (Camp et al., 2002). The self-configuration feature of such an infrastructure performed the communication in an efficient manner throughout the entire network. VANET is a part of the mobile ad-hoc networks. For research purposes, the vehicle ad-hoc networks seem to be the most prominent area of interest. For intent of communication several vehicles and their components (Chauhan et al., 2017) across the roadside are always linked to one another, so this infrastructure should also be self-configured in design. They need no static infrastructure for them (Rajamani et al., 2000). The routing structure of the network is modified whenever the system topology is disrupted, however, there are widely changing nodes or vehicles participating in the system generate very difficult situation to perform tasks. Again, for aim of forwarding packets between nodes, a greedy position-based routing technique known as edge node based greedy routing (EBGR) is often used. There are some nodes accessible on the edge of the source or forwarding node transmission range. Based on the potential score of its nearest node, the much more appropriate next hop is appointed (Mustafa et al., 2010). There is a minimization of the end-to-end latency of the packet propagation as compared with the existing VANET routing protocols. Protection of VANETs is violated by numerous forms of attack. The attacks happen once when one node keeps causation multiple messages to alternative nodes that are assumed to be from completely Varying identities. In certain instances, a Sybil attack always be feasible. Attacker pretends to be having several identities or nodes throughout this attack. A malicious node may behave as a real node even though it was more than one number of nodes, whether by impersonating various nodes or simply by claiming false identities (Rawat et al., 2012). The collision in the network begins which ends in inflicting Sybil attack within the network. The primary goal of DoS attack seems to be prevention of a legitimate user from using the resources as well as the services. Along with the network, the entire channel can also be jammed in this attack. This result is an inability of not being able to access the network by the authorized vehicles. The DDOS attack is more harmful than the DOS attack, because of its distributive nature. For the purpose of launching the attack, various types of locations are used. Various time slots can be used for the purpose of sending the messages where the natures of the message, as well as the time slot, are different for each vehicle. V2V and V2I can both have DDOS attack them. The unauthorized users use the reply attack which helps them to be a legitimate user or the road side unit (RSU) (Malla et al., 2013). The replaying of already generated frames is done to the new connection by the attacker. In wormhole attack, a packet is received by the malicious node in the network at a point which tunneled back to another point at another location where rests of the packets present within the network. Inside VANET, there are different routing protocols. In DGR (directional greedy routing), Via choosing the nodes that heads nearer target destination, the hop count reduced (Iqbal et al., 2009). A-STAR stands for Anchor-based-street and traffic aware routing. It is a position-based routing mechanism which uses city bus routes for the purpose of identifying an anchor path for packet delivery along with the high connectivity. GyTAR (Greedy Traffic-Aware Routing) is the proposed technique which used in city environments where the intersection kept as the base. The geographical routing protocol is used for this techniquc. E-GyTAR is the modified version of GyTAR routing protocol. The junction is selected dynamically because it based on vehicular traffic density. The AODV is an on-demand routing protocol, which described as a reactive routing protocol (Ahmad et al., 2010). Sometimes throughout the network, the source node needs to be routed to a particular destination then the routing establishment initialized by the route discovery process. Every ncighbor relay the RREQ packet on their own to the neighbors of the source node.

20.2 LITERATURE REVIEW

Rajamani et al. (2002) have proposed the spacing policies which help in the highway vehicle automation. Here, a spacing policy (a nonlinear function of speed) is discussed. The autonomously present information is utilized by the spacing policy. The guarantee of proving string stability by the spacing policy seen after the analytical calculations. Also, it ensures the traffic flow stability by keeping the smaller steady state spacing. Larger traffic flow capacities which in range of 20–65 m/h are provided through this mechanism. The adaptive cruise control (ACC) vehicles present on the highways of today can readily use the spacing policy.

Peel et al. (2001) have developed a design framework of intelligent transport systems which related to the aspects of road sweeping vehicle automation. The exchange of information from car to car as well as car to road is the main aim for the road condition information transferring module. The security issues of VANETs are discussed and their solutions to be derived. This is been made sure that solutions implemented under certain security patterns.

Camp et al. (2002) have reviewed mobility models for ad-hoc network. In this chapter, the researchers recommended security infrastructure for vehicular communication. The main aim of proposed structural design holds organization of identities. Further there should be the involvement of cryptography keys which ensure the security of communications and integrate the privacy enhancement technologies. The parameters which could be enriched such as to provide improved privacy and security protection for future that aimed to be achieved by the system proposed.

Wu et al. (2011) have explained a new technology which proposed for the communications being held at vehicles present within shorter range. Both the V2V and V2I communications are introduced under the highway scenario. The work discusses the network characteristics of the driving network environment through this chapter.

Kwag et al. (2006) have discussed the performance evaluation of the IEEE 802.11 ad-hoc networks in this chapter. This evaluation is performed in the vehicle to vehicle (V2V) communication for analyzing the IEEE 802.11 ad-hoc networks performance. The V2V communication which occur in vehicular background focuses on justification which is important for security correlated services. The flexibility effects are also considered here.

Haas et al. (2007) have devised performance measurements of simulations of VANETs. The information traces of vehicle movements which created by the traffic simulators are depend on traffic model theory. The research that archives movement of vehicles based on broad scale recordings discussed in this chapter.

20.3 SYBIL ATTACK IN VANETS

The attack occurs when a single node keeps sending multiple messages to other nodes which are pretended to be from varying identities. In certain instances, Sybil attack is possible. It can only be exempted from either the extreme conditions as well as possibilities of resource parity and entity coordination. A type of confusion occurs in the whole network when a single node starts sending multiple copies of its selves. There is a chance that all the fake, illegal ID's and the authority are claimed (Figure 20.1).

The collision within the network starts which causing Sybil attack in the network. Both internal and external attacks can be triggered by this type of attack. That being said, by establishing metrics of authentication, external attacks can indeed be prevented. For internal attacks, it is not feasible. There are one to one mapping of the identity and role inside a network. The Sybil points represented as A, B, C, D nodes in Figure 20.1, generate a false identity inside the network. Sybil attack has become such a crucial attack whereby multiple messages are sent from the attacker and then sent to various vehicles with multiple IDs every time. This leads all other nodes to get distracted in such a manner that the nodes think the messages are coming from certain nodes. Because of this, there is indeed a jam inside the network. This causes the vehicle to take another route and leave that path to something like the advantage for attacker. There are two types of Sybil attacks namely, sensitive, and throughput sensitive attacks.

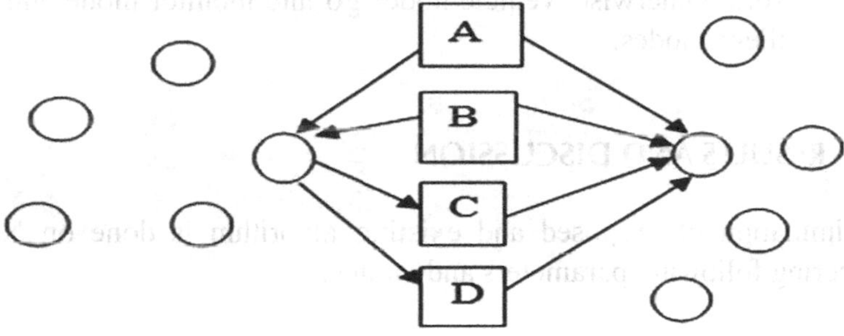

FIGURE 20.1 Sybil attack.

20.4 PROPOSED TECHNIQUE

In this work, technique has been discussed which detect and isolate malicious nodes that accountable to trigger Sybil attack from network. The suggested approach involves node strength methodology in monitor mode. In this technique, the RSUs flood ICMP messages in network. The vehicle nodes when obtain the ICMP messages will begin Trying to send towards nearest roadside units (rsus) its transmission strength rate All details will be obtained by the roadside infrastructure and the info will be shared with one another. The node of vehicle with multiple signal strength values will be detect as the node which Trigger an intrusion into the networks. The RSUs transmit the controlling information packets in the network to validate which node would be the fraudulent node and then once the control packets are received, nodes go out to monitor mode and continue monitoring their neighboring nodes. The node that's suspicious be detected and multiple path routing technique applied which isolate malicious nodes from the network.

20.4.1 WORKING STEPS FOR PROPOSED WORK

- Firstly, road side unit (RSU) save the information regarding deploy vehicles;

- If any new vehicle joins the network, then all information (network id, route, etc.), send to RSU;
- Identity changed by any vehicle send important message to all nearest RSU;
- RSU check the adjacent vehicles information based on signal strength;
- Finally, if the signal strength matched then communication start in the network. Otherwise, vehicle nodes go into monitor mode and detect the illegal nodes.

20.5 RESULTS AND DISCUSSION

The simulation of proposed and existing algorithm is done on NS2 by considering following parameters and values.

20.5.1 SIMULATION DESIGN

For simulation purpose, we construct a network, which comprises wireless ad-hoc network for the formation of simulation result. As we all know that VANET is performed in a real environment, hence for the better analysis of the results of our protocol we design a realistic scenario of road traffic. The network, which we configure is consists of vehicles, routes, and RSUs.

To test the performance of all above present methodologies, a program is written in TCL and tested in real environment. The snapshots of proposed work execution are shown in Figure 20.2.

```
prachi@prachi-Inspiron-5559: ~/codes
prachi@prachi-Inspiron-5559:~$ cd codes
prachi@prachi-Inspiron-5559:~/codes$ ns congestion.tcl
num_nodes is set 36
INITIALIZE THE LIST xListHead
using backward compatible Agent/CBR; use Application/Traffic/CBR instead
using backward compatible Agent/CBR; use Application/Traffic/CBR instead
using backward compatible Agent/CBR; use Application/Traffic/CBR instead
using backward compatible Agent/CBR; use Application/Traffic/CBR instead
using backward compatible Agent/CBR; use Application/Traffic/CBR instead
using backward compatible Agent/CBR; use Application/Traffic/CBR instead
using backward compatible Agent/CBR; use Application/Traffic/CBR instead
using backward compatible Agent/CBR; use Application/Traffic/CBR instead
using backward compatible Agent/CBR; use Application/Traffic/CBR instead
using backward compatible Agent/CBR; use Application/Traffic/CBR instead
using backward compatible Agent/CBR; use Application/Traffic/CBR instead
using backward compatible Agent/CBR; use Application/Traffic/CBR instead
using backward compatible Agent/CBR; use Application/Traffic/CBR instead
using backward compatible Agent/CBR; use Application/Traffic/CBR instead
using backward compatible Agent/CBR; use Application/Traffic/CBR instead
Start of simulation..
channel.cc:sendUp - Calc highestAntennaZ_ and distCST_
highestAntennaZ_ = 1.5, distCST_ = 550.0
SORTING LISTS ...DONE!
prachi@prachi-Inspiron-5559:~/codes$
```

FIGURE 20.2 Initialization of network.

In this work, at first finds the signal strength of each node, i.e., signal strength of every node and their neighbors. Figure 20.2 demonstrates the initialization of network.

In Figure 20.3, vehicular ad hoc network is installed with static number of road side sensor nodes and vehicles which travels freely on roads and V2V communication and vehicle to road side communication is also accessible.

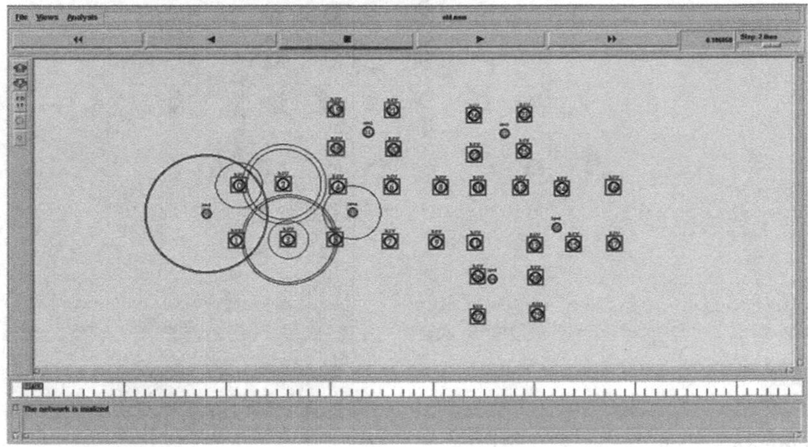

FIGURE 20.3 Network deployment.

In Figure 20.4, the smart cars are moving on roads and communicating with one another, i.e., car with identification number 1 communicating with car that registration number is 4.

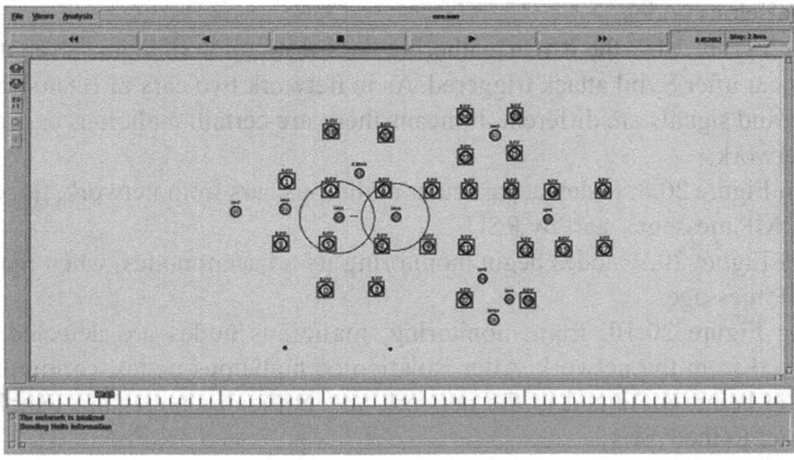

FIGURE 20.4 Communication between cars.

In Figure 20.5, new car gets its registration identity and this information flooded in the network.

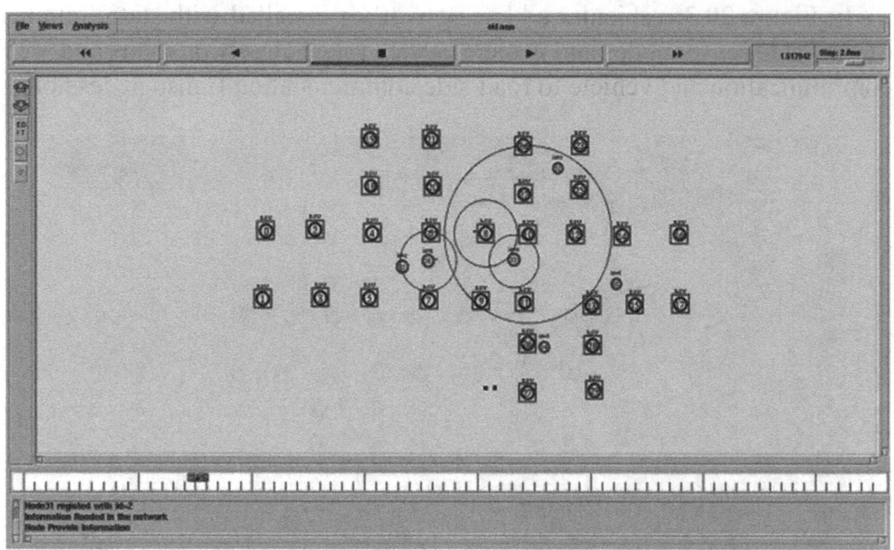

FIGURE 20.5 Allocation of new identity.

In Figure 20.6, newly registered nodes change their identity like 2 to 4 but a node with identity 4 already exists in the network. It means Sybil attack has been triggered in the network. The mischievous nodes have short distance from legitimate node. So, source node begins sending information to malicious nodes.

In Figure 20.7, the RSUs' gathering the information of signal strength of each car after Sybil attack triggered. As in network two cars of ID number 4 exits and signals are different. It means there are certain malicious cars exist in network.

In Figure 20.8, to detect particular malicious cars from network, flooding of ICMP messages start by RSU.

In Figure 20.9, nodes begin monitoring its adjacent nodes, when receive ICMP messages.

In Figure 20.10, from monitoring, malicious nodes are detected and isolated from the network. After isolation of malicious nodes, communication is being continued in the network and malicious nodes are unable to contact to the RSUs.

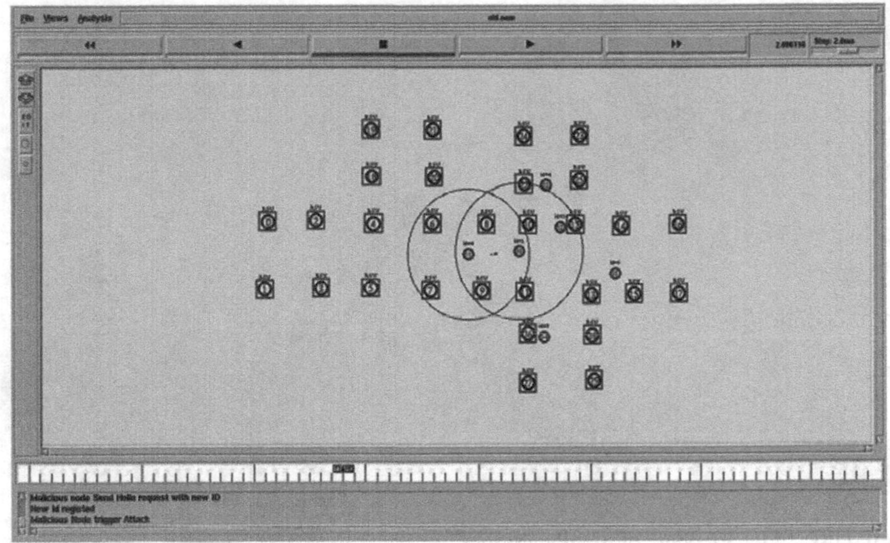

FIGURE 20.6 Attack triggered.

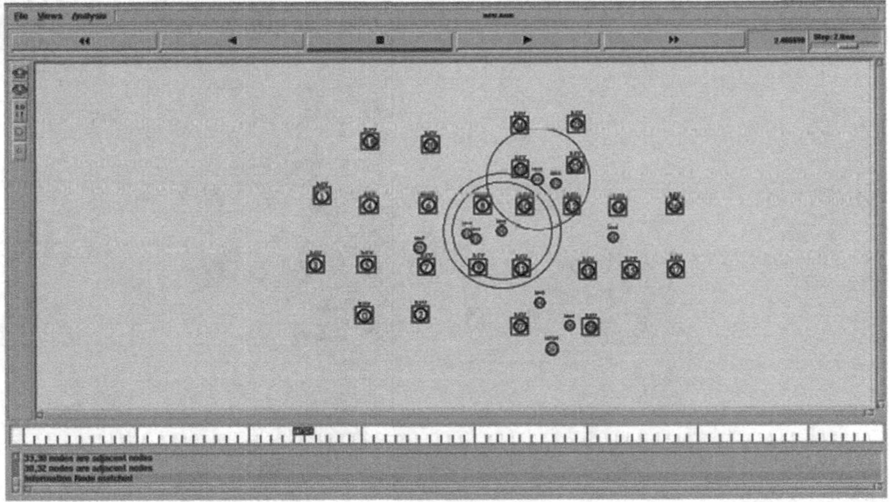

FIGURE 20.7 Getting signal strength information.

20.6 RESULT ANALYSIS

Throughput can be defined as the number of packet data received per unit time and it is represented in bps.

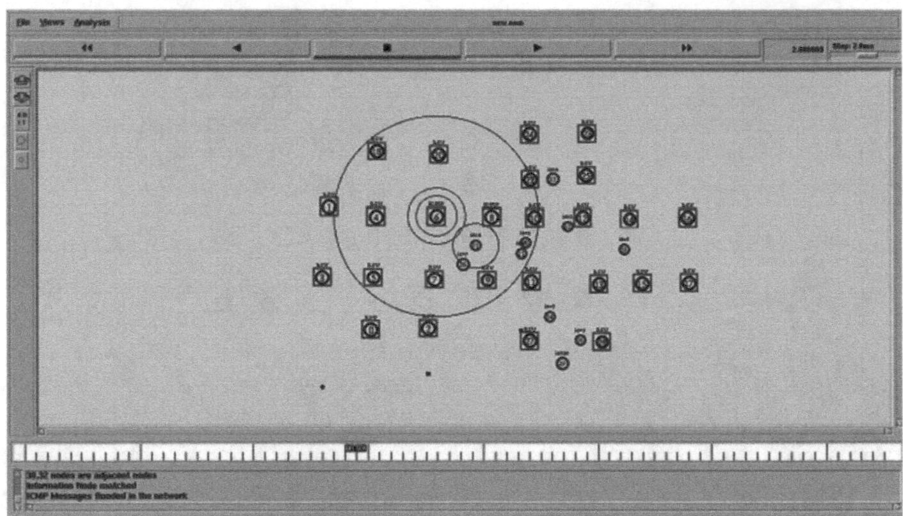

FIGURE 20.8 Flooding of ICMP messages.

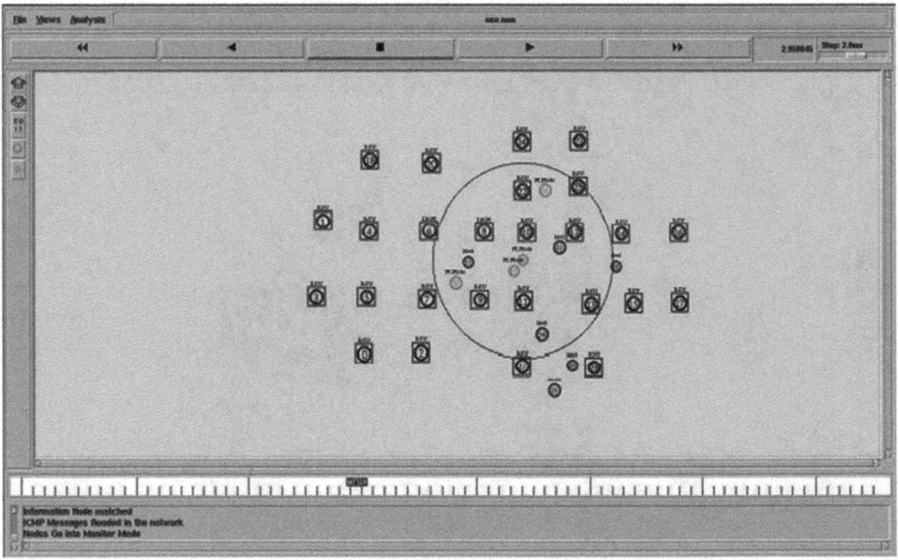

FIGURE 20.9 The monitoring of adjacent nodes.

$$T = RP / TT \tag{1}$$

where; T is the throughput; RP is the received packets; TT is the data transmission time (bps).

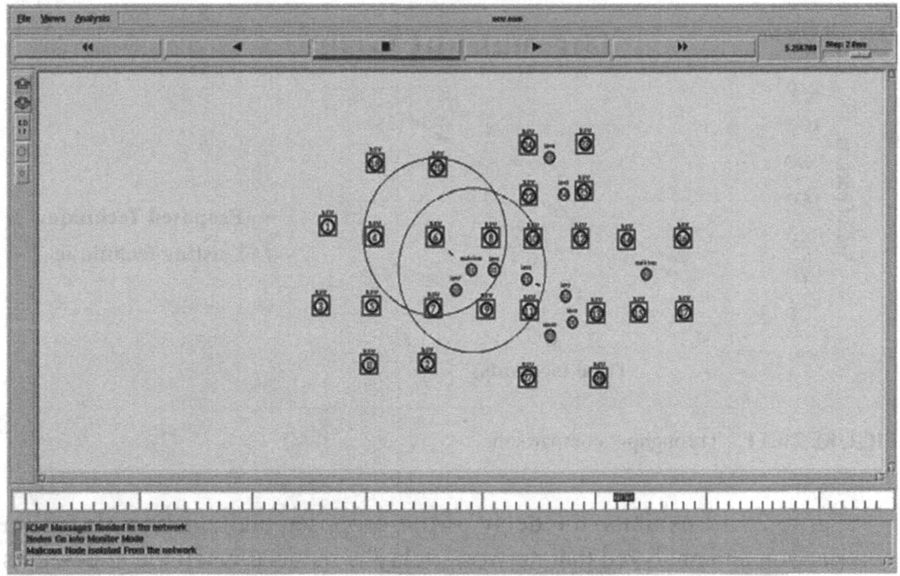

FIGURE 20.10 Detection and isolation of malicious nodes.

E2E (end-to-end) delay defined as the time taken between sending of a packet and it's receiving on the destination.

$$D = RT - ST \tag{2}$$

$$AD = TD / Number\ of\ packets \tag{3}$$

where; D is the delay; RT is the receiving time; ST is the sending time; AD is the average delay; TD is the total delay.

Packet loss in a communication is the difference between the generated and received packets.

$$PL = GP - RP \tag{4}$$

where; PL is the packet loss; GP is the generated packets; RP is the received packets.

Routing Overhead is the number of routing packets required for network communication.

$$Routing\ overhead = Rouring\ packet\ counts \tag{5}$$

Figure 20.11 exhibits, the throughput of the proposed and existing technique comparison and analyzed that the network throughput is increased at the steady rate after the malicious node isolation.

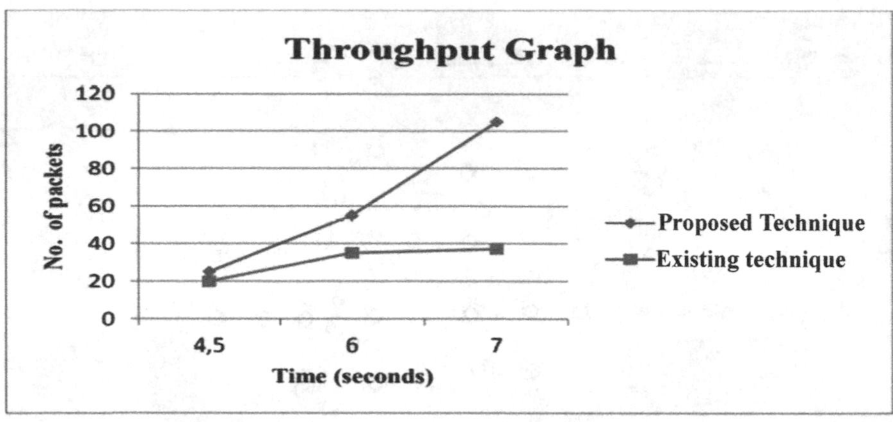

FIGURE 20.11 Throughput comparison.

Figure 20.12 exhibits, the delay of the proposed and existing algorithm comparison and analyzed that network delay is reduced when the attack gets isolated from the network.

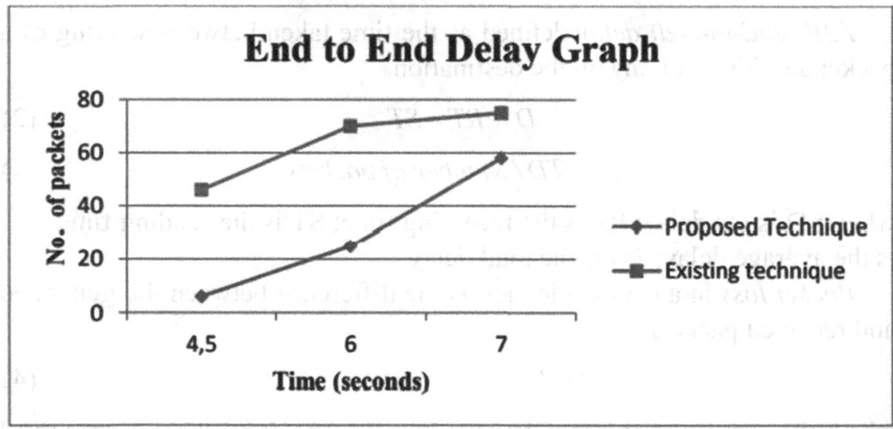

FIGURE 20.12 End to end delay comparison.

Figure 20.13 describe the packet loss of the proposed and existing technique comparison and analyzed that the network packet loss is reduced when Sybil attack gets isolated from the network. If the packet loss is less then number of packets drops also less.

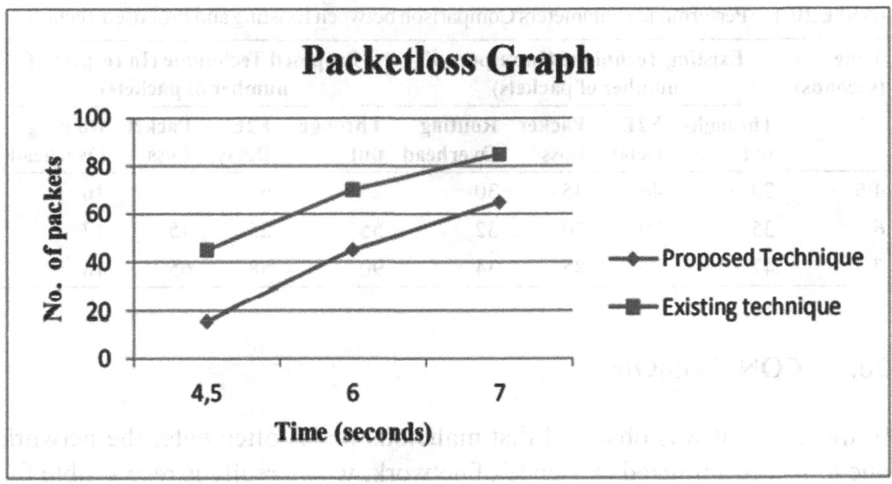

FIGURE 20.13 Packet loss comparison.

Figure 20.14 shows, the routing overhead are the parameter which measures the extra packets that transmitted in network. The routing overhead in network is reduced on detection and isolation of attack from the network. It means fuel emission decreased (Table 20.1).

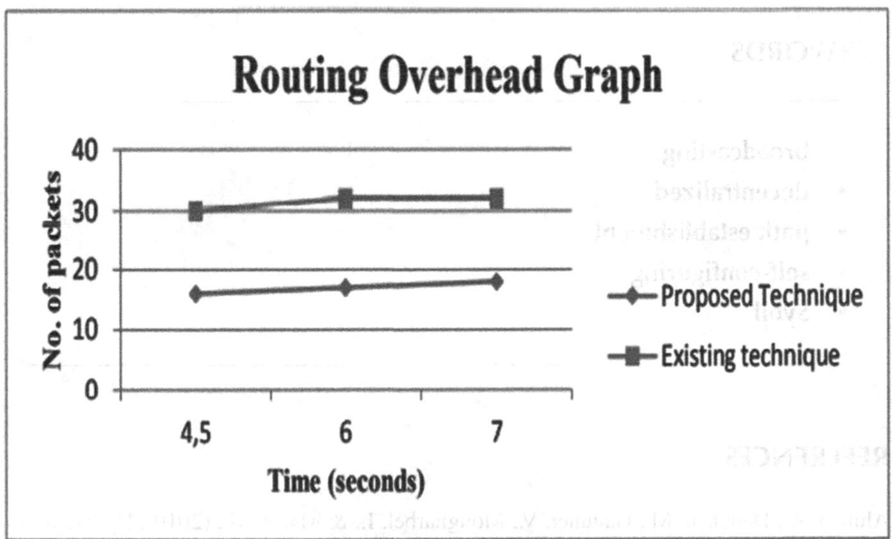

FIGURE 20.14 Routing overhead comparison.

TABLE 20.1 Performance Parameters Comparison between Existing and Proposed Technique

Time (seconds)	Existing Technique (In respect of number of packets)				Proposed Technique (In respect of number of packets)			
	Through-put	E2E Delay	Packet Loss	Routing Overhead	Through-put	E2E Delay	Packet Loss	Routing Overhead
4.5	20	46	45	30	25	6	16	16
6	35	70	70	32	55	25	45	17
7	47	75	85	34	90	58	65	18

20.7 CONCLUSION

In this study, it was observed that malicious nodes often enter the network due to its decentralized existence of network, which really is responsible for causing different forms of active and passive attacks. This analysis provides a justification for the identification of malicious nodes that responsible for triggering a Sybil attack on the network. The simulation of implemented technique has been done on NS2 and the result shows that overall performance increased of the network. Future work will include a further improvement of our technique by removing the chances of information spoofing in the network.

KEYWORDS

- **broadcasting**
- **decentralized**
- **path establishment**
- **self-configuring**
- **Sybil**

REFERENCES

Ahmad, A., Doughan, M., Gauthier, V., Mougharbel, I., & Marot, M., (2010). Hybrid multi-channel multi-hop MAC in VANETs. In: *Proceedings of the 8th International Conference on Advances in Mobile Computing and Multimedia* (pp. 353–357).

Balamaha, L. D., & Shankar, M. K. V., (2014). Sybil attack detection with reduced bandwidth overhead in urban vehicular networks. *International Journal of Engine ring Trends and Technology (IJETT), 12*, 578–584.

Camp, T., Boleng, J., & Davies, V., (2002). A survey of mobility models for ad hoc network research. *Wireless Communications and Mobile Computing, 2*(5), 483–502.

Chang, S., Qi, Y., Zhu, H., Zhao, J., & Shen, X., (2011). Footprint: Detecting Sybil attacks in urban vehicular networks. *IEEE Transactions on Parallel and Distributed Systems, 23*(6), 1103–1114.

Chauhan, P., & Mandoria, H. L., (2017). An empirical study of vehicles communication in vehicular ad-hoc network. In: *International Journal of Innovative Research in Science, Engineering and Technology* (Vol. 6, pp. 3763–3769).

Haas, J. J., Hu, Y. C., & Laberteaux, K. P., (2009). Real-world VANET security protocol performance. In: *GLOBECOM 2009–2009 IEEE Global Telecommunications Conference* (pp. 1–7). IEEE.

Hao, Y., Tang, J., & Cheng, Y., (2011). Cooperative sybil attack detection for position-based applications in privacy preserved VANETs. In: *2011 IEEE Global Telecommunications Conference- GLOBECOM 2011* (pp. 1–5). IEEE.

Iqbal, S., Chowdhury, S. R., Hyder, C. S., Vasilakos, A. V., & Wang, C. X., (2009). Vehicular communication: Protocol design, testbed implementation and performance analysis. In: *Proceedings of the 2009 International Conference on Wireless Communications and Mobile Computing: Connecting the World Wirelessly* (pp. 410–415).

Kwag, S. J., & Lee, S. S., (2006). Performance evaluation of IEEE 802.11 ad-hoc network in vehicle to vehicle communication. In: *Proceedings of the 3rd International Conference on Mobile Technology, Applications and Systems* (pp. 47-es).

Malla, A. M., & Sahu, R. K., (2013). Security attacks with an effective solution for dos attacks in VANET. *International Journal of Computer Applications, 66*(22).

Mustafa, B., & Raja, U. W., (2010). *Issues of Routing in VANET*. School of computing at Blekinge Institute of Technology.

Peel, G., Michielen, M., & Parker, G., (2001). Some aspects of road sweeping vehicle automation. In: *2001 IEEE/ASME International Conference on Advanced Intelligent Mechatronics. Proceedings (Cat. No. 01TH8556)* (Vol. 1, pp. 337–342). IEEE.

Rabieh, K. M., & Azer, M. A., (2011). Combating sybil attacks in vehicular ad hoc networks. In: *Recent Trends in Wireless and Mobile Networks* (pp. 65–72). Springer, Berlin, Heidelberg.

Rajamani, R., (2000). On spacing policies for highway vehicle automation. In: *American Control Conference*. Chicago, Illinois.

Rawat, A., Sharma, S., & Sushil, R., (2012). VANET: Security attacks and its possible solutions. *Journal of Information and Operations Management, 3*(1), 301.

Raya, M., & Hubaux, J. P., (2007). Securing vehicular ad hoc networks. *Journal of Computer Security, 15*(1), 39–68.

Uchhula, V., & Bhatt, B., (2010). Comparison of different ant colony-based routing algorithms. *IJCA Special Issue on MANETs,* (2), 97–101.

Wu, H., Palekar, M., Fujimoto, R., Guensler, R., Hunter, M., Lee, J., & Ko, J., (2005). An empirical study of short-range communications for vehicles. In: *Proceedings of the 2nd ACM International Workshop on Vehicular ad Hoc Networks* (pp. 83, 84).

Xiao, B., Yu, B., & Gao, C., (2006). Detection and localization of sybil nodes in VANETs. In: *Proceedings of the 2006 Workshop on Dependability Issues in Wireless Ad Hoc Networks and Sensor Networks* (pp. 1–8).

CHAPTER 21

SECURING THE INFORMATION USING COMBINED METHOD

BIJOY KUMAR MANDAL, SAPTARSHI ROYCHOWDHURY, PAYEL MAJUMDER, and ANWESA DAS

NSHM Knowledge Campus, Durgapurs, West Bengal, India, E-mail: writetobijoy@gmail.com (B. K. Mandal)

ABSTRACT

In this project, we propose an algorithm to hide data inside image using image steganography technique. Using image steganography, image can be secured that has to be sent over the internet. There are two images one is cover image and another one is secret image with a secret message that need to be transferred over the network. The whole process is divided into two parts first part is encryption that deals with the process of converting plain text into cipher text using AES algorithm and the second part deals with decryption using DES algorithm to get the original plain text from a cipher text.

21.1 INTRODUCTION

With the help of image processing, we have proposed an algorithm by combing the two method, cryptography, and steganography that is used to provide security to the secret message by hiding the way of communication (Fridrich, Goljan, and Soukal, 2014). Steganography is the process of hiding the relevant information over the network inside an image (Petitcolas, Anderson, and Kuhn, 2008).

Advances in Data Science and Computing Technology: Methodology and Applications. Suman Ghosal, Amitava Choudhury, Vikram Kumar Saxena, Arindam Biswas, & Prasenjit Chatterjee (Eds.)

Nowadays, technology is growing very fast. Everything is dependent on the internet and hence security is must more needed (Cheddad et al., 2010). So, steganography is used to make it more secure to transform the information over the network in more way that is convenient. Here we will use AES algorithm (which convert the plain text into the cipher text) to make It more secure and only the authorized user can access (Cheddad et al., 2009). Those data which made it safe from the cyber-crime and on another algorithm which is DES algorithm (which is reverse process of AES algorithm means extract the plain text from cipher text). Steganography is basically based on message hiding and image stitching which provide double security and intruders cannot access the useful information (Liu and Tsai, 2007). In the growing age of technology, where technology is used by a common man in day to day life, where each second millions of messages are transmitting across the world wide over network. Where the world is increasing with technology, the cyber-crime is also increasing, so providing the network security is not sufficient. We need extra security while transferring messages or images. Using image steganography, we proposed an algorithm to provide security to the image, which is being transferred rapidly in day-to-day life (Girling, 1987).

21.1.1 IMAGE PROCESSING

Image processing is a process of converting an image into digital form and some operation can be performed on it, to extract some useful information from the image. In the process of image, processing input is image say photograph, any video frame and output can be image, or it may be characteristic of that image (Manfred, 1989). In this age of technology, everything is performed digitally so image processing play a great role on it in many aspects. In Image processing system, images are treated as two-dimensional signals. Image processing includes this mention below steps:

- Import of an image through via image acquisition;
- Recognizing an image;
- The result can be altering image or report based on analysis of image.

Major reason why this encryption and decryption is used is security while we transferring the message or media through internet or other communication network (Wendzel and Keller, 2016). As lots of message and information is transformed from one place to another through internet media, which need security from hackers or unauthorized users. Proving this security.

21.1.2 ENCRYPTION

The process where the plain text is transformed into unreadable format. This is done by using an algorithm for unauthorized users. Here the message or the data is encoded for which unauthorized user cannot access. The types of encryption are:

1. **Symmetric Encryption:** In this encryption, the key that is used for encryption and decryption is same. Therefore, this use the same key for the message need to send.
2. **Public Key Encryption:** In this, the algorithm use both the keys, i.e., one for public that is called public key and other for authorized user, which is called secret key. In addition, these two keys are linked with each other.

21.1.3 DECRYPTION

Decryption is a process of converting the encrypted data into its original form. It is opposite process of encryption. Only authorized user can decode this encrypted data because the help of secret key. Only can do decryption those user can decrypt the information that have a valid key for that encrypted information (Krzysztof, 2010; Rivest, Shamir, and Adleman, 1978).

21.2 STEGANOGRAPHY

Steganography has vast effect on nowadays world and introduced few new technologies. It also provides the security during messaging in mobile. This Steganography word came from two Greek words, one is steganos and other is graptos. The first steganos means Covered whereas graptos means writing, which makes the meaning "cover writing." Steganography means hiding of message in specific medium like audio, video, and many more. It uses multimedia as cover image as we know in steganography, we use one message which in kept inside an image (known as cover image) and the encrypted image is called stego image (Junod, 2001). People send lots of private and legal message through internet or the other communication network and security is key term for them, for those purpose; steganography is used where actual purpose is to protect the content or the information from the attack of unauthorized user (hacker) without altering the actual message.

Cryptography and steganography are two terms where cryptography is use to protect the data and steganography is use for hiding the data/message/information. There are different types of steganography:

1. **Image Steganography:** It is used to hide secret message inside an image. It uses LSB algorithm to hide the message.
2. **Audio Steganography:** It is used to hide secret message in the audio using digital representation.
3. **Video Steganography:** It is combination of image steganography and audio steganography. Video Steganography is to hide large amount of data.
4. **Text Steganography:** It is most difficult steganography technique because texts have less redundancy compare to audio or image.

21.3 CRYPTOGRAPHY

Cryptography is an important term that is used to provide network security. It is mainly associated with the converting the plain text into non-readable text and vice-versa. Cryptography is used for data security as well as for user authentication while cryptanalysis is the process where cryptosystem is break. Either the algorithm is alike to each other or you can say two sides of coin. Steganography does not alter the structure of original message whereas in cryptography it affects the original secret message (Bijoy, Debnath, and Xiao-Zhi, 2019). In steganography, the communication of message is hidden while in cryptography, encryption method is done (Figure 21.1).

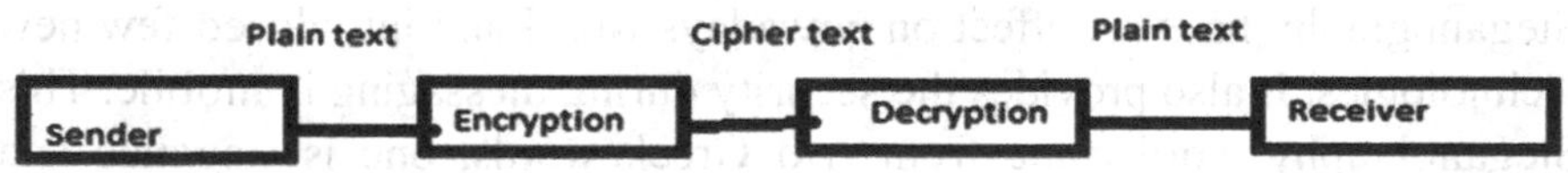

FIGURE 21.1 Process of cryptography.

21.3.1 SYMMETRIC KEY CRYPTOGRAPHY

Symmetric key cryptography is also known as private key cryptography. In this cryptography, a single key is used for both encryption of plain text and decryption of cipher text. The process of key exchange (means the key itself is exchanged between sender and receiver) constitutes a problem. In simple words if the key is revealed then beyond the sender and receiver than there is a chance to decrypt the message.

21.3.2 ASYMMETRIC KEY CRYPTOGRAPHY

Asymmetric key cryptography is also known as public key cryptography. In this cryptography, both public as well as private key are used to decrypt the data. Here public key is known by everyone and the private key is kept as secret. Either of the key can be used to encrypt the message, the opposite key is used for decrypt the message.

21.3.3 HASHING

Hashing is the method of cryptography where one form of data is converted into a unique string of text. Hashing can be done with any size or type.

21.4 ALGORITHM

1. **Input Image:** Here the secret message is hidden in selected file. The message can be in text form also.
2. **Image Quality Selection:** The quality of output image is selected. The more quality of image the less data can be hidden.
3. **AES Encryption:** The secret image is encrypted through the AES Algorithm. It provides the security id by mistake if unauthorized user decodes the image file it can't access or encrypt the information encrypted by AES.
4. **LSB Algorithm:** LSB can hide a huge amount of information to the lower bits of data/information.
5. **Encoded and Encrypted Image:** Here, the message which is hidden in image is ready to send to the user. But the user must have that secret key to decrypt or to open the algorithm (AES).

21.4.1 ON SENDER SIDE

- **Step 1:** At first, take the inputs the message and any image as image1. and another image as image2.
- **Step 2:** Message and image1 need to be broken into n number of equal parts.
- **Step 3:** Message or information are encrypted using AES (keys 512) Algorithm.

- **Step 4:** Do random dynamic mapping between the sub-parts of image1 and the sub-part of encrypted message.
- **Step 5:** After the mapping, sub-part of the encrypted message is hidden with the sub-part of the image1 using LSB algorithm.
- **Step 6:** The hidden information is again encrypted using AES algorithm.
- **Step 7:** Now, all sub-part are hide into the image 2 using LSB algorithm. After that all the sub-part hide into image 2.
- **Step 8:** Again, image2 is broken into n number of equal parts.
- **Step 9:** Each part then separately sent to the recipient.

21.4.2 ON RECEIVER SIDE

- **Step 1:** On the receiver's end, the sub parts are received.
- **Step 2:** Now, the parts image need to be merged based on their index values to regain original cover image.
- **Step 3:** Process of decryption of secret image is done using AES algorithm same as encryption, but in reverse order.
- **Step 4:** Now, get original message using LSB algorithm.
- **Step 5:** Apply AES algorithm for decryption.
- **Step 6:** Message and image need to be merged.
- **Step 7:** We obtain the secret image on receiver end.

21.4.3 LEAST SIGNIFICANT BIT (LSB)

1. **Embedding of Text in Image:**
   i. Calculating the pixel (IMAGE) need to encrypt;
   ii. Make a loop through the pixels of the image to be encrypted;
   iii. In each passing round we get the RGB Value of pixel image;
   iv. Make each RGB value to zero;
   v. Get or extract the character from the hidden binary form and hide the 8 bit binary code in pixel (LSB) of image.
2. **Extracting the Embed Message from the Image:**
   i. Calculate the pixel;
   ii. Loop till one finds 8 consecutive zeros in pixel of image;
   iii. Convert the picked LSB from pixel element to character.

21.5 RESULTS

The results are shown in Figures 21.2–21.7.

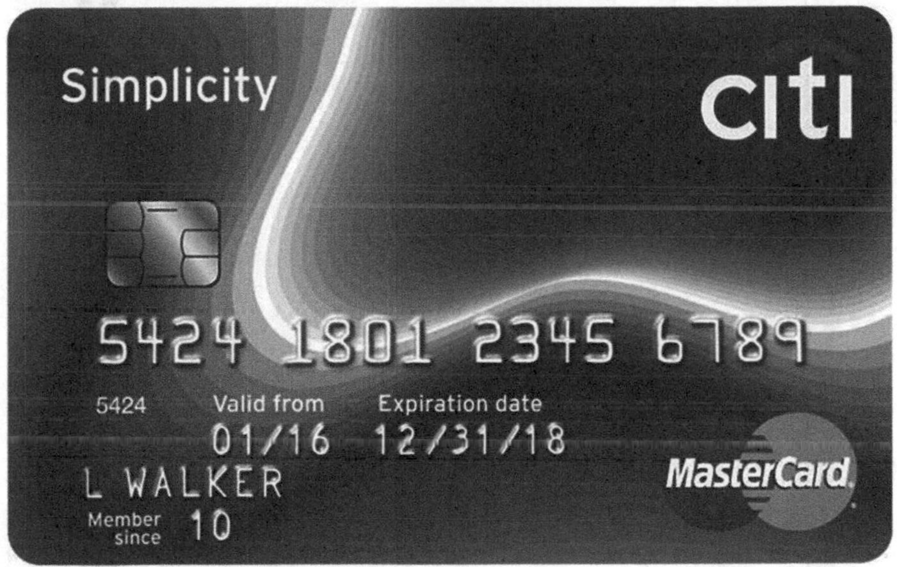

FIGURE 21.2 Secrete image for input.

FIGURE 21.3 Cover image for input.

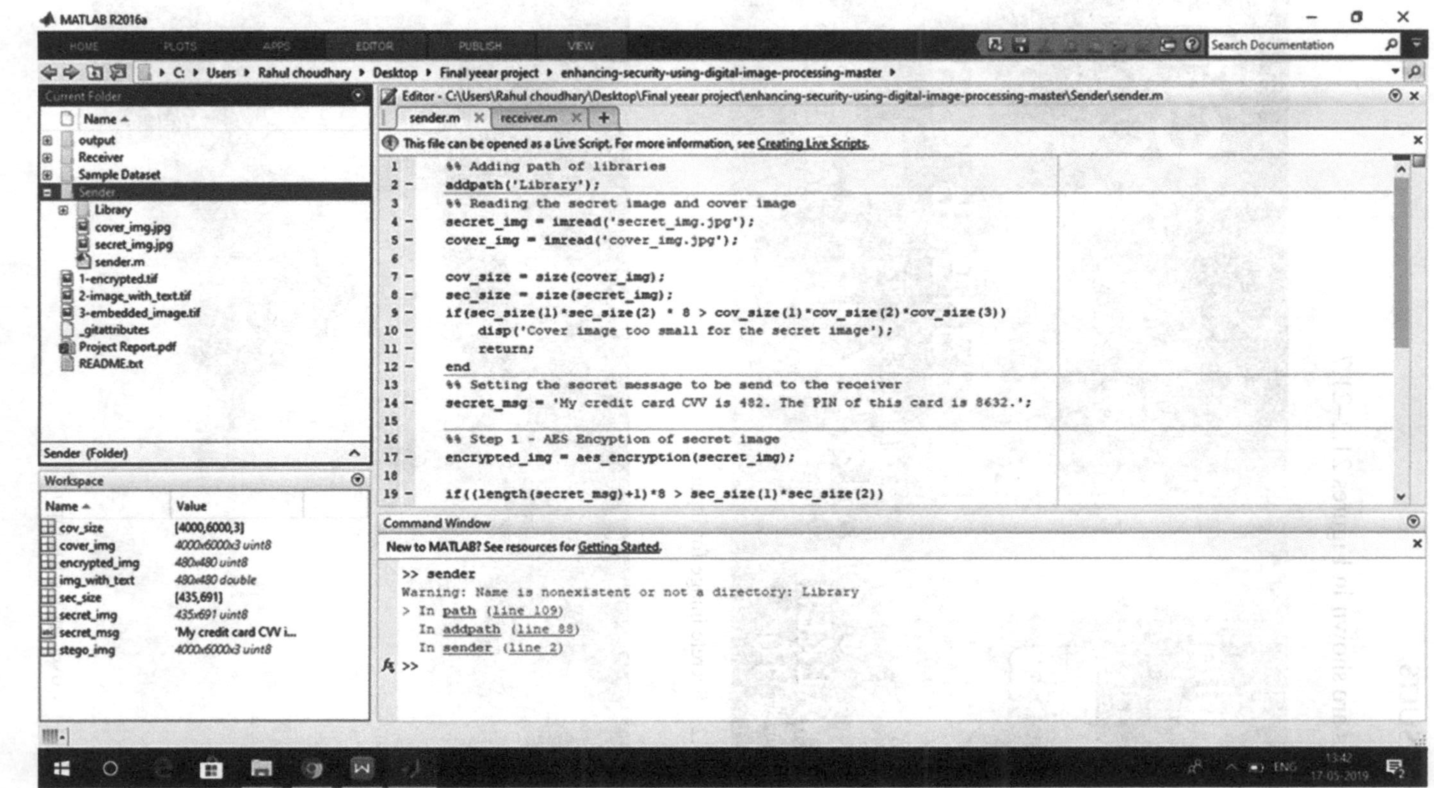

FIGURE 21.4 Sender side.

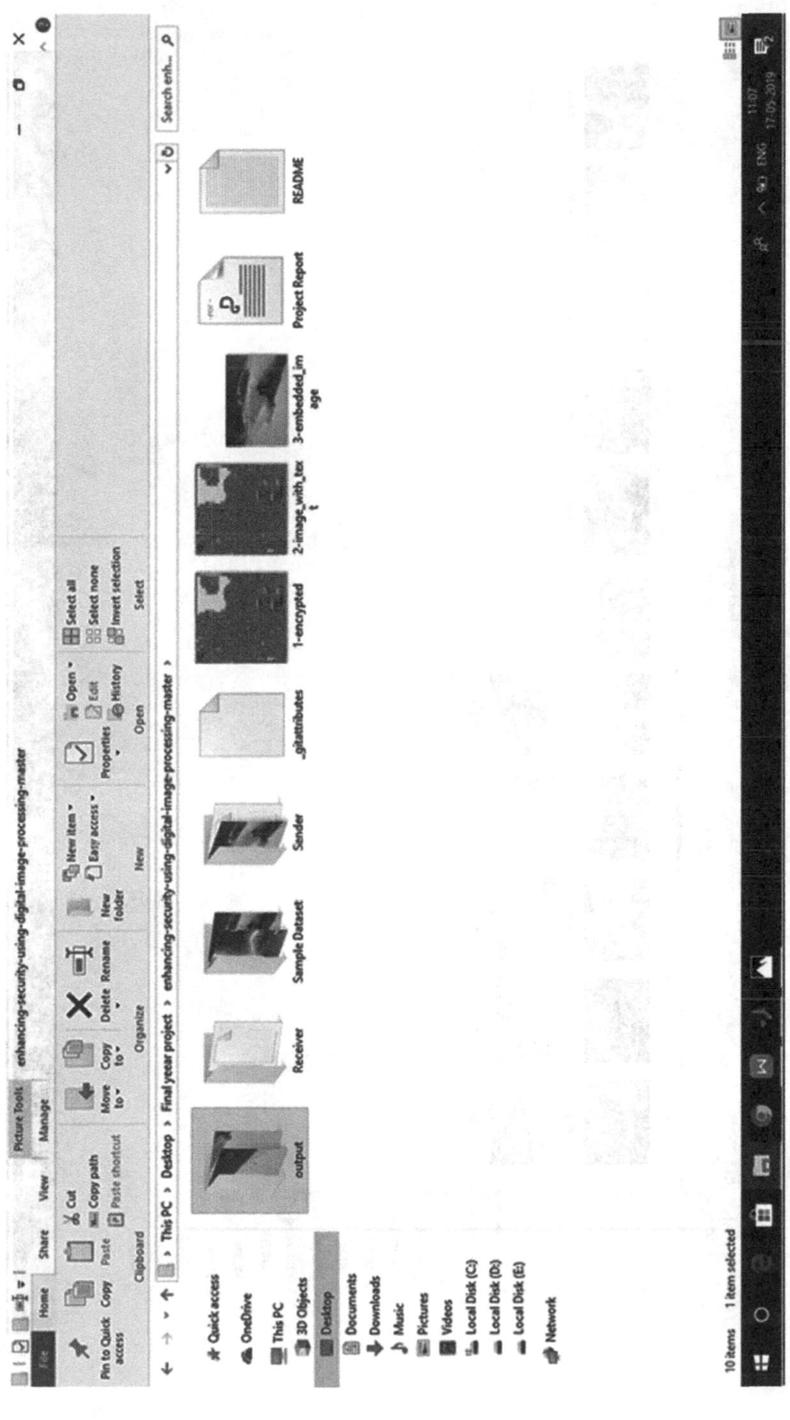

FIGURE 21.5 Encrypted image.

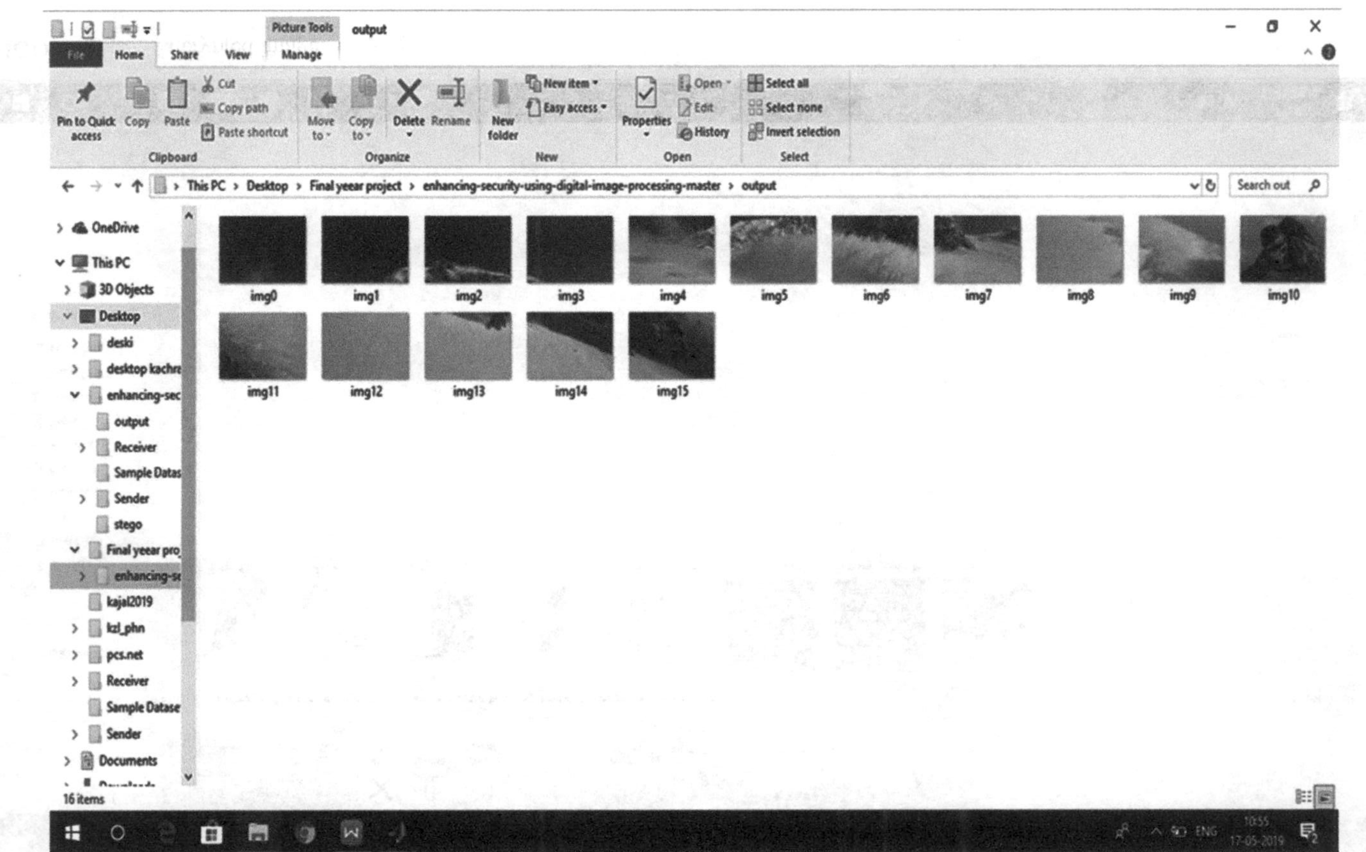

FIGURE 21.6 Sender side out after encryption process.

FIGURE 21.7 Receiver side output after final process.

21.6 CONCLUSION

Nowadays, security is major aspect so we have done this project. In this project we used steganography technique which is called image steganography. The data is hidden inside the image which is called stego image. The main aim of this project to enhance security. In future, to provide greater security to the data we can use different keys for encryption and decryption. Audio can also embed in carrier media. Size of images and message can be different. It will provide an effective way to select an image to hide a message. A secret key is needed to decode the secret message. Steganography has much use in future. This is double checker security for transfer of message through internet. Its application is in use of protection of data alteration and storing confidential data. Accessing of digital media/data that is stored for authorized user can also be control by steganography.

KEYWORDS

- **AES**
- **cipher text**
- **data encryption standard (DES)**
- **LSB**
- **steganography**

REFERENCES

Bijoy, K. M., Debnath, B., & Xiao-Zhi, G., (2019). "Data security using 512 bits symmetric key based cryptography in cloud computing system." In: *International Journal of Advanced Science and Technology* (Vol. 129, pp. 1–10).

Cheddad, A., Condell, J., Curran, K., & Mc Kevitt, P., (2009). "A skin tone detection algorithm for an adaptive approach to steganography." In: *Signal Processing* (Vol. 89, No. 12, pp. 2465–2478).

Cheddad, A., Condell, J., Curran, K., &. Mc Kevitt, P., (2010). "Digital image steganography: Survey and analyses of current methods." In: *Signal Processing* (Vol. 90, No. 3, pp. 727–752).

Fridrich, J., Goljan, M., & Soukal, D., (2014). "Searching for the stego key." *Proc. SPIE, Electronic Imaging, Security, Steganography, and Watermarking of Multimedia Contents*, 70–82.

Girling, C. G., (1987). "Covert channels in LAN's." *IEEE Transactions on Software Engineering, 13*(2), 292–296.

Junod, P., (2001). "On the complexity of Matsui's attack." *Selected Areas in Cryptography* (Vol. 2259, pp. 199–211). Lecture Notes in Computer Science.

Krzysztof, S., (2010). "HICCUPS: Hidden communication system for corrupted networks." In: *Proc. of the Tenth International Multi-Conference on Advanced Computer Systems ACS'2003* (pp. 31–40).

Liu, T. Y., & Tsai, W. H., (2007). "A new steganographic method for data hiding in Microsoft word documents by a change tracking technique." In: *IEEE Transactions on Information Forensics and Security* (Vol. 2, No. 1, pp. 24–30).

Manfred, W., (1989). "Covert channels in LAN protocols." Local area network security. *Lecture Notes in Computer Science, 396*, 89–101.

Petitcolas, F. A. P., Anderson, R. J., & Kuhn, M. G., (2008). "Information hiding: A survey." *Proceedings of the IEEE (special Issue), 87*(7), 1062–1078.

Rivest, R. L., Shamir, A., & Adleman, L., (1978). "A method for obtaining digital signatures and public-key cryptosystems." *Communications of the ACM, 21*(2), 120–126.

Wendzel, S., & Keller, J., (2016). Low-attention forwarding for mobile network covert channels. *12th Joint IFIP TC6 and TC11 Conference on Communications and Multimedia Security (CMS)* (Vol. 7025, pp. 122–133). Lecture Notes in Computer Science.

CHAPTER 22

IMPLEMENTATION OF DNA CRYPTOGRAPHY IN IoT USING CHINESE REMAINDER THEOREM, ARITHMETIC ENCODING, AND ASYMMETRIC KEY CRYPTOGRAPHY

ANANYA SATPATI (DAS), SOUMYA PAUL, and PAYEL MAJUMDER

Department of Computer Science and Engineering, NSHM Knowledge Campus, Durgapur, Arrah Shibtala Via Muchipara, Durgapur–713212, West Bengal, India, E-mail: ananya.das@nshm.com (A. Satpati)

ABSTRACT

In today's world security is an important issue where internet plays an important role. Network security helps us by protecting our valuable data from unauthorized access, harmful cryptanalyst, etc. It also confirms that shared data is kept secure. Stack of protection layer of network security help us to create a complex cipher. Generally, an intruder focuses on the file, when it is available in a transmission medium, instead of before transmitting the file, when it only available to the sending device (like IoT device). But if we simultaneously increase the security of sending/receiving devices (like IoT devices) to create more complex cipher with the security of the transmission process, then our system will be more strong and, in this chapter, emphasize on this matter. In this chapter, we use a robust security system that can help any IoT device to create a complex cipher data. In this chapter encryption process done by knapsack algorithm and public key,

---

Advances in Data Science and Computing Technology: Methodology and Applications. Suman Ghosal, Amitava Choudhury, Vikram Kumar Saxena, Arindam Biswas, & Prasenjit Chatterjee (Eds.)

where public key is generated by Chinese remainder theorem. Our system is a public key crypt system, where spread is generally slower than a private key crypto system. But here, we can overcome this problem. For more confusing encrypted data creation we have taken the help of DNA encoding and Rail Fence technique. After that to compress it, we have used arithmetic encoding technique.

22.1 INTRODUCTION

In 21st century advancement of technology, storage of big data, fast transmission of huge data through network, embedded system all are the example of a new thing known is called Internet of Thinking (IoT) (Zeadally, Das, and Sklavos, n.d.). The internet of things (IoT) is established on smart and self-conviction nodes contained in a dynamic and global environment (Raj, Bashar, and Ramson, 2020). It is one of the most developing technologies, sanctioning the things to interact world. Recent research has considered IoT indicates smart and interconnected things capable of sharing their perceptions through the Internet. Deoxyribonucleic acid (DNA) encoding is a very popular topic in cryptography domain based research area. DNA is actually a biological molecules, which having the capacity to store, process, and transmit information, these capacities are the source of idea of DNA cryptography (Review on DNA Cryptography, 2019). We generally take the chemical characteristics of biological DNA sequences in traditional cryptography system to hide the structure of plain text from an intruder, or in other word we can say to create more confusion on plain text. DNA computing is the heart of this innovative method. The methodologies of DNA cryptography are not coded mathematically, thus, to crack a DNA cryptosystem is very difficult (Review on DNA Cryptography, 2019). Based on the techniques of encryption and decryption we can classify cryptosystem in two ways, there are Symmetric Key Encryption and Asymmetric Key Encryption (Review on DNA Cryptography, 2019). We all are known Symmetric key encryption is faster than Asymmetric key encryption, but if we compare based key transmission process, then we will see strength of security level of asymmetric key encryption is much more higher than symmetric key encryption, because every time security is our first priority in a data transmission process. Hence in this chapter we use asymmetric key DNA encryption technique. In this chapter the modulo operation of Chinese Remainder Theorem use for key generation purpose.

22.2 PRELIMINARIES

22.2.1 CHINESE REMAINDER THEOREM (FOROUZAN AND MUKHOPADHYAY, 2003; JIA ET AL., 2019)

In our cryptography system in many cases, we use the concept congruence to map Z to Z_n, it not a one to one mapping, where infinite number of Z map to one member of Z_n. As example we can say 2 mod 10 = 2, 12 mod 10 = 2, 22 mod 2 = 2, and so on. Hence according to modulo arithmetic we can say 2, 12 and 22 are called congruent mod 10. But in Chinese remainder theorem among all congruent equation variable is only one but it works on different modulo (Forouzan and Mukhopadhyay, 2003). According to Chinese remainder theorem we can write, (Jia et al., 2019) if p_1, p_2. p_t are pairwise co-prime positive integer, then the system of congruence.

$$x \equiv S_1 \,(mod\, p_1)$$
$$x \equiv S_2 \,(mod\, p_2)$$
$$x \equiv S_t \,(mod\, p_k)$$

has a unique solution:

$$x \equiv \sum_{i=1}^{t} \frac{P}{p_i} .\, Si.y_i \;\; mod\; P,$$

where; $P = \prod_{i=1}^{t} p_i, \frac{P}{p_i} .\, S_i.y_i \; mod\; P$, and "≡" denotes congruence (Jia et al., 2019).

22.2.2 DNA ENCODING (LUE ET AL., 2007; BISWAS ET AL., 2019; PELLETIER AND WEIMERSKIRCH, 2002; PAUL ET AL., 2017; NASKAR ET AL., 2019)

The concept DNA sequence comes from biological. By using different characteristics of biological DNA like capacity to store, process, and transmit information, etc., in cryptography field, we can relate these two different fields. Characteristics of DNA cryptography is the merging form of biological and computational characteristics, which plays an important role to increase the security level in a cryptography process (Lue et al., 2007). Strength of security level of DNA cryptography is much higher than traditional cryptosystems, hence it is very popular in modern cryptosystem (Biswas et al., 2019; Pelletier and Weimerskirch, 2002). A DNA sequence is a combination of four nucleic acids: A (adenine), T (thymine), C (cytosine) and G (guanine). Depend on base pairing rules, the purine-adenine (A) and the pyrimidine-thymine (T) create one pair, and the pyrimidine-cytosine (C) and the purine-guanine (G) create another pair. After following various characteristics of different nucleic acids, we can say A and T are complementary, as well as the characteristics of G and C are also complemented, where {00, 01, 10, and 11} are the representation of {A, T, G, C} respectively (Biswas et al., 2019). Simple DNA structure shown in FIGURE 1 and corresponding all values of four nucleic acids base on encryption and decryption map rules for DNA sequence shown in Table 1.

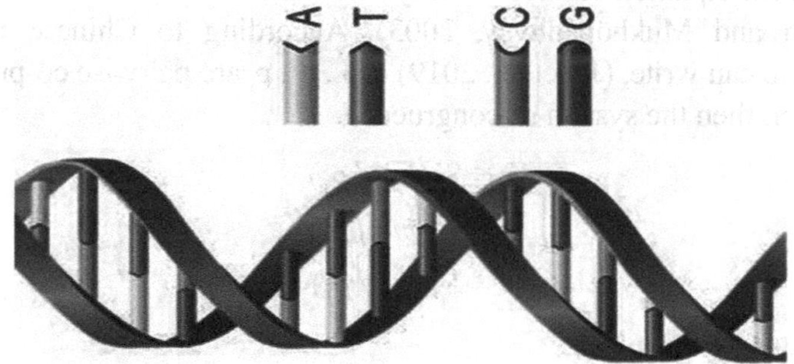

FIGURE 22.1 Simple DNA structure (Paul et al., 2017; Naskar et al., 2019).

TABLE 22.1 Encryption and Decryption Map Rules for DNA Sequences (Paul et al., 2017; Naskar et al., 2019)

	A	T	C	G
Rule 1	00	11	10	01
Rule 2	00	11	01	10
Rule 3	11	00	10	01
Rule 4	11	00	01	10
Rule 5	10	01	00	11
Rule 6	01	10	00	11
Rule 7	10	01	11	00
Rule 8	01	10	11	00

22.2.3 XOR ALGEBRAIC OPERATION FOR DNA SEQUENCE

XOR operation of DNA sequence is very useful in encryption process of cryptography. XOR operation for DNA sequences means bit wise XOR operation between corresponding binary values of nucleic acids of DNA sequence. Resultant value of XOR operation between various DNA components, which are used in the encryption system is shown in Table 22.2.

TABLE 22.2 XOR Operation for DNA Sequences (Paul et al., 2017; Naskar et al., 2019)

⊕	**A**	**T**	**C**	**G**
A	A	T	C	G
T	T	A	G	C
C	C	G	A	T
G	G	C	T	A

22.2.4 ASYMMETRIC KEY CRYPTOSYSTEM (KAHATE, 2003; ASYMMETRIC CRYPTOGRAPHY), N.D.)

In Asymmetric key cryptography two different keys are used, one for encryption another for decryption (Kahate, 2003). There are public key and private key, hence asymmetric key cryptosystem also called public key cryptography. The general idea of asymmetric key cryptosystem shown in FIGURE 2. The recipient public key use for encryption purpose and recipient private key use for decryption purpose.

- **Advantages [Asymmetric Cryptography (Public Key Cryptography), n.d.]:**
  - In asymmetric key no key exchange is required, no key distribution problem arise here.
  - Strength of security is higher than symmetric, as private key never disclose among to all.
  - In asymmetric key cryptography sender has no scope to gainsay regarding sending message, as non-repudiation characteristics supported by asymmetric key cryptography.
- **Disadvantages [Asymmetric Cryptography (Public Key Cryptography), n.d.]:**
  - In long size massage decryption, generally asymmetric key algorithm is not used, because asymmetric key process is comparatively slower than symmetric key algorithm.
  - If receiver loses his/her private key then message decryption never possible.
  - If private key of a person is revealed by an unauthorized person, then unauthorized person can also read all secret message's owner.

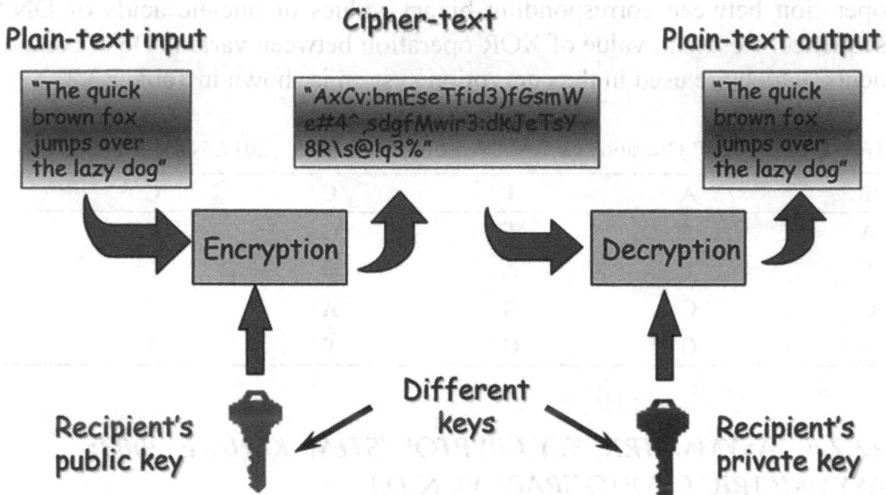

FIGURE 22.2 Asymmetric cryptography.

22.2.5 RAIL FENCE TECHNIQUE (KAHATE, 2003)

Plain text to cipher text transmission have two technique, one is substitution and another is transposition technique (Kahate, 2003). Rail Fence technique is one of the popular transposition techniques, it also called "zigzag cipher." The Rail Fence technique algorithm (Kahate, 2003) is:

- Write down the plain text as a diagonal order;
- After writing diagonal order, read the plain text in row wise.

The pictorial representation of Rail Fence Technique shown in bellow:

Plaintext T H I S I S A S E C R E T M E S S A G E

Rail Fence Encoding key = 4

T						A						T						G	
	H				S		S				E		M				A		E
		I		I				E		R				E		S			
			S						C						S				

Ciphertext T A T G H S S E M A E I I E R E S S C S

Here "key" value define the number of rows.

22.3 PROPOSED SCHEME

In our scheme system takes plan text and encrypt it using asymmetric key cryptographic scheme. At first in the sender uses a public key of receiver which is created by Chinese Remainder Theorem and Knapsack algorithm. Similarly, Knapsack algorithm again use at the last step of decryption. The Knapsack algorithm is described in bellow:

- We have used key pair of public key and private key in the knapsack algorithm.
- The public key is used only for encryption process and will be known to everyone and whereas the private key is used during decryption process (Asymmetric Cryptography (Public Key Cryptography, n.d.)).
- The elements which are from a predefined set of numbers are in knapsack are provided, if the sum is Known then we can find out the sum of the numbers quickly of the knapsack, rather it is not an easy job to find those elements which has participated in the knapsack to generate the sum.

Following steps are to followed for knapsack algorithm:

Let N = {N_1, N_2…………, N_n} be a predefined set and

I = (I_1, I_2…………., I_n) be a solution set in which I_i is either 0 or 1.

The elements' sum in knapsack can be calculated as:

S = knapsack Sum (N, I) = N_1I_1 + N_2I_2+……………+N_nI_n.

- Objective is to find the solution set (I1, I2,………………., In).
- When knapsack Sum(N,I) is given then it is not difficult to find S but if S&N is given then finding the value of X is difficult which is the case of inverse of knapsack Sum(N,I).

➢ **Super-Increasing Sets:**

- It is easy to find knapsack Sum and its inverse if the sets S is super increasing.
- Anon empty set is called super-increasing if every n^{th} term element >= the sum of all (n–1) elements previous to it.

$$\text{i.e., } N\_i >= \sum_{i=1}^{N-1} N\_i$$

- The algorithm for knapsack and its inverse is given as follows:

➢ **Algorithm:**

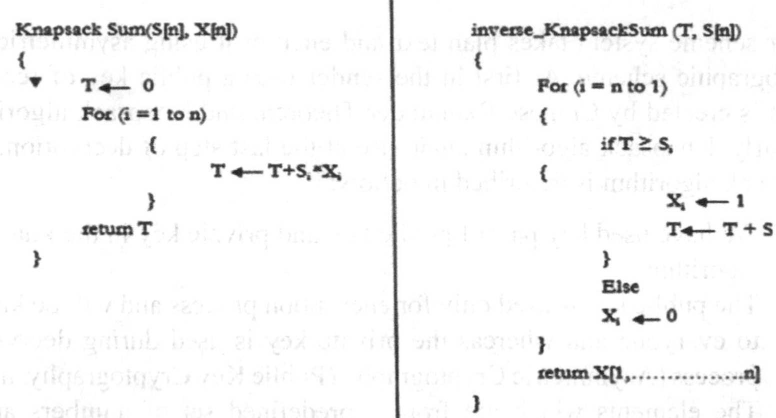

Knapsack Sum(S[n], X[n])
{
T ← 0
For (i =1 to n)
{
T ← T+S_i*X_i
}
return T
}
inverse_KnapsackSum (T, S[n])
{
For (i = n to 1)
{
if T ≥ S_i
{
X_i ← 1
T ← T + S
}
Else
X_i ← 0
}
return X[1,............n]
}

➢ ***Example:*** Let us consider N = [3, 7, 12, 30, 60, 115] and S = 82. To find the vector I?

i	Ni	S	S ≥ Ni	Ii	S,S – Ni × Ii
6	115	82	No	0	S←82 – 115 × 0 = 82
5	60	82	Yes	1	S←82 – 60 × 1 = 22
4	30	22	S	0	S←22 – 30 × 0 = 22
3	12	22	Yes	1	S←22 – 12 × 1 = 10
2	7	10	Yes	1	S←10 – 7 × 1 = 3
1	3	3	Yes	1	S← 3 – 3 × 1 = 0

So, the solution set is I = [1,1,1,0,1,0].

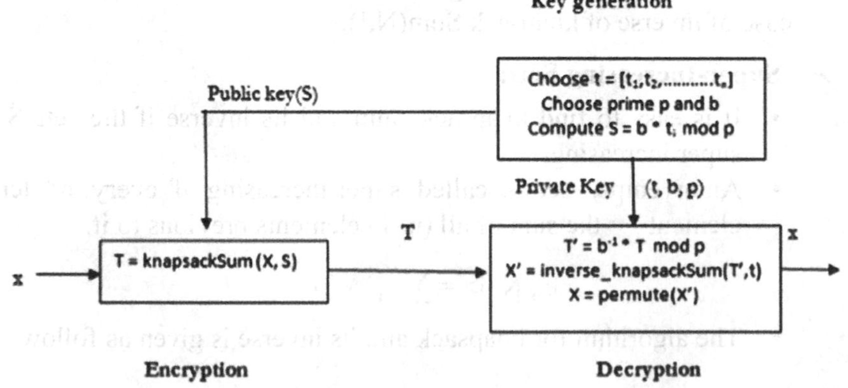

FIGURE 22.3 Merkel-Hellman Knapsack cryptosystem.

Consider Vivek and Arun want to communicate with each other.

1. **Generation of Key:**
   i. Let I= I_1, I_2,.........I_n be a super-increasing set of values.
   ii. Choose a prime p such that p > I_1 + I_2 +......+ I_n and a number q with 1 <q < p.
   iii. Arun calculates his public set of items, N_i = q × I_i mod p.
   iv. The public key is N and private key is I,q and p.
2. **Encryption Process:**
   i. Vivek encrypts the M using S= knapsack_Sum(M,N) algorithm.
   ii. Vivek then sends S as ciphertext to Arun.
3. **Decryption Process:**
   When Arun receives S from Vivek, he does the following:
   i. Calculates S' = q^{−1} × S mod p;
   ii. Uses M' = inverse_knapsackSum(S,'I);
   iii. Permutes M' to find M which is original plaintext.

In our proposed algorithm encryption happens in two phases. In our 1st phase after getting the plain Text in binary form we encrypt the data using knapsack algorithm but difference is that we use Chinese remainder theorem to generate the pair of keys instead of modular transformations (Asymmetric Cryptography (Public Key Cryptography), n.d.), in our 2nd phase of encryption we have used DNA encoding technique of cryptography followed by Rail Fence technique which creates more confusion in data to be send and then we finally compress the encrypted data using Arithmetic encoding algorithm then we send this to receiver.

In decryption end we have applied reverse process of encryption.

22.4 RESULT AND DISCUSSION

The design of our proposed cryptographic system is discussed in detail by using Knapsack followed by DNA cryptography is shown in Figure 22.1. The recipient broadcast the Knapsack public key to the sender. The sender encrypts the message using Knapsack algorithm and uses receiver's knapsack public key (using Chinese remainder theorem) in the 1st phase of cryptography and in 2nd phase of cryptography we have used DNA cryptography followed by rail fence technique then the encrypted data goes through a compression technique by using arithmetic encoding technique and generates the final cipher Text and then it is sent to the receiver.

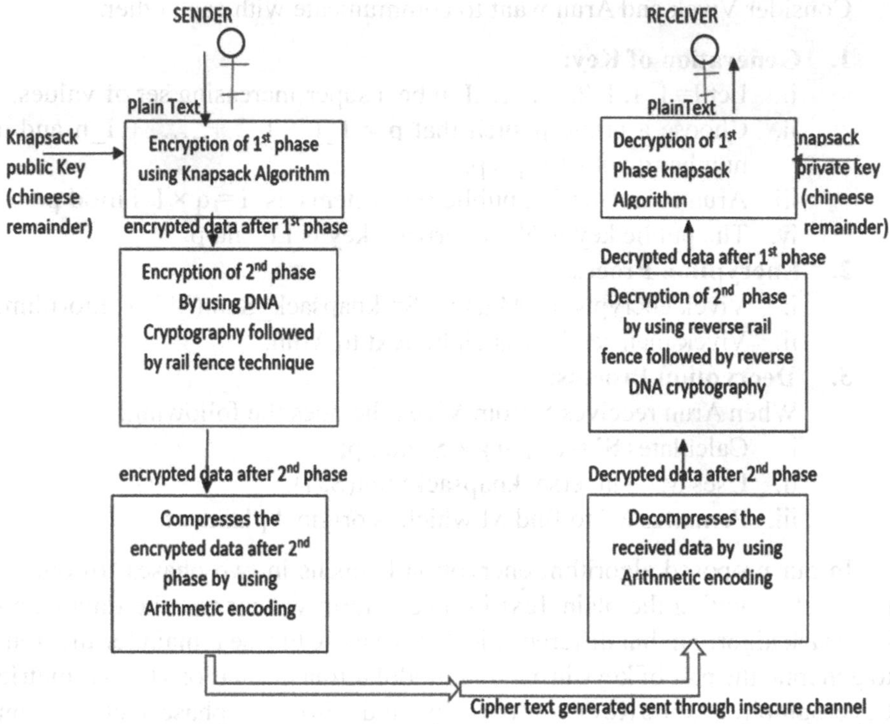

FIGURE 22.4 Process diagram.

After receiver receives the ciphertext it first decompresses the received data then decrypts the decompressed data in just reverse order of two phases of cryptography where in decryption process of knapsack it uses first reverse order of rail fence and DNA cryptography and then it uses Knapsack private key of recipient generated by using Chinese remainder theorem to reveal the plain text back.

In our proposed algorithm we tried to test the confidentiality and Integrity of the original message content. So, in our algorithm one testing is done for checking the encryption and decryption of messages (data files). And the second test is for the generation of key of knapsack algorithm. Each test has conducted has successfully generated plain text to cipher text and cipher text back to plain text. Our Key generation have also been done successfully by generating public key and private key encryption and decryption process, respectively.

In the following table, it can be seen that this test successfully returns the cipher text to plaintext.

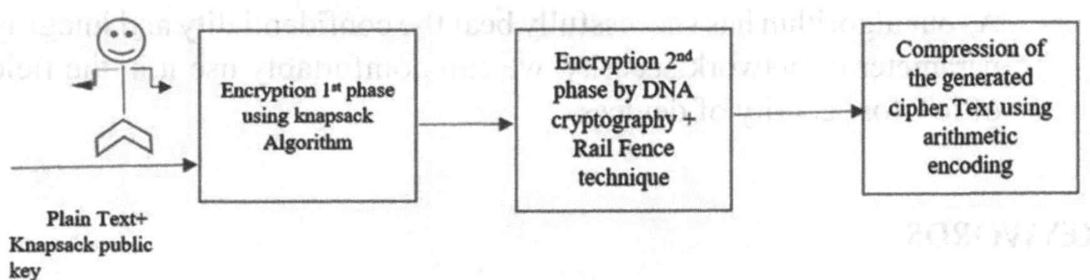

Encryption followed by compression of data

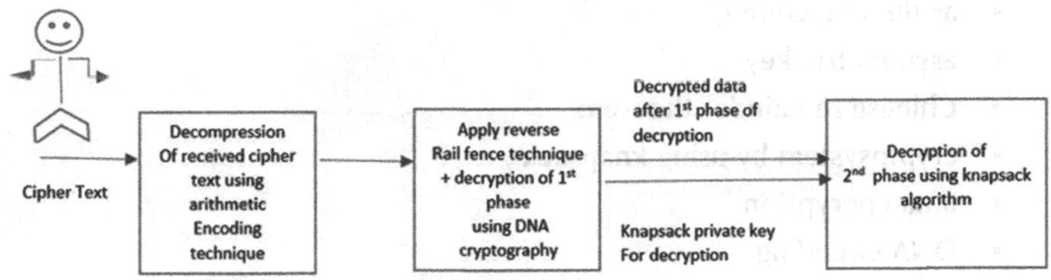

Decompression followed by decryption Of data

Encryption and Decryption of the Message

Encryption	Decryption
1. Plaintext: 100101111	Cipher text: 5 18 25
Knapsack Public Key: 1	Knapsack Private Key: {5,7,13}
Ciphertext-5 18 25	Plaintext: 100101111
2. Plaintext: 111100101	Cipher text: 23 5 16
Knapsack Public Key: 36	Knapsack Private Key:{5,7,11}
Ciphertext-23 5 16	Plaintext: 111100101
3. Plaintext: 101111100	Cipher text: 58 69 7
Knapsack Public Key: 1223	Knapsack Private Key:{7,5,51}
Ciphertext-58 69 7	Plaintext: 101111100

22.5 CONCLUSIONS

Following conclusions can be drawn from the analysis and obtained result of our project is:

- Our proposed algorithm using Knapsack clubbed with DNA cryptography will give new class of cryptography where the two phases of cryptography will contribute more towards the secure transmission of data and will be less prone to different types of attacks.

- As our algorithm has successfully beat the confidentiality and integrity parameter of network security, we can comfortably use it in the field of IoT for security of devices.

KEYWORDS

- **arithmetic coding**
- **asymmetric key**
- **Chinese remainder theorem**
- **cryptosystem by using knapsack**
- **data encryption**
- **DNA encoding**
- **rail fence technique**

REFERENCES

Biswas, M. R., Alam, K. M. R., Tamura, S., & Morimoto, Y., (2019). A technique for DNA cryptography based on dynamic mechanisms. *Journal of Information Security and Application, 48*, 102363.

Forouzan, B. A., & Mukhopadhyay, D., (2003). *Cryptography and Network Security* (3rd edn.). McGraw-Hill; Chennai.

Jia, X., Wang, D., Nie, D., Luo, X., & Sun, J. Z., (2019). A new threshold changeable secret sharing scheme based on the Chinese remainder theorem. *Information Sciences, 473*, 13–30.

Kahate, A., (2003). *Cryptography and Network Security* (1stedn.). McGraw-Hill; New Delhi.

Kate Brush (2020). *Asymmetric Cryptography (Public Key Cryptography).* [Online] https://searchsecurity.techtarget.com/definition/asymmetric-cryptography (accessed on 08 December 2021).

Lue, M., Lai, X., Xion, G., & Qin, L., (2007). Symmetric-key cryptosystem with DNA technology. *Sci. China Ser. F., 50*(3), 324–333.

Mandrita Mondal, & Kumar S. Ray (2019). *Review on DNA Cryptography.* [Online] https://arxiv.org/ftp/arxiv/papers/1904/1904.05528.pdf (accessed on 08 December 2021).

Naskar, P. K., Paul, S., Nandy, D., & Chaudhuri, A., (2019). "DNA encoding and channel shuffling for secured encryption of audio data." *Multimedia Tools and Applications, 78*(17), 25019–25042. Springer Nature.

Paul, S., Dasgupta, P., Naskar, P. K., & Chaudhuri, A., (2017). "Secured image encryption scheme based on DNA encoding &chaotic map." *Review of Computer Engineering Studies, IIETA, 4*(2), 70–75.

Pelletier, O., & Weimerskirch, A., (2002). Algorithmic self-assembly of DNA tiles and its application to cryptanalysis. In: *Proc. of 4th Annual Conf. on Generic and Evolutionary Computation* (pp. 139–146). Morgan Kaufmann Publishers Inc.

Raj, S. J., Bashar, A., & Ramson, S. R. J., (2020). Elliptical curve cryptography based access control solution for IoT based WSN. *International Conference on Innovative Data Communication Technologies and Application* (pp. 742–749). Springer Nature.

Zeadally, S., Das, A. K., & Sklavos, N. (2019). *Cryptographic Technologies and Protocol Standards for Internet of Things, Internet of Things*. https://doi.org/10.1016/j.iot.2019.100075.

CHAPTER 23

ROBUST TECHNIQUE TO OVERCOME THWARTING COMMUNICATION UNDER NARROWBAND JAMMING IN EXTREMELY LOW FREQUENCY OF CDMA-DSSS SYSTEM ON CODING-BASED MATLAB PLATFORM

TANAJIT MANNA,[1] RAHIT BASAK,[2] MAINAK DAS,[2] and ALOK KOLE[3]

[1]*Electronics and Communication Engineering Department, Pailan College of Management and Technology, Kolkata–700104, West Bengal, India, E-mail: tanajitmanna@gmail.com*

[2]*Computer Science and Engineering Department, Pailan College of Management and Technology, Kolkata–700104, West Bengal, India*

[3]*Electrical Engineering Department, RCC Institute of Information Technology, Kolkata–700015, West Bengal, India*

ABSTRACT

This chapter illustrates an approach for smooth communication to mitigate the jamming under narrow band in code division multiple access (CDMA) system in case of direct sequence spread spectrum (DSSS) single user system. In security sector improvement based communication under narrowband frequency is highly demandable. An analytical as well as experimental comparison of power spectral density (PSD) of CDMA under narrowband frequency with and without jamming condition is illustrated in

Advances in Data Science and Computing Technology: Methodology and Applications. Suman Ghosal, Amitava Choudhury, Vikram Kumar Saxena, Arindam Biswas, & Prasenjit Chatterjee (Eds.)

the chapter. By variation of signal power and Barker code, an optimized system is illustrated supported by bit error rate (BER) and signal to noise ratio (Eb/N) curve. Experimental fluctuation of signal power provides us low narrowband jamming as well as jamming free PSD. Variations of Spreading factor supported by Barker Code illustrate the results in favor of optimized model of the system under narrow band jamming condition. The abovementioned unique proposal of CDMA-DSSS system designing, modeling, and performance analysis are implemented using coding based MATLAB® platform.

23.1 INTRODUCTION

Nowadays, the need of optimized performance in DS-CDMA communication system transmitter and receiver both are real active interest as the application areas for the DSSS communication system is increasing day by day because of the advantages of such system in real time practice world. In the presence of narrowband jamming environment, the modeling is an approach toward the goal of improving the performance of a DSSS-CDMA communication system. In this work it has shown clearly that the performance of a DSSS-CDMA receiver system in under the narrowband jamming can be significantly boosted by modifying the power of the signal (Abimoussa and Landry, 2000; Lichuan and Hongya, 2005; Raju, Ristaniemi, Karhunen, and Oja, 2002). Variation of spreading factor with specific security or barker code is illustrated in the optimization method.

23.2 RELATED WORK

Jamming reduction under very low frequency band in DS-CDMA system is very cursed job. Anti-jamming authorization in CDMA (Abimoussa and Landry, 2000) was first blueprint in concerned field since the year around 2000. The employment was flourished in the tangible domain by adding particular subspace jut model (Lichuan and Hongya, 2005). By the arsenal of uncommitted component analysis (Ristaniemi, Raju, and Karhunen, 2002) this research job was clutch in CDMA domain sharply. In the meantime, this jamming notion was offer in FM technology for abdication of particular band (Lichuan and Hongya, 2003). Most cabalistic work done in DS-CDMA detection technique under near-far field (Manna and Kole, 2016).

23.3 SYSTEM DESIGN AND MODELLING

In this method we introduced a MATLAB® programming base modeling and jamming mitigate technique. To ingredient the problem we account a spreading factor with specific value. Depends upon the spreading factor model generate a barker code. For generate the barker code we depend on following formula and algorithm:

$$(Cv)^{n} = \sum_{j=0}^{N-v+1} aj + v^{n} aj^{n} \tag{1}$$

Where, $(Cv)^n \leq 1$ for all $1 \leq v \leq N+1$. The above equation based on auto-correlation function.

For Barker generation following model was account (Figure 23.1).

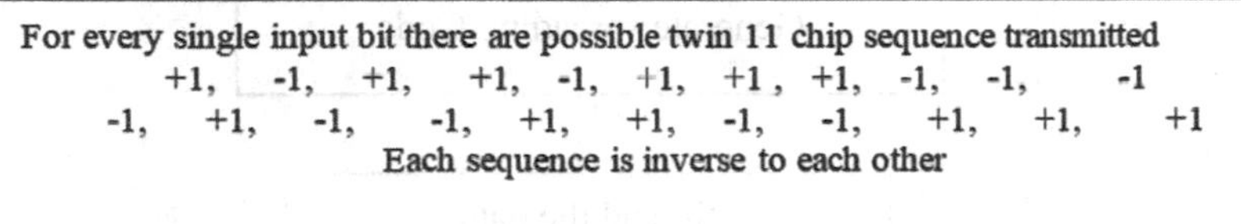

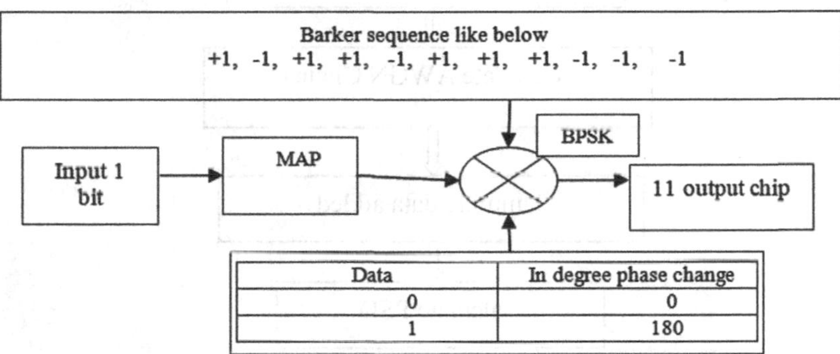

FIGURE 23.1 Barker code generation.

After creation of barker code, the modulation technique like QPSK is selected for operation. After that we create jamming data follows by channel. Depend upon the spreading code we spread the data. Generation of additive white gaussian noise (AWGN) channel grade up the noise in the system significantly, as well as the projected system almost removed the noise very sharply. In the time of detection, the signal de spread with same chip. In the receiver edge we nabbing the PSD for analyze the jamming amount.

Depending on Barker Code we design the system on proposed algorithm (Figure 23.2).

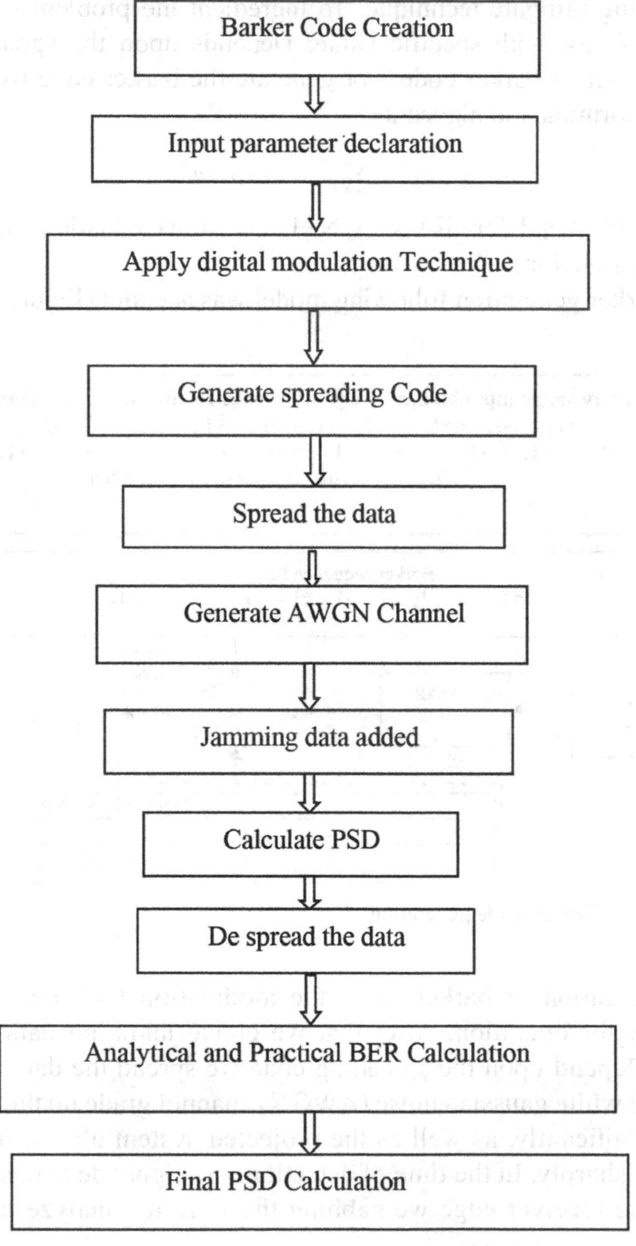

FIGURE 23.2 Algorithm for proposed model.

To achieve our goal, power spectral density (PSD) and bit error rate (BER) calculation like below. For PSD we used following formula.

$$Sx(f) = \lim_{T\to\infty}\left(\frac{mean\left[x(f)^2\right]}{T}\right) \quad (2)$$

T depict time period.

MATLAB® programming we have to modify the Eqn. (2) like below:

$$Sx(f) = \lim_{T\to\infty}\tfrac{1}{T}\int\int_{T/2}^{T/2} C(t_1)C(t_2)mean\left(n(t)_1 n(t)_2\right)e^{-j2\pi f(t_2-t_1)}dt_1 dt_2 \quad (3)$$

In the Eq. (3), product of mean n(t) and C(t) treated as noise. So, the intermediate PSD treated as noise which makes the both System more secure. In the above equation C(t) is the deterministic signal and n(t) is Gaussian with zero mean.

BER define as:

$$P_b = Q\left(\frac{\beta_{\max}}{2}\right) \quad (4)$$

$$\beta_{\max} = \sqrt{\left[\frac{2}{N}\int^{T_b}[p(t)-q(t)]\,dt\right]} \quad (5)$$

Here; P_b accounted as BER. Filter matched pulse described by difference of p (t) and q(t). T_b is time period.

23.4 RESULTS AND DISCUSSION

For mitigate the jamming we accounted 4 variations of experimental data. All the stages we consider 13 as spreading factor value. For first part we consider 5dB SIR value. As per Figure 23.3 no jamming data reflected in PSD. If we analysis the Figure 23.4, in 0 to 2.5 Hz frequency jamming reflected.

In 13 spreading factor value and 5 dB SIR are not enough for removing narrowband jamming which reflected in Figure 23.5. In Figure 23.5 smooth line represent the analytical BER and line with bubble represented BER with jamming data.

In 13 spreading factor value and 5 dB SIR are not enough for removing narrowband jamming which reflected in Figure 23.5. In Figure 23.5 smooth line represent the analytical BER and line with bubble represented BER with jamming data.

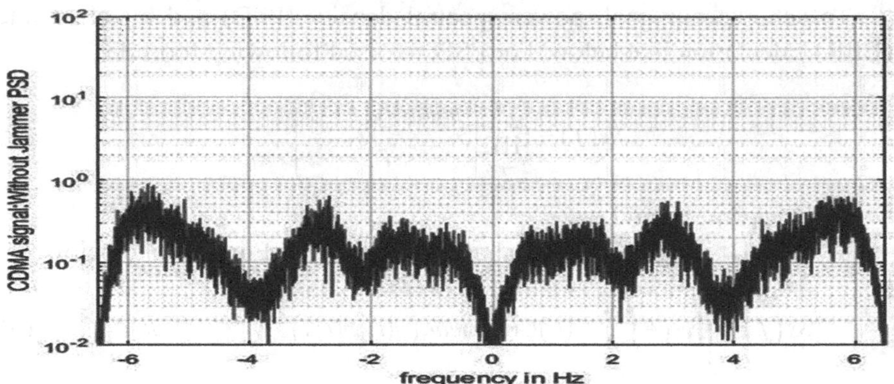

FIGURE 23.3 DS-CDMA PSD with no jamming.

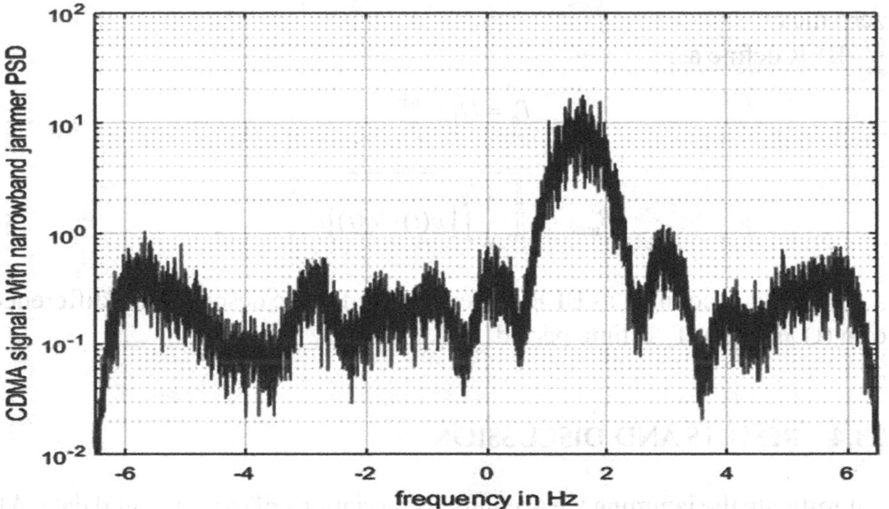

FIGURE 23.4 DS-CDMA PSD with jamming.

In second stage signal to interference ratio is 8 dB. In this condition results reflected in Figures 23.6, 23.7 and 23.8. Figure 23.6 delimitate no jamming PSD, whereas Figure 23.7 delimitate the jamming concept with PSD. In Figure 23.8, the jamming value minimized as per BER representation. The BER partially minimized compared to Figure 23.5.

Figures 23.9–23.11 represent the results with SIR value 13 dB. Figure 23.9 no data jamming is remains same. In Figure 23.10 jamming mitigation reflected positively. Here also PSD plot swing in between positive and negative half express the distortion in the performance. But the distortion not in practically, it depends on resolution. In Figure 23.11 it is clear that jamming and no jamming BER is closed to each other.

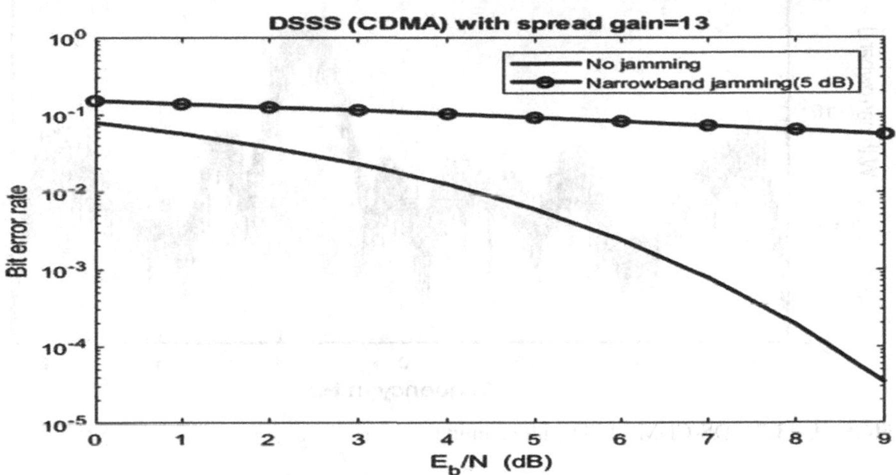

FIGURE 23.5 Bit error rate with and without jamming.

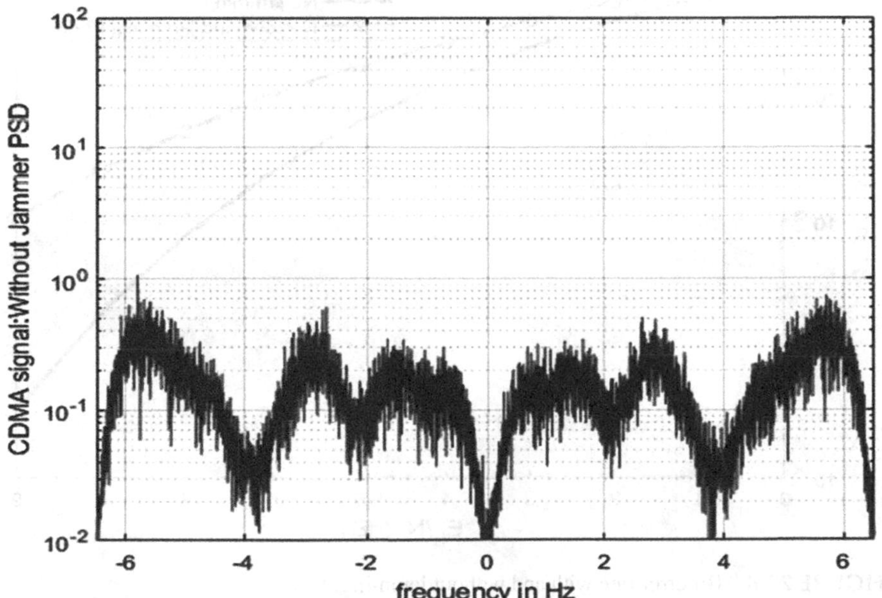

FIGURE 23.6 DS-CDMA PSD with no jamming.

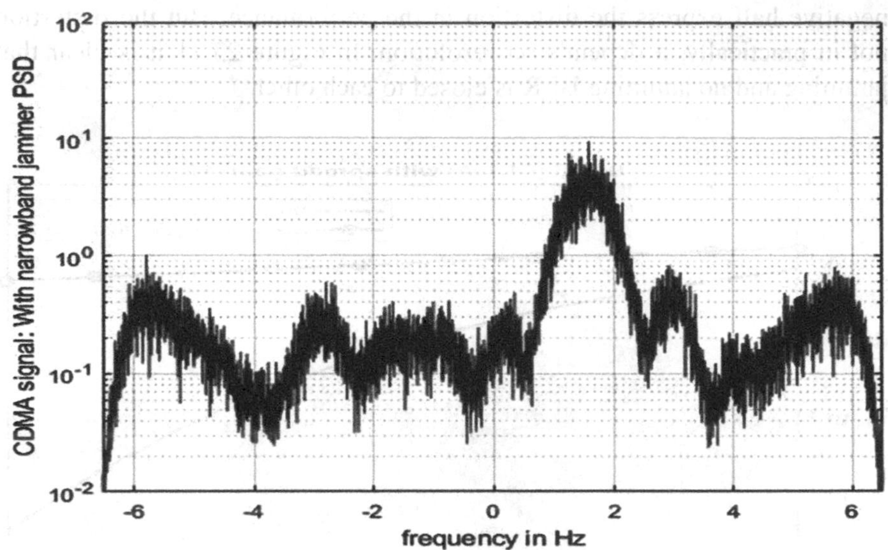

FIGURE 23.7 DS-CDMA PSD with jamming.

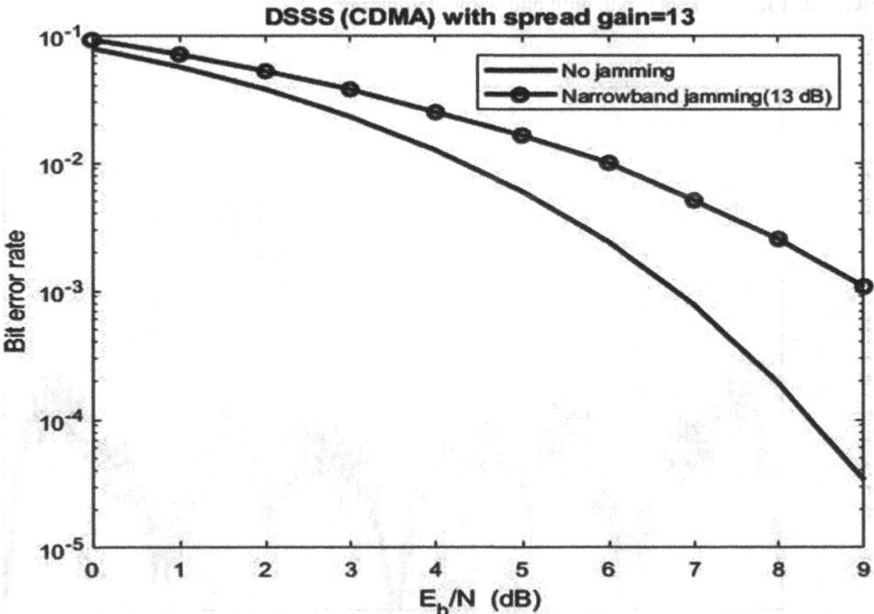

FIGURE 23.8 Bit error rate with and without jamming.

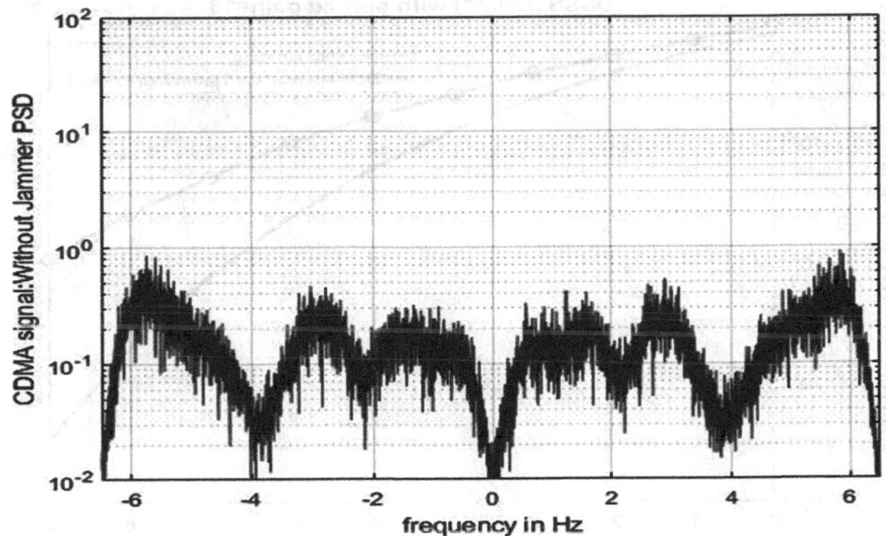

FIGURE 23.9 DS-CDMA PSD without jamming.

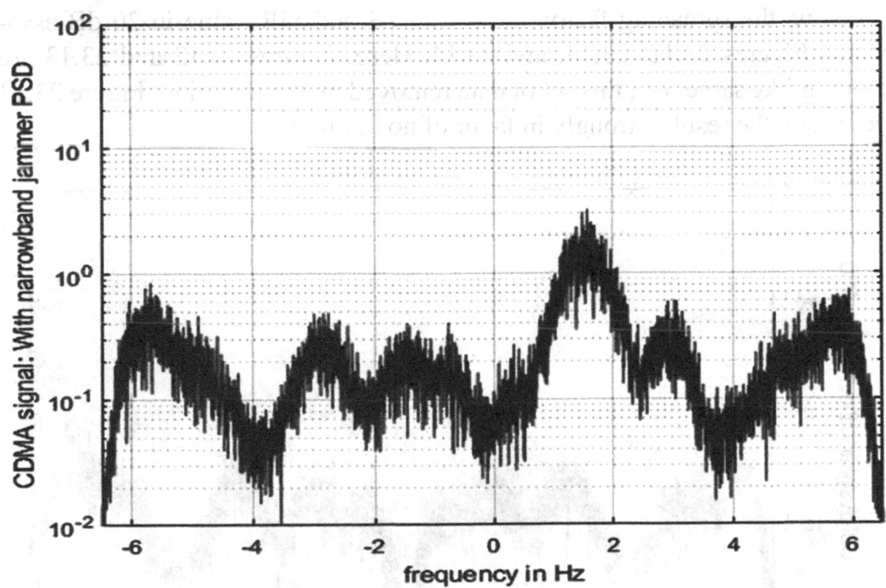

FIGURE 23.10 DS-CDMA PSD with jamming.

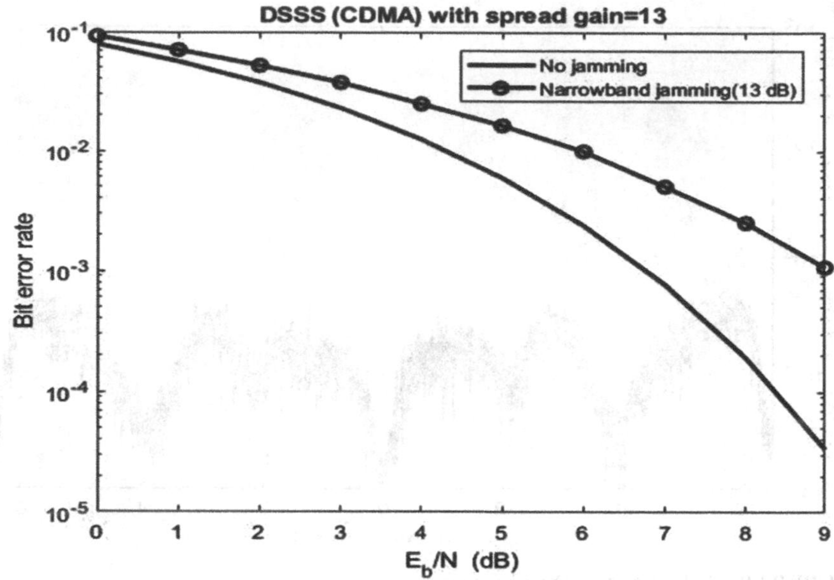

FIGURE 23.11 Bit error rate with and without jamming.

Now, the spreading factor value with 13 and SIR value in 20 dB associated Figures 23.12, 23.13, and 23.14. Here, Figures 23.12 and 23.13 are looking like same, i.e., this set of data removed all the jamming. Figure 23.14 described the results strongly in favor of no jamming.

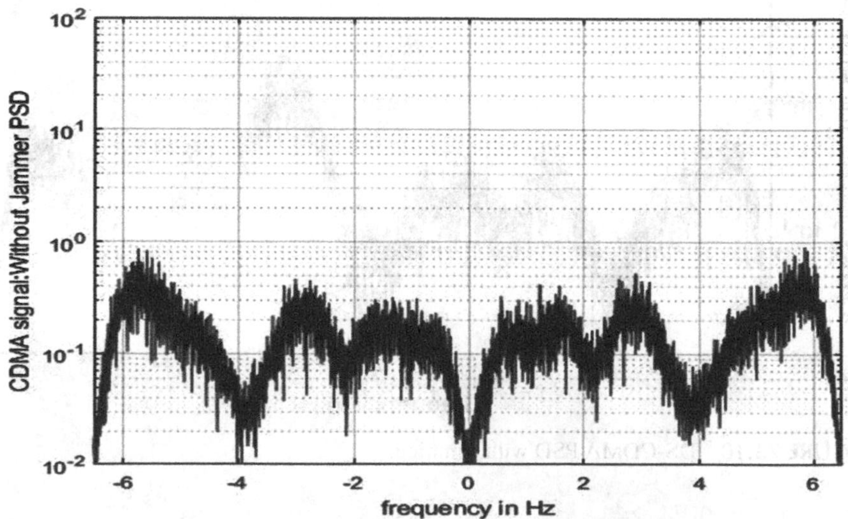

FIGURE 23.12 DS-CDMA PSD without jamming.

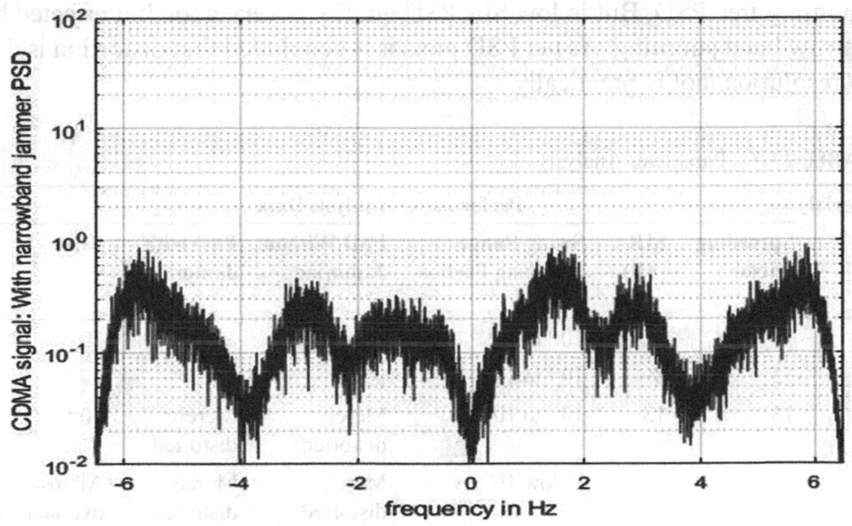

FIGURE 23.13 DS-CDMA PSD with jamming.

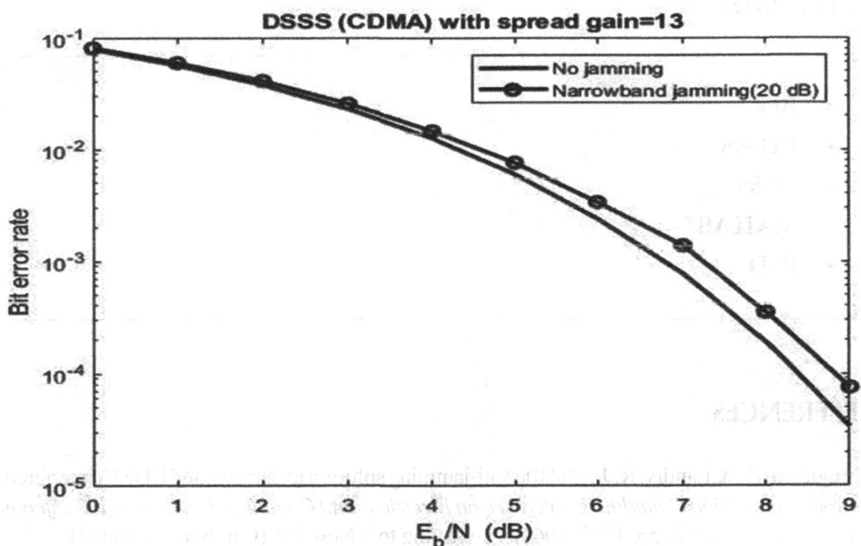

FIGURE 23.14 Bit error rate with and without jamming.

23.5 CONCLUSION AND FUTURE DEVELOPMENT

As per Table 23.1, we calculate all the parameters under spreading factor 13. It are clearly visible that in same spreading factor high SIR gives up almost jamming free PSD. But in low SIR PSD quality is very good but effected by narrow band jamming. As per PSD plots it is concluded that distortion is due to resolution, not is practically.

TABLE 23.1 Parameter Analysis

Serial No.	Performance Analysis Data					
	Spreading Factor	**SIR (dB)**	**Swing Range of Jamming Portion of PSD**	**PSD Without Jamming**	**PSD with Jamming**	**BER**
1	13	5	10^{-1} to 10^{1}	good	good	10^{-3}
2	13	8	10^{-1} to $10^{0.7}$	good	good	10^{-2}
3	13	13	10^{-5} to 10^{+5}	Merely distorted	Merely distorted	10^{-1}
4	13	20	Below 10^{0}	Merely distorted	Merely distorted	Almost overlapped

KEYWORDS

- **BER**
- **CDMA**
- **DSSS**
- **MATLAB®**
- **PSD**

REFERENCES

Abimoussa, R., & Landry, R. J., (2000). Anti-jamming solution to narrowband CDMA interference problem. In: *2000 Canadian Conference on Electrical and Computer Engineering. Conference Proceedings* (Vol. 2, pp. 1057–1062). Navigating to a New Era (Cat. No.00TH8492).

Fang-Biau, U., Jun-Da, C., & Sheng-Han, C., (2004). Smart antennas for multiuser DS/CDMA communications in multipath fading channels. *Eighth IEEE International Symposium on Spread Spectrum Techniques and Applications - Program and Book of Abstracts* (IEEE Cat. No.04TH8738) (pp. 400–404).

Lichuan, L., & Hongya, G., (2003). Time-varying AR modeling and subspace projection for FM jammer suppression in DS/SS-CDMA systems. *The Thirty-Seventh Asilomar Conference on Signals, Systems & Computers, 1*, 623–627.

Lichuan, L., & Hongya, G., (2005). Subspace projection and time- varying AR modeling for anti-jamming DS-CDMA communications. In: *14th Annual International Conference on Wireless and Optical Communications, 2005* (p. 104). WOCC.

Lu, M., Cheng, F., Wei, S., & Gang, Q., (2017). Comparison of detection methods for non-cooperative underwater acoustic DSSS signals. *Signal Processing Communications and Computing (ICSPCC) 2017 IEEE International Conference*, 1–5.

Lu, M., Cheng, F., Wei, S., & Gang, Q., (2018). *A Real-Time Detection System for Non-Cooperative Communicates Electronic and Automation Control Conference (IMCEC) 2018 2nd IEEE* (p. 1331). Underwater acoustic DSSS signals based on LabVIEW, advanced information management 1335.

Manna, T., & Kole, A., (2016). Performance analysis of secure DSSS multiuser detection under near far environment. *International Conference on Intelligent Control, Power and Instrumentation 2016 (IEEE Conference)*. doi: 10.1109/ICICPI.2016.7859713 IEEE ISBN No.:978-1-5090- 2636-4. IEEE Xplore.

Raju, K., Ristaniemi, T., Karhunen, J., & Oja, A., (2002). Suppression of bit-pulsed jammer signals in DS-CDMA array system using independent component analysis. In: *2002 IEEE International Symposium on Circuits and Systems* (Vol. 1). Proceedings (Cat. No.02CH37353).

Rida, T., Saqib, E., Sobia, J., & Saqib, A., (2019). Performance analysis of multi-user polar coded CDMA system. *Applied Sciences and Technology (IBCAST) 2019 16th International Bhurban. Conference* (pp. 1000–1005).

Ristaniemi, T., Raju, K., & Karhunen, J., (2002). Jammer mitigation in DS-CDMA array system using independent component analysis. In: *2002 IEEE International Conference on Communications Conference Proceedings* (Vol. 1, pp. 232–236). ICC 2002 (Cat. No.02CH37333).

Seyedi, A., & Saulnier, G. J., (2001). A sub-channel selective orthogonal frequency division multiplexing spread spectrum system. In: *2001 MILCOM Proceedings Communications for Network - Centric Operations Creating the Information Force (Cat. No.01CH37277)* (Vol. 2 pp. 1370–1374).

Sriyananda, M. G. S., Joutsensalo, J., & Hämäläinen, T., (2013). Interference cancellation schemes for spread spectrum systems with blind principles. In: *2013 IEEE 27th International Conference on Advanced Information Networking and Applications (AINA)* (pp. 1078–1082).

Tsung-Chi, L., Chia-Cheng, H., Cheng-Yuan, C., Guu-Chang, Y., & Wing, C. K., ((2007). Study of MFSK/FH-CDMA wireless communication systems without symbol-synchronous assumption. *IEEE Sarnoff Symposium*, 1–5.

PART VII

Miscellaneous Topics in Computing

CHAPTER 24

THE UTILITY OF REGIONAL LANGUAGE FACILITY IN ACCOUNTING SOFTWARE FOR RETAIL MARKET OF WEST BENGAL

MADHU AGARWAL AGNIHOTRI and ARKAJYOTI PANDIT

Assistant Professor, Department of Commerce,
St. Xavier's College (Autonomous), Kolkata, West Bengal, India,
E-mail: madhu.cal@gmail.com

ABSTRACT

The role of accounting software in the realm of business practice is immense as they provide information that forms the primary reference for a stakeholder's decision making process and portfolio management. Most of the process in accounting software as well as the financial statements reproduced by it is most often in English. However, in a country like India where only 10% of the total population is well versed with the English language such accounting process and information become cumbersome. Moreover, with linguistic diversity present in every 15 to 20 kms of the country there is an absolute chance of English being not known to all the accounting professionals and stakeholders. Added to this the Census Report of 2011 confirms that only 28% of the total Indian population receive their education in English language which justifies the preference of regional language in accounting process. In the backdrop of such a scenario the current research chapter tries to assess the utility of regional language facility in accounting software through opinion collected from 90 accounting professionals and stakeholders

Advances in Data Science and Computing Technology: Methodology and Applications. Suman Ghosal, Amitava Choudhury, Vikram Kumar Saxena, Arindam Biswas, & Prasenjit Chatterjee (Eds.)

in West Bengal, India. The research also tries to explore the effect of regional language facility in accounting software on intra business communication at managerial level. At the end the study seeks to understand the role played by multilanguage software in meeting customer satisfaction.

24.1 INTRODUCTION

The Indian retail market is at present valued at a whooping Figure 24.1 trillion US Dollars with 40 million participants directly connected with the selling and purchasing process (Economic Times, 2017). The accounting process of such a huge market is bound to be complex and cumbersome not only owing to its size but the linguistic diversity present in the stakeholders (Aryan, Alrabei, and Haija, 2016). In such a scenario, the role of accounting software in fastening and simplifying the accounting process is immensely important. It is also utterly necessary that the Accounting software and the financial statements produced thereon must have adequate clarity, worthy to be understood by all concerned.

In India, the accounting process and outcomes are mostly developed in English. However, it must be noted that only 10% of the entire Indian population are fluent with the English language (Census Report of India, 2011). Moreover, the country boosts of having 22 official languages with linguistic diversities seen in every 15 to 20 kms (Linguistic Diversity of India, 1969). In the backdrop of such a varied linguistic mélange, processing, and preparation of accounting information in regional language not only makes the use of accounting software more convenient for the user but helps the stakeholders of business have a better knowledge of the accounting information. Any information in a known vernacular posse more coherence than the same conveyed in a less known language (Linguistic Diversity of India, 1969). Accounting information being highly relevant to the market participants cannot be an exception. Such a milieu calls for understanding the utility of regional language facility in accounting software.

In any business sector irrespective of the country it pertains to, the stakeholders are liable to get full-fledged accounting information pertaining to the business or market they operate in. According to Professor Laithy Aryan, the quality and clarity of accounting information plays a vital role in investment decisions, portfolio management and liquidity analysis (Aryan, 2017). In India as most of the population is better versed in their regional language over and above English, it is a dire need to interpret the utility of regional language facility in accounting software from the point of view of all the users and makers of accounting information. The present research chapter analyzes the utility of regional language facility in accounting software and if the presence of the concerned facility helps in meeting customer satisfaction. Alongside, the research assesses the effect of presence of 'regional language facility in accounting software' on intra business communication at managerial level.

24.2 REVIEW OF LITERATURE

Researches pertaining to the role of accounting software in business practice have been conducted multiple times. However, studies related to the relevance of regional language in such accounting software are scanty. Among the few researches conducted, the one that deserves mention at the foremost is 'Multilingual Information Management' by Ximo Granell (2015; Garnell, n.d.). The study on the importance of Accounting Software in small business by Walton (2019) highlights both the significance and the limitations of current day accounting software in the business atmosphere. The report of Centre for Development of Advanced Computing (2019) on 'Multilingual computing and heritage computing' (2019) clearly defines the role of multi-language facility in various government sectors including the Rajya Sabha [the upper house of the Indian Parliament]. Danet and Herring (2007) in their book have detailed the use of multilingual facility in communication over the internet. The research of Public Utility Accounting by Farber and McDonnell (2012) highlights the need for accounting lucid enough to be understood by the lay man. A mention must be made of the laudable study on the impact of accounting software for business performance by Wickramasinghe (2017) which vividly describes the significant position of accounting software in business performance. These research chapters provide an evidence for the importance of multilingual facility in business perspectives.

24.3 RESEARCH GAP

Studies in the past have successfully pointed out the utility of accounting software in the progress of business. The researches have also been ardent in assessing the significance of accounting information in business decision making. However, it must be understood that the accounting information generated by the accounting software must be lucid and should have enough clarity to be understood by all concerned. Researchers however, have confirmed that most of the financial information produced by accounting software is in English, a language practiced fluently by only 10% of the total Indian population. Moreover, the Indian subcontinent is home to more than 700 languages with linguistic diversity in every 15 to 20 kms. Under such a scenario it is natural that accounting information in English may appear complicated for the non-English speakers of the Indian population. But that does not mean the respective persons are not entitled to clarified accounting information. In this scenario, if the accounting information can be generated in vernacular language, then much of the problem regarding the lucidity of such information to the users who are not well versed in English is solved. Though a lot of studies proclaim the utility of accounting software in Business practice but none of them defines the clarity that financial information could produce if prepared in regional languages spoken by the majority of stake holders. This forms the core area for assessing the utility of regional language facility in accounting software.

Another important aspect lies in the fact that almost 79% of the accounting professionals had their education mostly in vernacular language. These professionals are bound to be more convenient in accounting in a language they have been taught all throughout their academic life. This necessity is also a reason to determine the utility of the presence of regional language facility in accounting software.

24.4 RESEARCH OBJECTIVES

- To explore the utility of regional language facility in accounting software;
- To assess the effect of regional language facility in accounting software on intra business communication at managerial level;
- To understand the role played by Multilanguage software in meeting customer satisfaction.

24.5 RESEARCH METHODOLOGY

The research takes into account the primary data collected from a population of 90 accounting professionals and users of accounting information across the state of West Bengal. The selection of the respondents was limited to only those persons who have been using accounting software for at least 6 months. The population was selected using snow ball sampling technique. From the period of February 2020 to March 2020 the respondents were asked

20 questions mostly through online and verbal interview. The first part of the questionnaire contained 10 questions about the respondent's personal information [e.g., Educational qualification, medium of education, etc.]. While in the second part, the respondents were asked to convey their degree of agreeability towards a particular question using Likert's 5-point scale. The second part had nine questions which conclusively have formed the variables for the research. At the end to have a better clarity of research the respondents were also asked about any potential problems resulting from the use of regional language in accounting and accounting information generation. The opinion of the respondents was noted down and the descriptive statistics was used for the analysis. For the convenience of research, the respondents were divided into three different groups based on their work experience. The mean score of the responses of each category were considered to determine the degree of agreeability of the category for a particular question. The respective groups with the number of respondents in each one of them is shown in Table 24.1.

TABLE 24.1 Categories of Respondents and Number of Respondents in Each Category

Work Experience	Number of Respondents
6 months to 5 years	38
Above 5 years to 10 years	27
Above 10 years	25
Total	**90**

24.6 ANALYSIS

The opinion of the respondent categories for each of the variables was recorded in a data sheet. The respondent categories which had a mean score higher than the overall average score signifies a greater degree of agreeability towards a particular variable than the rest. The entire findings are shown in Table 24.2.

The numerical analysis as presented in Table 24.2 is graphically shown in Figure 24.1.

24.6.1 INFERENCE FROM TABLE 24.2 AND FIGURE 24.1

The Numerical Analysis of the mean score obtained by the respondents as represented through Table 24.2 and Figure 24.1 vividly proclaims the

TABLE 24.2 Mean Score Obtained by Each Category of Respondents and Overall Mean Score for Each Variable

SL. No.	Variables	Work Experience of Respondents			Overall Mean Score for Each Variable
		6 Months to 5 Years	5 Years Above to 10 Years	10 Years Above	
Vr.1	Presence of regional language facility in accounting software makes its use more convenient.	3.69	4.85	4.35	4.30
Vr.2	Presence of regional language in accounting software fastens the accounting process	3.6	4.62	4.4	4.21
Vr.3	Preference of regional language over English language for better internal communication	3.43	4.57	4.21	4.07
Vr.4	Preference of regional language over English language for better external communication	4.06	4.21	3.95	4.07
Vr.5	Presence of regional language facility in accounting software helps auditor to retrieve classified information	4.08	4.31	4.05	4.15
Vr.6	Business documents prepared in regional language have a positive impact on business communication with supplier.	4.12	4.45	4.2	4.25
Vr.7	Invoices prepared in regional language is more acceptable to customers over invoices prepared in English	3.33	4.37	4	3.9
Vr.8	Quotations prepared in regional language is more desirable to customers over quotations prepared in English	4.30	4.37	4.3	4.32
Vr.9	Information about customer reward, payback terms or any incentive given in regional language has more coverage of customers than if it is mentioned in English.	3.75	4.74	4.2	4.23

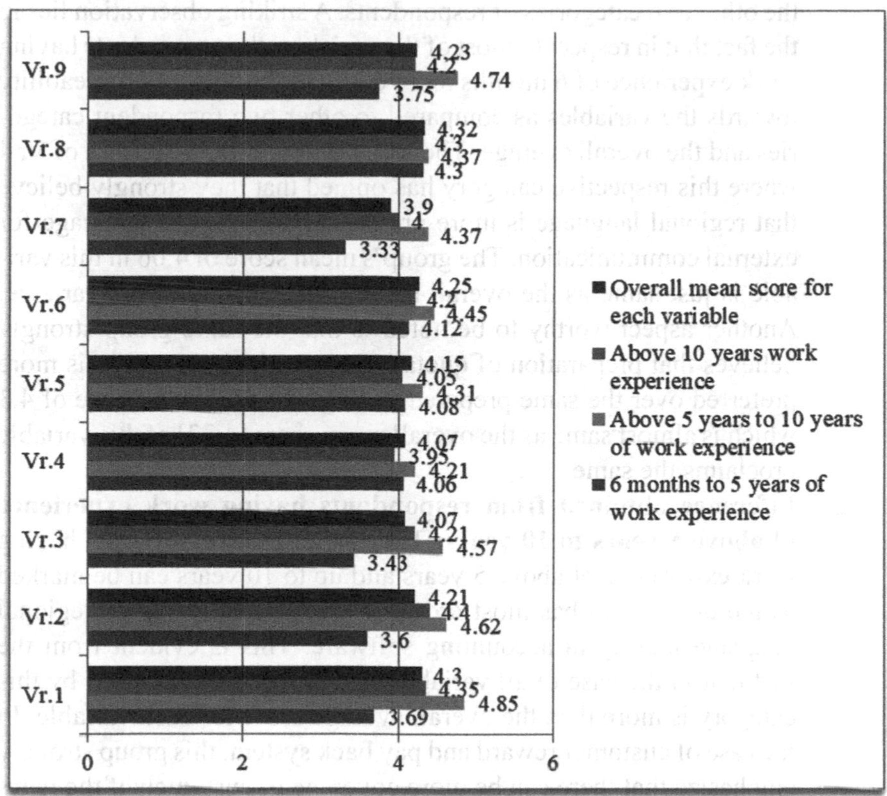

FIGURE 24.1 Mean score obtained by each category of respondents and overall mean score for each variable.

favorability of every respondent category towards the utility of regional language facility in accounting software. The mean score pertaining to each variable is found to be approx. 4 in Likert scale of 1 to 5. This statistic is a vivid representation of the agreeability for utility of regional language in accounting software. A deeper insight regarding the opinion of each category of respondents as mentioned in Table 24.1 is summarized as follows:

1. **Inference obtained from respondents having work experience of 6 months to 5 years:** The respondents having work experience of 6 months to 5 years have opined that most of them believe that presence of regional language in accounting software makes the accounting procedure convenient for them. However, the degree of agreeability of this category of respondents [3.69] regarding this issue is less as compared to the overall average [4.3] as well as from

the other two categories of respondents. A striking observation lies in the fact that in respect to most of the variables the respondents having work experience of 6 months to 5 years have shown less agreeability towards the variables as compared to other two respondent categories and the overall average. The exception is only in the case of Vr.4 where this respective category has opined that they strongly believe that regional language is more preferred over English language for external communication. The group's mean score of 4.06 in this variable is just same as the overall mean score of 4.07 of the variable. Another aspect worthy to be noted is that the same group strongly believes that preparation of quotations in regional language is more preferred over the same prepared in English. The mean score of 4.3 which is almost same as the overall mean score [4.32] of this variable proclaims the same.

2. **Inference obtained from respondents having work experience of above 5 years to 10 years:** The category of respondents having work experience of above 5 years and up to 10 years can be marked as the group who has most ardently favored the utility of regional language facility in accounting software. This is evident from the fact that in the case of all variables the mean score obtained by this category is more than the overall average score for each variable. In the case of customer reward and pay back system, this group strongly emphasize that there can be more coverage of customers if the news about such rewards and its processing are done in regional language. The mean score of 4.74 which is almost 0.51 points more than the overall average proclaims the same.
3. **Inference obtained from respondents having work experience of above 10 years:** The most experienced working group of the research also strongly agrees that there is high utility of regional language in accounting software. The mean scores of such respondents are in almost parlance with the overall mean score for each response.

24.7 FINDINGS

To put it in a phrase, the utility of regional language facility in accounting software is immense as almost all the respondents have strongly agreed to the viability of the variables. However, a fair amount of diversity in responses is shown by the respondent categories. To highlight the reason for such diversity the findings have been categorized for each category of respondents:

1. **Findings pertaining to respondents having work experience of 6 months to 5 years:** It is evident from the numerical analysis that the respondents having work experience between 6 months and 5 years agree to a less degree to all the factors relating to the utility of regional language facility in accounting software in comparison to the other two categories. The primary reason for such result can be attributed to the fact that most of the respondents in this category had been taught in English during their academic life and are well versed in the language. They feel more convenient in using the accounting software in English because they are habituated with accounting terms in English language and find it difficult to prepare bills and quotations in regional language. In their feedback regarding the potential problems in using regional language while accounting they have opined that understanding business documents in regional language becomes complicated as most of them are not aware of the accounting terms in vernacular. However, when the question of external communication and preparation of quotations and invoices in regional language arise, even this group consider that regional language is more preferred. This is owing to the fact that even though the users of the accounting software in this category are comfortable in English, but the persons studying the accounting statements thus produced are more comfortable if such statements are reproduced in regional language.
2. **Findings pertaining to respondents having work experience of above 5 years to 10 years:** The respondents having work experience of above 5 years and up to 10 years strongly agree to the utility of regional language in accounting software. This category is more experienced than the previous one and has naturally dealt with more customers. Most of the respondents of this category had their medium of education in regional language and are less tech-savvy as compared to the earlier category. They not only find English language as complex but the entire process of computerized accounting is inconvenient for them. The main reason for such a scenario is because of the fact that they had less access to computer and computing terms during their school and college life. This category naturally works faster if regional language facility in accounting software is offered to them and both internal and external communication in regional language becomes easier. The group also believes that customers when explained accounting information in a regional language are more likely to favor the company and develop reliability towards

the same. This group believes that regional language facility in accounting software not only increases customer base but also helps in the company's brand creation and development by gaining customer loyalty.

3. **Findings pertaining to respondents having work experience of above 10 years:** The most experienced group of respondents also shows a high affinity towards presence of regional language facility in accounting software. Their opinion greatly matters as they have handled more clients than the previous one and have a greater degree of understanding of the nuances of accounting owing to their work experience. This group strongly believes the necessity of regional language facility in accounting software but in no way underestimates English language. They have handled foreign clients in addition to domestic clients and believe that both English and regional language facility in accounting software is necessary as that would make it regionally viable and internationally acceptable. The reason for such a perspective is due to the fact that this group of respondents have become well versed in English language during their tenure of work as well as possess the knowledge of regional language. The group's client and colleague base has more diversity than the previous two groups. To them business communication and customer satisfaction can be done both by regional language and English depending on the sphere of accounting concerned.

24.8 RECOMMENDATION

The role of regional language facility in accounting software is huge as all the respondents have strongly agreed upon its viability. However, for software to sustain in the global business scenario, accounting only in regional language cannot suffice. For operational efficiency presence of regional language facility is a must but for global expansion English which is the most widely practiced professional language is required.

The feedback of the respondents indicates that professional accounting terms [like Balance Sheet, print, etc.], are better understood in English than in vernacular. Under such a background it is highly desirable that accounting software have mixed language facility where professional accounting terms are present in English and the same has been transliterated in regional language.

Owing to the perspective of business expansion on an international basis, the accounting software demand the presence of both English and regional language facility simultaneously. Such facility will increase operational efficiency as well as will provide impetus for global outreach. Accounting information as produced by the accounting professionals are often found to be difficult to decipher for the stake holders who does not have much of accounting knowledge. To tackle the situation, it is an urgent need that the annual reports and other statements be prepared both in English and regional language to have more clarified information.

Regional language facility helps in better internal communication pertaining to break up of accounting process. This need vividly promotes the utility of regional language facility in accounting software. On a conclusive note, the research reveals that in order to have a more clarified accounting process regional language facility is absolute necessity but must be precluded by the presence of English language in the same.

The entire recommendation can be summarized through the following model of existing and proposed dimensions of accounting software as shown in Figure 24.2.

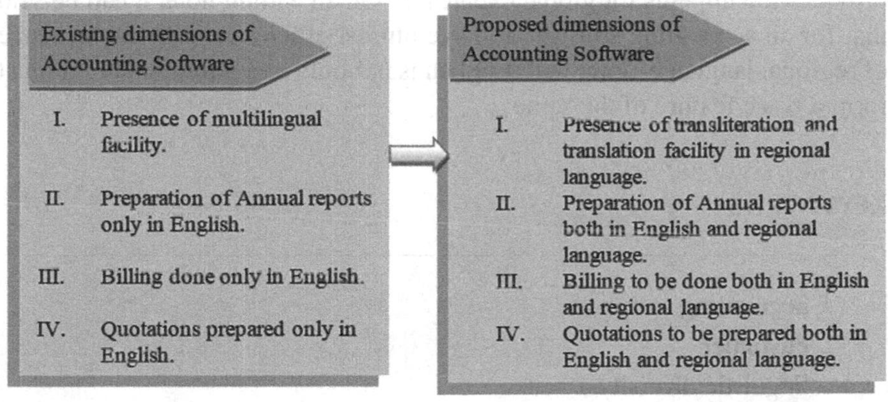

FIGURE 24.2 Model for existing and proposed dimensions of accounting software.

24.9 CONCLUSION

The accounting scenario of India is a massive area with the presence of a varied level of linguistic diversity. Added to this it must be understood that the Indian sub-continent is home to 700 languages with linguistic diversity present in every 15 to 20 kms. For all the stakeholders of the accounting process interpretation of financial information in the single language of English is indeed ungainly considering the fact that only 10% of the total

Indian population is fluent with the language. This proclaims the need for regional language facility in accounting software. However, it must be understood that professional accounting terms are better understood in English than in regional language. This very reason is worthy enough to convey that not only regional translation but regional transliteration is equally important in accounting software.

For any stake holder of accounting process, the financial information is very relevant and is the best reference for decision making. This rationale intensely justifies the clarity of accounting information. For a non-English speaker the clarity of accounting information will be more when such accounting information is reproduced in vernacular. In the light of such significant reason the utility of regional language facility in accounting software is noteworthy. However, it must be kept in mind that in the necessity of the presence of regional language facility in accounting software the importance of English language in the same cannot be underestimated. Operational efficiency in a multi linguistic country like India obviously demands regional language facility but the presence of English alongside provides the impetus for global expansion. On an ending note, it can be said that for an accounting software to have utmost efficiency both the presence of regional language along with English is not only a requirement but it must form a basic feature of the same.

KEYWORDS

- **accounting software**
- **customer**
- **linguistic diversity**
- **multi-language**
- **regional language**
- **retail market**

REFERENCES

Aryan, L., (2017). *The Role of Accounting Information Quality in Management*. ResearchGate.

Aryan, L., Alrabei, A., & Haija, A., (2016). *The Role of Accounting Information Quality in Enhancing Cost Accounting Objectives in Jordanian Industrial Companies*. ResearchGate.

Census Report of India, (2011). https://censusindia.gov.in/2011-Common/Archive.html.

Danet, B., & Herring, S., (2013). *The Multilingual Internet: Language, Culture, and Communication Online*. Oxford University Press, ISBN: 978-0195304794.

Economic Times, (2017). *Indian Retail Market Expected to Reach $1 Trillion by 2020*. https://economictimes.indiatimes.com/industry/services/retail/indian-retail-market-expected-to-reach-1-trillion-by-2020/articleshow/61662729.cms.

Farber, B., & McDonnell, (2012). *Public Utility Accounting*/ American Public Power Association.

Garnell, X. (2014). *Multilingual Information Management* (1st edn.). Elsevier, ISBN 97218433 47712.

https://www.cdac.in/index.aspx?id=mlingual_heritage (2019). *Centre for Development of Advanced Computing, Multilingual Computing and Heritage Computing.*

Linguistic Diversity of India, (1969). Sodhganga, Chapter III, pp. 92–122.

Walton, A., (2019). *The Importance of Accounting Software*. Chron.

Wickramasinghe, M., (2017). Impact of accounting software for business performance. *Imperial Journal of interdisciplinary Research, 3*(5). ISSN 24541362.

CHAPTER 25

COMPARATIVE ANALYSIS OF STOREY DRIFT OF A G+9 RC-STRUCTURED BUILDING DUE TO SEISMIC LOADS IN VARIOUS SEISMIC REGIONS

BUDHADITYA DUTTA,[1] ADITYA NARAYAN CHAKRABORTY,[1] AYUSHI SHAH,[1] DEBARKA BRAHMA,[1] AMIT DEB,[2] and TUSHAR KANTI DE[3]

[1]*Undergraduate Student, Department of Civil Engineering, Amity University Kolkata, Kolkata, West Bengal–700135, India, E-mail: bdneil3103@gmail (B. Dutta)*

[2]*Assistant Professor, Department of Civil Engineering, Amity University Kolkata, Kolkata, West Bengal–700135, India*

[3]*Associate Professor and Head of Department, Department of Civil Engineering, Budge Budge Institute of Technology, Budge Budge, West Bengal–700137, India*

ABSTRACT

Seismic loads are one of the important loads that are to be considered while designing any structure, like, buildings, power plants, towers, bridges, etc., irrespective of using a reinforced-concrete frame or steel frame. The Indian subcontinent has a history of highly devastating earthquakes over the years. The main reason behind the high frequency and intensity of earthquakes is the driving of the Indian plate into the Eurasian plate at a rate of 47 mm/year. As per geographical statistics, about 54% of the landmass of

Advances in Data Science and Computing Technology: Methodology and Applications. Suman Ghosal, Amitava Choudhury, Vikram Kumar Saxena, Arindam Biswas, & Prasenjit Chatterjee (Eds.)

India is vulnerable to earthquakes. The Indian Standard code for designing structures against seismic loads is 1893 (Part 1): 2016. As per IS 1893, India can be divided into 4 seismic zone: Zone II, Zone III, Zone IV, Zone V, with Zone V experiencing the highest level of seismicity and Zone II the least. The object of this chapter is to analyze and compare the story drift of a G+9 RC-structured residential building against seismic loads in the 4 different seismic regions. The software used to carry out the analysis is ETABS 2016.

25.1 INTRODUCTION

Seismic tremors cause ground movements in an arbitrary design in all directions having critical flat and vertical ground increasing speeds as function of time. Designs exposed to ground movements react in a vibratory style. The most extreme reaction acceleration during the elastic stage relies upon the normal recurrence of vibration of the design and the scale of the damping. The most extreme inertia loads following up on a structure during a tremor is dictated by duplicating the mass by the acceleration (Bulusu, 2016).

The distinctive parameters like intensity, duration, etc., of seismic ground vibrations depend upon the degree of the earthquake, its depth of focus, distance from epicenter, properties of soil or medium through which the seismic waves travel and the soil strata when the structure stands. The random earthquake motions can be reduced in any three mutually perpendicular directions (Bulusu, 2016). The horizontal direction is normally the prominent, while vertical acceleration is considered in large-span structures (Jangid, n.d.). The response of a structure to the ground depends on the nature of foundation soil, form, material, size, and mode of construction of structures and the duration and characteristics of ground motion.

The following assumptions are considered while designing earthquake-resisting structures (IS 1893 (Part 1), 2016):

- Imprudent ground movements of a tremor are complicated, sporadic in character, changing period and time and of brief length. They may not cause resonance as pictured under consistent sinusoidal excitations, besides in tall constructions established on profound delicate soil.
- During the time of the earthquake, any wind, flood, or waves will not occur.
- For static analysis, elastic modulus of the materials shall be considered, unless otherwise mentioned.

The components in reinforced concrete casing structures that are significant under seismic stacking are the joints of sections and bars or floor slabs, structural walls, spliced areas, shear walls, and coupling beams. All in all, exceptional consideration ought to be taken for cyclic loadings in any place where underlying discontinuity exists. Basic regions for such cases can likewise be the area where there is a chance of plastic hinge formation (Necdet, 2004).

25.2 DEFINITIONS

As per IS 1893 (Part 1): 2016 (IS 1893 (Part 1), 2016):

1. *Design horizontal seismic coefficient,* A_h (Clause 6.4.2) (IS 1893 (Part 1), 2016):

$$A_h = \frac{\frac{Z}{2} * \frac{Sa}{g}}{\frac{R}{I}}$$

   where; Z is the seismic zone factor (IS 1893 (Part 1), 2016); I is the importance factor (IS 1893 (Part 1), 2016); R is the response reduction factor (IS 1893 (Part 1), 2016); $\frac{Sa}{g}$ is the design acceleration coefficient for different soil types, normalized with peak ground acceleration, corresponding to natural period T of structure (IS 1893 (Part 1), 2016)

2. *Design lateral force,* V_B (Clausc 7.2.1) (IS 1893 (Part 1), 2016)

$$V_B = A_h \times W$$

   where; A_h is the design horizontal seismic load (IS 1893 (Part 1), 2016); W is the seismic height of the building (IS 1893 (Part 1), 2016).

25.3 MODEL ANALYSIS

The ETABS 2016 software has been used for 3D model analysis. As per Figure 25.1 – Seismic Zones of India (IS 1893 (Part 1): 2016), the topography of India can be divided into four seismic zones (IS 1893 (Part 1), 2016):

1. **Zone II:** Regions which fall under Low Damage Risk Zone and have Seismic zone factor, Z = 0.1 (IS 1893 (Part 1), 2016).

2. **Zone III:** Regions which fall under Moderate Damage Risk Zone and have Z = 0.16 (IS 1893 (Part 1), 2016).
3. **Zone IV:** Regions which fall under High Damage Risk Zone and have Z = 0.24 (IS 1893 (Part 1), 2016).
4. **Zone V:** Regions which experience highest intensity of earthquakes and have Z = 0.36 (IS 1893 (Part 1), 2016).

A model G+9 RC-structured building is placed at each of these 4 zones keeping the material properties and structural members unchanged, unless needed based on design check results (Table 25.1).

TABLE 25.1 Sites Chosen for Analysis from Different Seismic Zones

Location	Seismic Zone	Z	Soil Type	I	R
Bengaluru	II	0.10	II	1	5
Kolkata	III	0.16	II	1	5
Delhi	IV	0.24	II	1	5
Guwahati	V	0.36	II	1	5

Model characteristics:

- Base of building structure: 20 m × 20 m;
- Number of floors: 10;
- Height of each floor: 3.3 m;
- Height of base floor: 3.5 m;
- Total height of building structure: 33.2 m;
- Aspect ratio: 1.6:1.

Initially, the structure is designed using M25 concrete with 250×400 beams, 450×500 columns and 125 mm thick two-way slabs. Further changes are made as per requirements, based on the stability of beams and columns after the design check, so that the building can attain the minimum stable structure required to resist the seismic loads. The RC designs are made in accordance with IS 456:2000 (IS 456, 2000).

The loads applied are:

- Dead load (DL) – Wall load, floor finish load, self-load (IS 456, 2000);
- Live Load (LL) (IS 456, 2000);
- Earthquake Loads along X-direction (EL_X) and Y-direction (EL_Y) (IS 456, 2000).

As per assumptions, wind load is neglected.

Soil type (taken from Table 4 – IS 1893 (Part 1): 2016).

- **Type I:** Rock or hard soil (IS 1893 (Part 1), 2016);
- **Type II:** Medium soil (IS 1893 (Part 1), 2016);
- **Type III:** Soft soil (IS 1893 (Part 1), 2016).

Storey drift is defined as the difference between the displacement of two consecutive storys (Jean-Marie, Nicolas, and Michael, 2010). Storey displacement is the absolute value of the displacement of the story under the action of lateral forces (Shariff et al., 2019). Clause 7.11.1.1 of IS 1893 (Part 1): 2016 states that story drift shall not exceed 0.004 times the story height (Sanyogita, 2019). Therefore, the maximum story drift permissible is less than (0.004 × 3,300) or 13.2 mm for storys 2 to 10, while, for story 1, it should be (0.004 × 3,500) or 14 mm.

The rotational reaction of building structures during solid ground movements has ended up being the fundamental driver of fractional or complete collapse. As of late various experiments have been done to show the seismic weakness because of building imbalance and mass or firmness abnormality (Figure 25.2) (Georgoussis, Tsompanos, and Makarios, 2015).

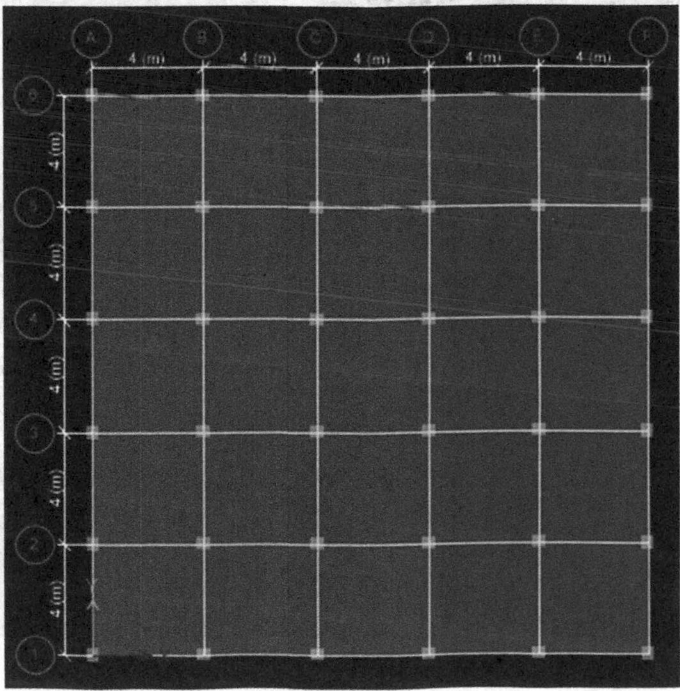

FIGURE 25.1 Plan of model structure

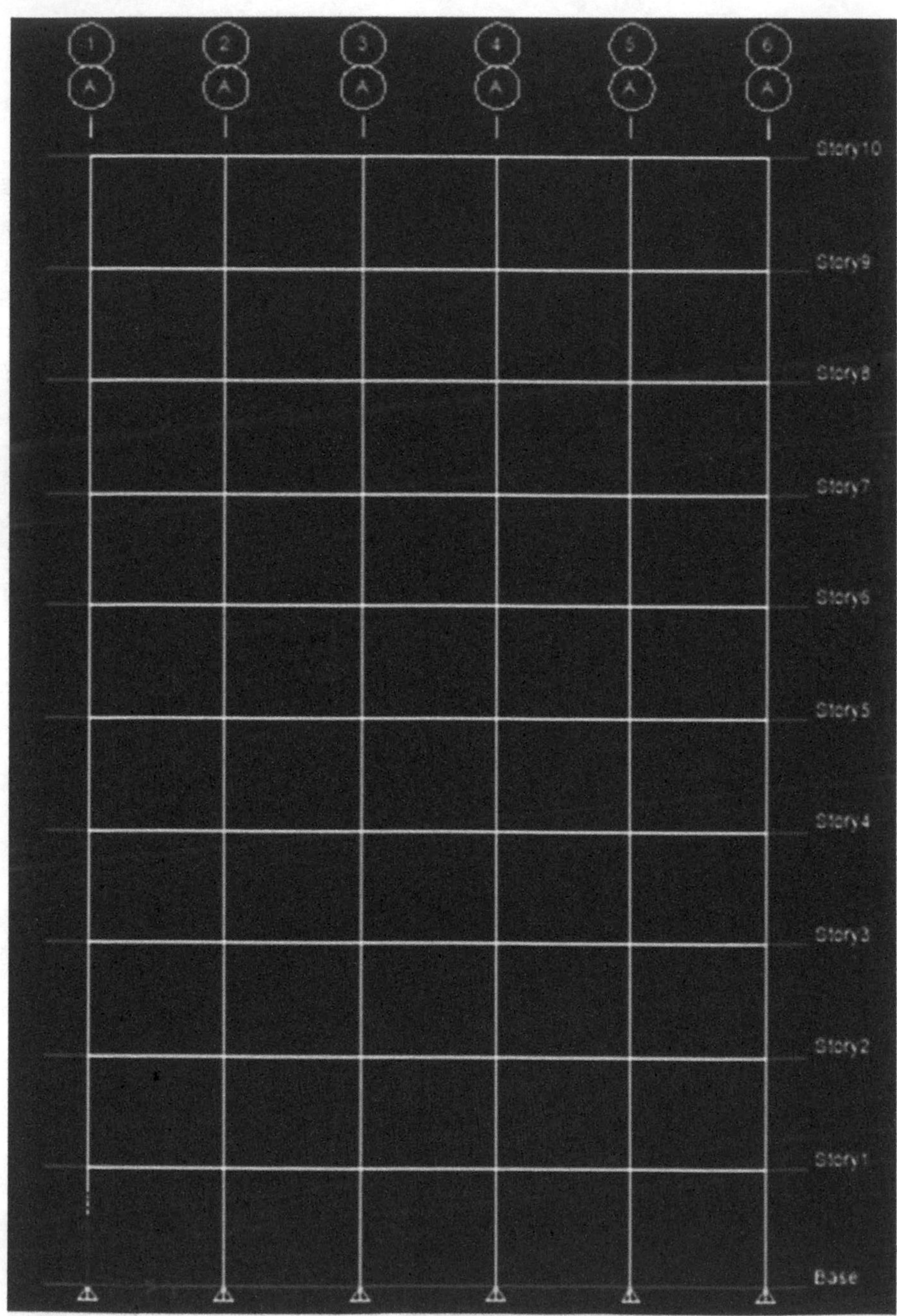

FIGURE 25.2 Elevation of model structure.

25.4 LOAD COMBINATIONS

As per Clause 6.3.1 of IS 1893 (Part 1) 2016, the following load combinations are to be considered for designing a seismic structure (IS 1893 (Part 1), 2016):

- 1.2[DL + IL ± (EL_X ± 0.3EL_Y)] and 1.2[DL + IL ± (EL_Y ± 0.3EL_X)] (IS 1893 (Part 1), 2016);
- 1.5[DL ± (EL_X ± 0.3EL_Y)] and 1.5[DL± (EL_Y ± 0.3EL_X)] (IS 1893 (Part 1), 2016);
- 0.9[DL ± (EL_X ± 0.3EL_Y)] and 0.9[DL± (EL_Y ± 0.3EL_X)] (IS 1893 (Part 1), 2016).

25.5 RESULTS OF THE ANALYSIS

1. Location: Bengaluru (Figures 25.3 and 25.4)

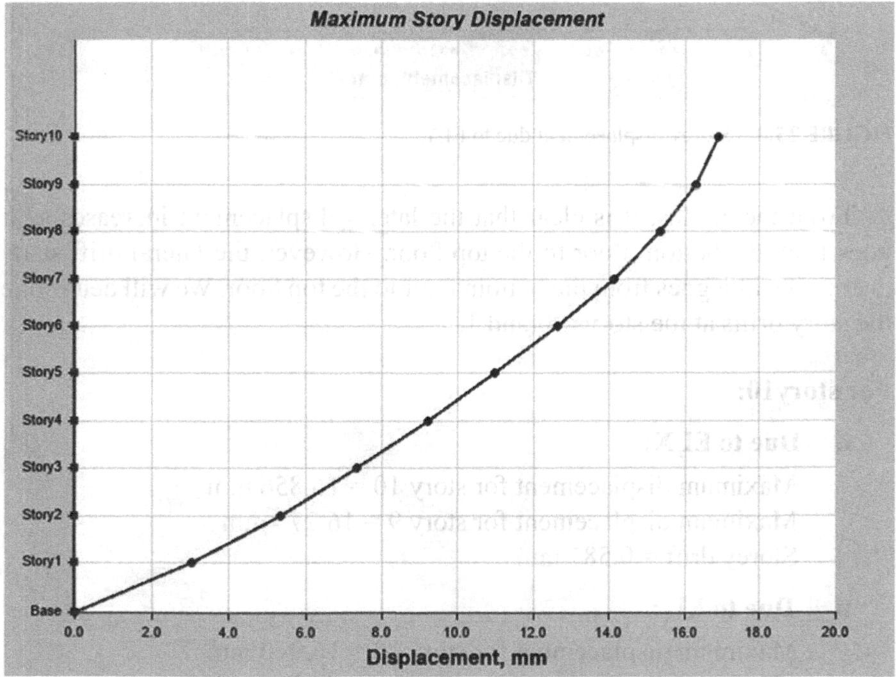

FIGURE 25.3 Storey displacement due to EL_X.

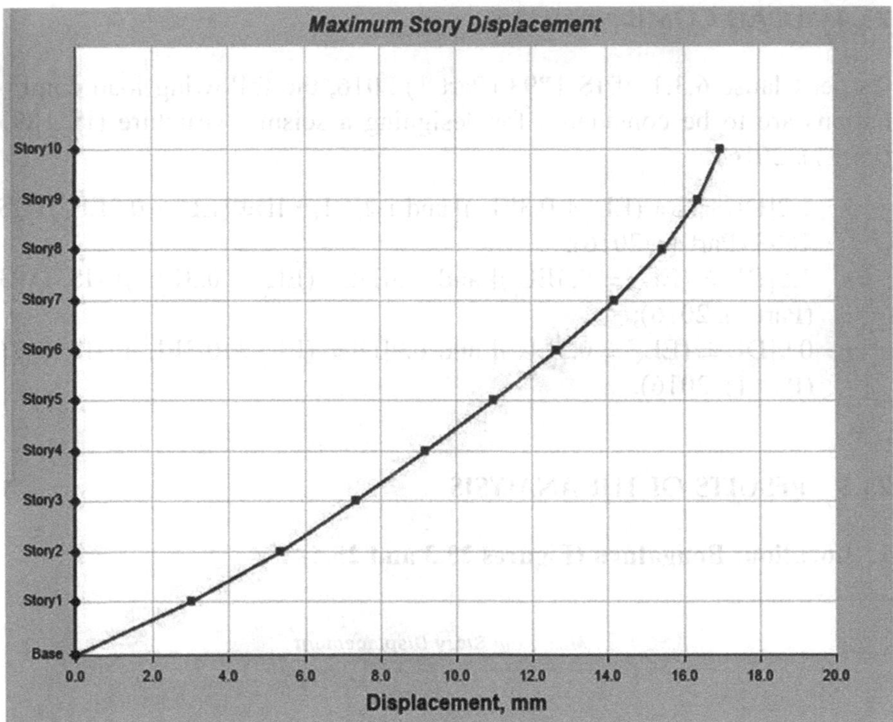

FIGURE 25.4 Storey displacement due to ELY.

From the graphs, it is clear that the lateral displacement increases as it goes from the bottom floor to the top floor. However, the lateral drift starts decreasing as it goes from the bottom floor to the top floor. We will determine the story drifts at the storys 10 and 1.

For story10:

i. **Due to ELX:**

Maximum displacement for story 10 = 16.856 mm
Maximum displacement for story 9 = 16.275 mm
Storey drift = 0.581 mm

ii. **Due to EL_Y:**

Maximum displacement for story 10 = 16.970 mm
Maximum displacement for story 9 = 16.381 mm
Storey drift = 0.589 mm

For story 1:

i. **Due to EL_X:**

Maximum displacement for story 1 = 3.051 mm
Maximum displacement for base = 0 mm
Storey drift = 3.051 mm

ii. **Due to EL_Y:**

Maximum displacement for story 1 = 3.042 mm
Maximum displacement for base = 0 mm
Storey drift = 3.042 mm

As, the story drifts in both the floors, are less the maximum permissible drift, for loads acting from both directions, we can consider the structure to be safe.

2. Location: Kolkata (Figures 25.5 and 25.6)

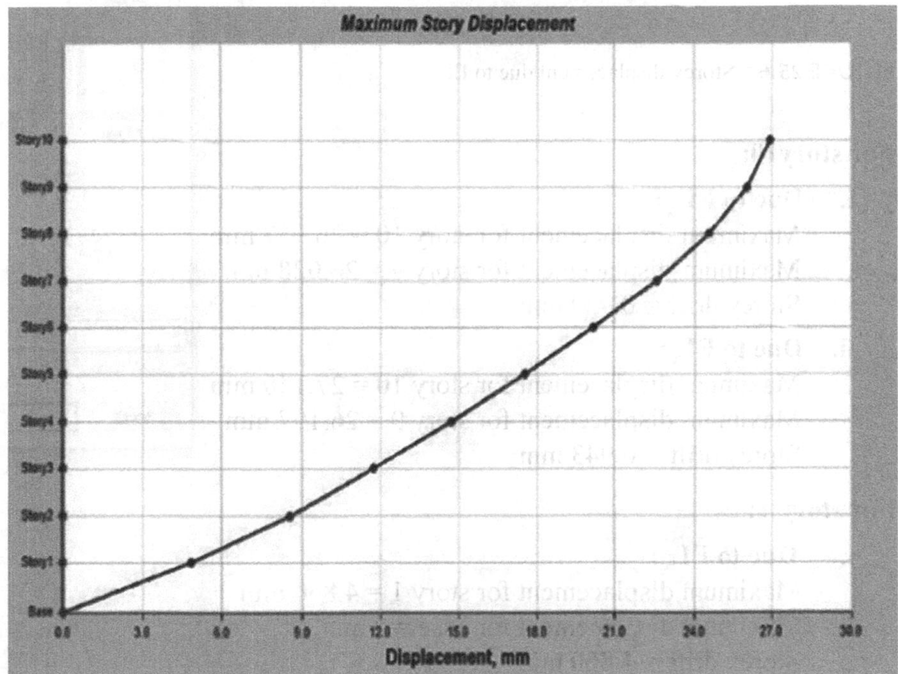

FIGURE 25.5 Storey displacement due to EL_X.

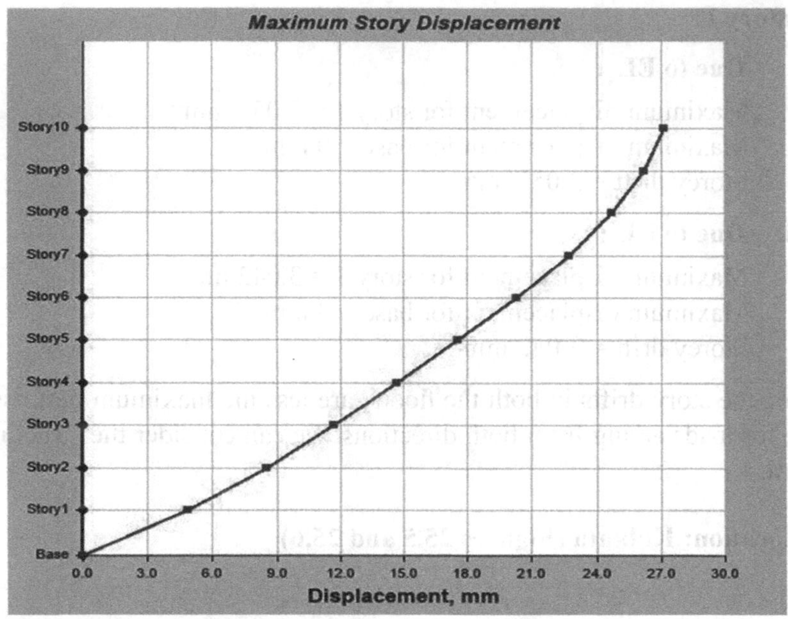

FIGURE 25.6 Storey displacement due to EL_Y.

For story10:

i. **Due to EL_X:**
Maximum displacement for story 10 = 26.957 mm
Maximum displacement for story 9 = 26.028 mm
Storey drift = 0.929 mm

ii. **Due to EL_Y:**
Maximum displacement for story 10 = 27.140 mm
Maximum displacement for story 9 = 26.197 mm
Storey drift = 0.943 mm

For story 1:

i. **Due to EL_X:**
Maximum displacement for story 1 = 4.860 mm
Maximum displacement for base = 0 mm
Storey drift = 4.860 mm

ii. **Due to EL_Y:**
Maximum displacement for story 1 = 4.846 mm
Maximum displacement for base = 0 mm
Storey drift = 4.846 mm

As, the story drifts in both the floors, are less the maximum permissible drift, for loads acting from both directions, we can consider the structure to be safe.

3. Location: Delhi (Figures 25.7 and 25.8)

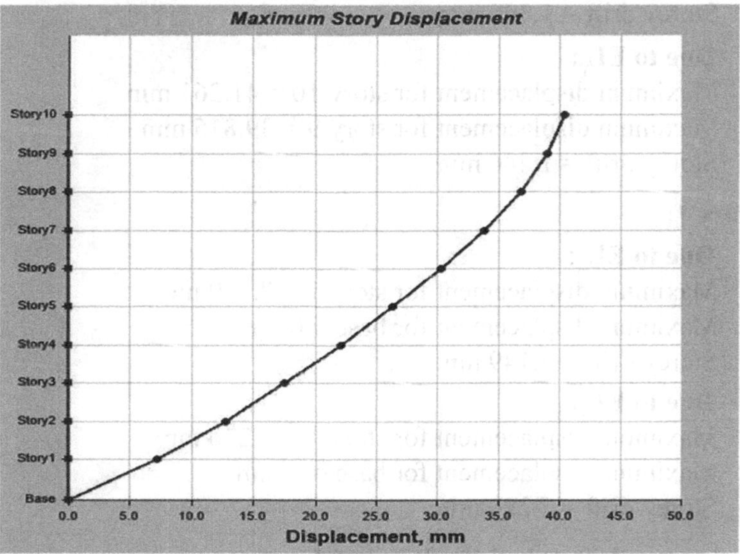

FIGURE 25.7 Storey displacement due to EL_X.

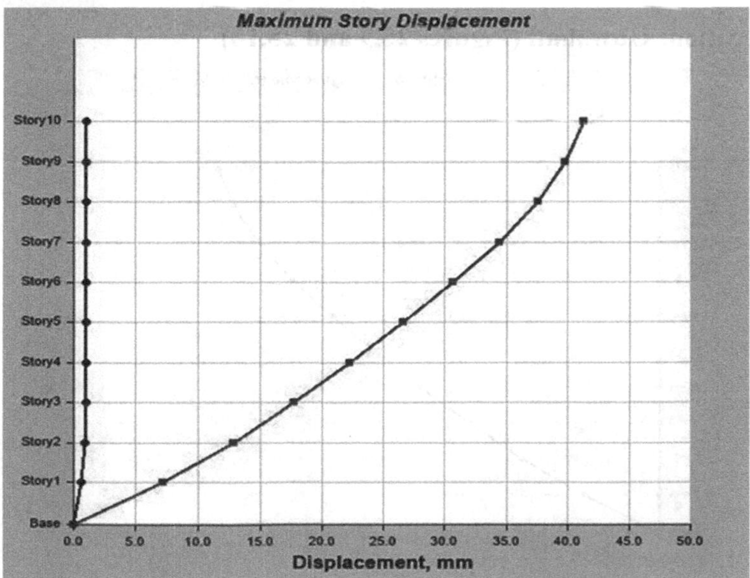

FIGURE 25.8 Storey displacement due to EL_Y.

For story 10:

i. **Due to EL_X:**
Maximum displacement for story 10 = 40.410 mm
Maximum displacement for story 9 = 39.011 mm
Storey drift = 1.399 mm

ii. **Due to EL_Y:**
Maximum displacement for story 10 = 41.261 mm
Maximum displacement for story 9 = 39.815 mm
Storey drift = 1.446 mm

For story 1:

i. **Due to EL_X:**
Maximum displacement for story 1 = 7.149 mm
Maximum displacement for base = 0 mm
Storey drift = 7.149 mm

ii. **Due to EL_Y:**
Maximum displacement for story 1 = 7.220 mm
Maximum displacement for base = 0 mm
Storey drift = 7.220 mm

As, the story drifts in both the floors, are less the maximum permissible drift, for loads acting from both directions, we can consider the structure to be safe.

4. Location: Guwahati (Figures 25.9 and 25.10)

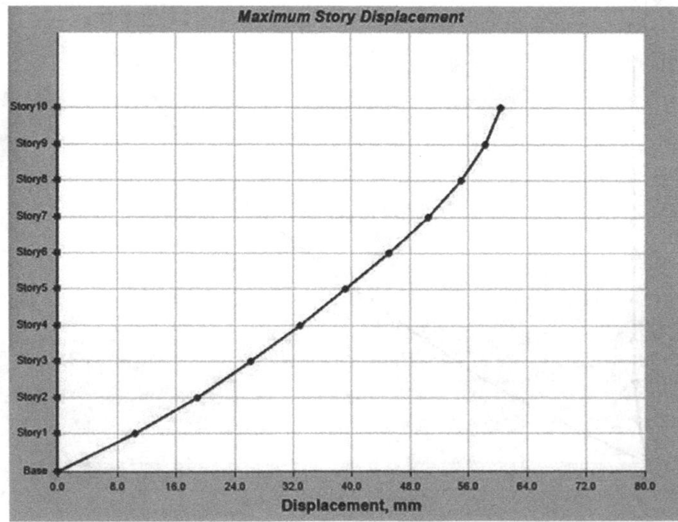

FIGURE 25.9 Storey displacement due to EL_X.

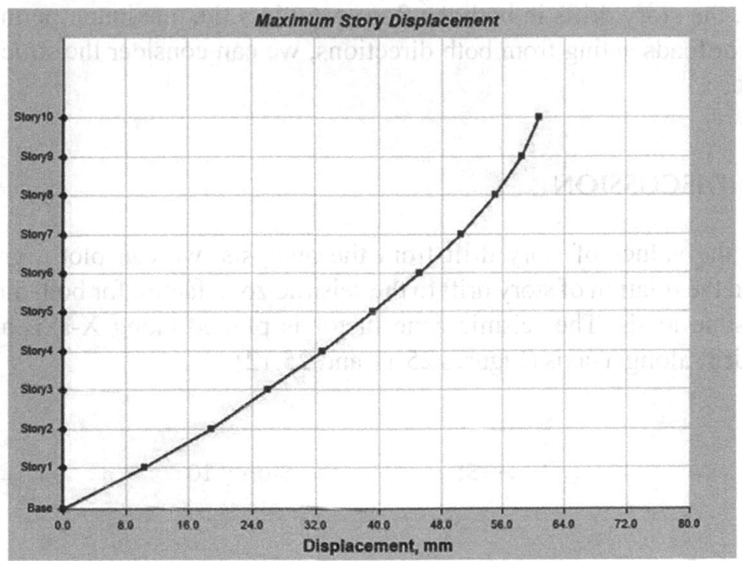

FIGURE 25.10 Storey displacement due to EL_Y.

For story 10:

i. **Due to EL_X:**

Maximum displacement for story 10 = 60.465 mm
Maximum displacement for story 9 = 58.358 mm
Storey drift = 2.107 mm

ii. **Due to EL_Y:**

Maximum displacement for story 10 = 60.662 mm
Maximum displacement for story 9 = 58.520 mm
Storey drift = 2.142 mm

For story 1:

i. **Due to EL_X:**

Maximum displacement for story 1 = 10.430 mm
Maximum displacement for base = 0 mm
Storey drift = 10.430 mm

ii. **Due to EL_Y:**

Maximum displacement for story 1 = 10.423 mm
Maximum displacement for base = 0 mm
Storey drift = 10.423 mm

As, the story drifts in both the floors, are less the maximum permissible drift, for loads acting from both directions, we can consider the structure to be safe.

25.6 DISCUSSION

Using the values of story drift from the analysis, we can plot a graph to explain the relation of story drift to the seismic zone factor, for both direction of seismic loads. The seismic zone factor is plotted along X-axis and the story drift along Y-axis (Figures 25.11 and 25.12).

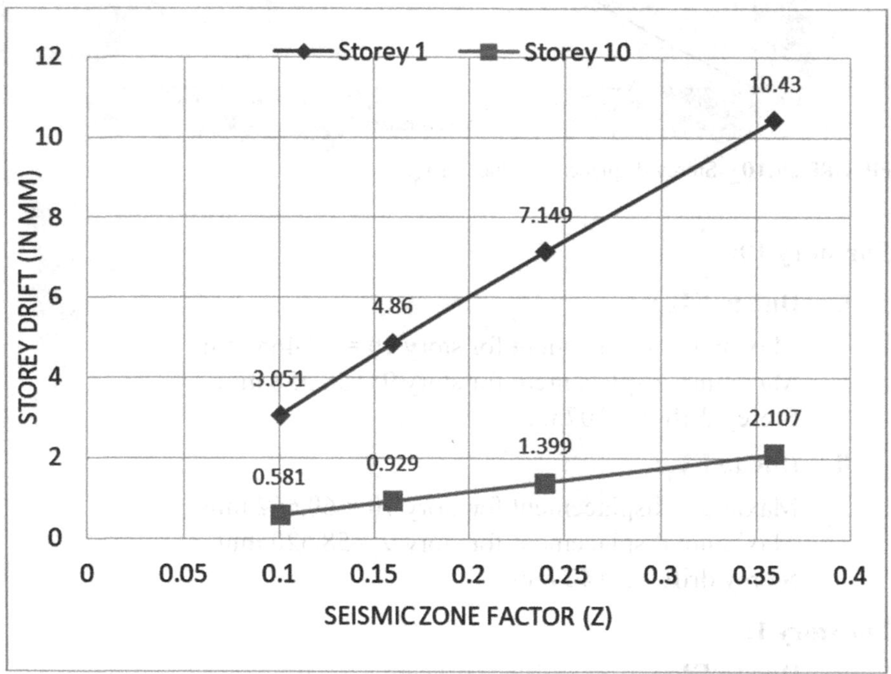

FIGURE 25.11 Seismic zone factor v/s story drift graph for storys 1 and 10 due to seismic loads along X-direction.

From the graphs it is observed that the slope for story drift curve for story 1 is much greater than that for story 10. This explains that the inertia forces acting along the upper part of the structure is greater than that acting along the lower part. This can cause shearing of the structure, which may result in concentration of stresses along the walls or joints, leading to failure or perhaps total collapse. Hence, measures should be taken to reduce the impact of vibration on the lower floors due to an earthquake.

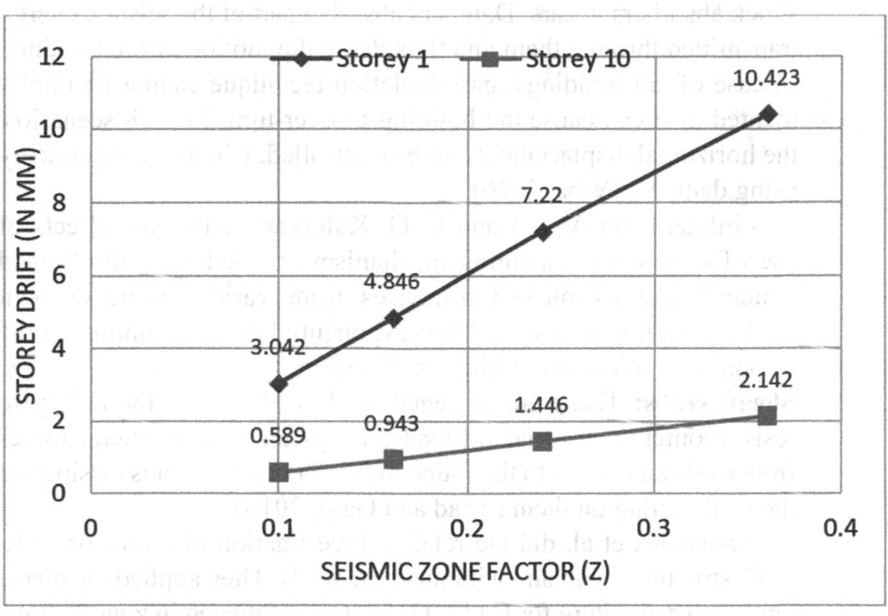

FIGURE 25.12 Seismic zone factor v/s story drift graph for storys 1 and 10 due to seismic loads along Y-direction.

Measures to reduce the impact of seismic vibration:

1. **Base-Isolation:** Base isolation is perhaps the most well-known methods for securing a construction against quake vibration. It is an assortment of primary components which ought to significantly decouple a superstructure from its base, laying on a shaking ground, accordingly ensuring the underlying uprightness. Base isolation is quite possibly the most significant assets of tremor designing relating to the passive underlying vibration control advances. The seclusion can be obtained by the utilization of different procedures, e.g., rubber bearings, friction bearings, ball bearings, spring frameworks and similar different methods. A versatile base separation framework incorporates a tunable isolator that can change its properties dependent on the contribution to limit the moved vibration (Behrooz, Wang, and Gordaninejad, 2014; Seismic Base Isolation, n.d.).

2. **Seismic Dampers:** Seismic performance of buildings can also be improved by installing seismic dampers in place of structural elements, such as diagonal braces. These dampers act like hydraulic shock absorbers in cars. Dampers absorb a part of the seismic energy transmitted through them and thus damp the motion of the building. In case of tall buildings, base-isolation technique cannot be implemented as it can cause the building to over-turn. In such scenarios, the horizontal displacement can be controlled, which can be done by using dampers (Vrba, 2020).

   Firdous Patel Anjum and L. G. Kalurkar studied the effects of use of exponential damping mechanism coupled with RC-framed structure and established responses from various finite element models and it was observed that exponential dampers improved the responses (Firdous and Kalurkar, 2018).
3. **Shear Walls:** These are stiffened walls which have the ability to resist moment due to lateral loads. They can transfer lateral forces from roofs and floors to the foundation of a building, thus dissipating the loads acting on them (Azad and Gani, 2016).

   Aainawala et al. did the relative investigation of multistoried R. C. C. structures with and without shear walls. They applied the quake burden to a structure for G+12, G+25, G+38 situated in zone II, zone III, zone IV, and zone V for various instances of shear wall position. They determined the lateral uprooting and story drift in all the cases. It was seen that multistoried R. C. C. structures with shear wall is prudent when contrasted with the one without shear wall. According to examination, it was concluded that displacement at various levels in multistoried structure with shear walls is similarly lesser unlike the R. C. C. building without shear wall (Aainawala and Pajgade, 2014).
4. **Moment-Resisting Frames:** The column-beam joints in a moment-resisting frame is capable of handling both shear and moment. Hence, they eliminate the space limitations of solid shear walls or braced frames. The column-beam joints are designed, taking into consideration stiffness, yet allow for some deformation for energy dissipation, taking advantage of ductility of steel.
5. **Energy-Dissipating Devices:** Energy-dissipating devices are used to reduce shaking of the building structure, which occurs when the building is made more resistive and may damage the vital portions of the building. Energy dissipates when ductile materials deform in a controlled way (Necdet, 2004).

Structural and architectural detailing, along with construction quality control are important factors to guarantee ductility and natural damping. They also limit the damages to a repairable range. The prospect of structural and non-structural damage is not avoidable without the practical use of energy-dissipating devices.

6. **Fiber-Reinforced Concrete:** This can be instrumental in reducing bond degradation within such critical regions of a reinforced concrete frame. Shear strength of reinforced concrete members can be increased by the addition of fiber confinement. There is a strong case for considering fiber reinforced concrete as an alternative to conventional concrete in reinforced concrete frame structures located in severe seismic zones (Patnaik and Adhikari, 2012).

   An increase of 30% in the punching resistance of slab-column connection as well as improved ductility of the connection was recorded by Alexander and Simmonds (Alexander and Simmonds, 1992) due to the addition of corrugated steel fibers.

25.7 CONCLUSION

From the story drift plots, it can be observed that, taking the conditions of soil type, importance factor and response reduction factor as constant, for a structure studied at different seismic zones, the drift decreases from the bottom floors to the top. This is opposite to the trend exhibited by story displacement. Storey displacement is maximum at the top floor, decreases towards the lower floors and ultimately becomes null at base or ground level. Hence, we can conclude that story displacement is directly proportional to the height of a building, while story drift is inversely proportional to the height. Storey drift is greater at the bottom as the bottom floors undergo lesser forces of inertia acting on them, than the higher floors, due to seismic vibration. Measures should be taken to keep the impact of seismic loads on the lower floors in check, in order to set up a stable structure. The various approaches that can be taken to reduce seismic impact are base-isolation technique, installation of seismic dampers, setting up shear walls, using moment-resisting frame, energy-dissipating devices, etc. The story drifts of the storys 1 and 10 of the model structure do not exceed the maximum drift limit in all 4 seismic zones, as specified by IS 1893 (Part 1): 2016. Hence the structures can be considered safe.

The critical bending moment of irregular frames is more than the customary casing for all structure statures. This is because of abatement in firmness of building outlines because of setbacks. Subsequently, there is need for giving greater fortification to irregular frames (Avantika, Vinayak, and Valsson, 2019).

KEYWORDS

- **reinforced concrete frame**
- **seismic load**
- **seismicity**
- **story drift**

REFERENCES

Aainawala, M. S., & Pajgade, P. S., (2014). "Design of multistoried R.C.C. buildings with and without shear walls." *International Journal of Engineering Sciences & Research Technology, 7*(3), 498–510. ISSN: 2277-9655.

Alexander, S. D. B., & Simmonds, S. H., (1992). Punching shear tests of concrete slab-column joints containing fiber reinforcement. *ACI Structural Journal.*

Avantika, H. D., Vinayak, D. V., & Valsson, V., (2019). "Effect of stiffness of different shapes of building to seismic zones." *International Journal of Scientific Research and Engineering Development, 2*(2).

Azad, Md. S., & Gani, S., (2016). "Comparative study of seismic analysis of multistory buildings with shear walls and bracing systems." *International Journal of Advanced Structures and Geotechnical Engineering, 5*, 72–77.

Behrooz, M., Wang, X., & Gordaninejad, F., (2014). Performance of a new magnetorheological elastomer isolation system. *Smart Materials and Structures, Smart Materials and Structures, 23*(4).

Bulusu, U., (2016). "*Earthquake Resistant Design of Structures.*" https://www.linkedin.com/pulse/earthquake-resistant-design-structures-udaya-bhaaskar-bulusu-1 (accessed on 08 December 2021).

Firdous, P. A., & Kalurkar, L. G., (2018). "Performance of R.C. frames coupled with exponential dampers under seismically triggered condition." *International Journal of Research in Engineering and Technology.*

Georgoussis, G., Tsompanos, A., & Makarios, T., (2015). Approximate seismic analysis of multi-story buildings with mass and stiffness irregularities. *Procedia Engineering, 125*, 959–966. 10.1016/j.proeng.2015.11.147.

IS 1893 (Part 1), (2016). *Criteria for Earthquake Resistant Design of structures – General Provisions and Buildings (Sixth Revision).*

IS 456, (2000). Indian Standard Code of Practice for Plain and Reinforced Concrete (Fourth Revision).

Jangid, R. S. (2013). *Lecture Series on Earthquake Engineering.* Department of Civil Engineering, IIT Bombay.

Jean-Marie, R., & Nicolas, I., & Michael, B., (2010). Inelastic seismic analysis of the SPEAR test building. *European Journal of Environmental and Civil Engineering, 14*, 855–867. 10.1080/19648189.2010.9693266.

Necdet, T., (2004). "Seismic isolation and energy dissipating systems in earthquake resistant design." In: *13th World Conference on Earthquake Engineering.* Vancouver, B.C. Canada.

Patnaik, A., & Adhikari, S., (2012). Potential applications of steel fiber reinforced concrete to improve seismic response of frame structures. *NED University Journal of Research.*

Sanyogita, S. B., (2019). Seismic analysis of vertical irregularities in buildings. In: Singh, H., Garg, P., & Kaur, I., (eds.), *Proceedings of the 1st International Conference on Sustainable Waste Management through Design. ICSWMD 2018* (Vol. 21). Lecture Notes in Civil Engineering.

Seismic Base Isolation." (2010). en.wikipedia.org. https://en.wikipedia.org/wiki/Seismic_base_isolation (accessed on 08 December 2021).

Shariff, M., Owais, M., Rachana, C., Vinu, S., & AsishDubay, B., (2019). "Seismic analysis of multistorey building with bracings using ETABS." *International Journal of Innovative Research in Science, Engineering and Technology, 8*(5).

Vrba, M., (2020). *Hydraulic Shock Absorber in Vehicles and Other Applications* [online]. Brno, [cit. 2021-01-29]. http://hdl.handle.net/11012/191756. Bachelor thesis. Technical University Brno. Faculty of Mechanical Engineering. Department of Automotive and Transportation Engineering.

CHAPTER 26

A PROPOSAL FOR DISINFECTION OF CONTAMINATED SUBJECT AREA THROUGH LIGHT

ANUBRATA MONDAL[1] and KAMALIKA GHOSH[2]

[1]*Research Scholar, School of Illumination Science Engineering and Design, (Jadavpur University, Kolkata) and Assistant Professor, Department of Electrical Engineering, Global Institute of Management and Technology, West Bengal, India*

[2]*Director, School of Illumination Science, Engineering and Design, Jadavpur University, Kolkata, West Bengal, India, E-mail: kamalikaghosh4@gmail.com*

ABSTRACT

At present world is passing through a pandemic situation due to sudden ingress of Novel Corona virus. People (including all ages) can be affected by this new type of virus. Till time there is neither vaccination has been invented nor arrested by antibiotics due its unknown characteristics. Many researchers are trying to develop medicines against it. At present self-protection is very much essential to combat with Corona. That is why disinfection/sanitization is one important aspect too. In most places liquid sanitizer like Sodium hypo chloride are in use. But in indoor application liquid sanitizer can affect or deformed the materials. From this point of view one can utilize photo biological effect of light. UV range of the light source has established its ability to destroy bacteria and virus too. Experimentation had been carried out in Illumination Engineering Laboratory of Jadavpur University with several light sources with

Advances in Data Science and Computing Technology: Methodology and Applications. Suman Ghosal, Amitava Choudhury, Vikram Kumar Saxena, Arindam Biswas, & Prasenjit Chatterjee (Eds.)

different intensities of UV contents from light sources to reduce germicidal effects. The process shows a successful result to reduce bacteria and viruses. Same may be utilized for sterilization of indoor areas. But UV has some deleterious effect on human being thus presence of them must be avoided. For earlier studies on other viruses, like H_1N_1, Ebola, etc. UV with proper chemical agents as a catalyst has found positive effect to combat viruses. The above two factors have been discussed in details in this chapter.

26.1 INTRODUCTION

Light is a popular form of energy resource which primarily provides visibility. If an energy resource can be harnessed properly same can be utilized as a weapon for good or bad purpose. At present entire world is in panicky situation due to sudden and wide spread attack of COVID-19 (Zaria, 2020), corona virus. Since it is an airborne virus, entire localized region must be sanitized properly by breaking / weakening the RNA structure of the virus. Chemicals are introducing through spray to combat the same. But in some confined places it is not possible to use the same as it may create damage of the surrounding material, making it wet, at the same time bring permanent health hazard to living beings. This calls for a device which may be used without contaminating the surrounding other living and nonliving objects. In this work effort has been made to develop a system to prevent the environment from the virus as well as a way of survival from the disease too.

26.2 PROPOSED METHODOLOGY

Light is an electromagnetic wave. It is accompanied by ultraviolet ray (UV) (Bolton, 2010) and infrared ray (IR) in Electromagnetic spectrum. Everybody gets scared by the word UV. But as said earlier proper knowledge helps us to harness as well as divert the same in proper way. The above waves have been further classified according to its characteristics and analysis (Table 26.1).

- VUV (Vacuum UV) light is harmful due to its ability of immediate retort organic molecules even at low doses and with oxygen atoms.
- UV-C light has biocide effects and usually utilized as germicidal agent. UVC directly affects deoxyribonucleic acid (DNA) and ribonucleic acid (RNA) by inducing molecular transformation (i.e., producing photoproducts in the genetic material).

- UV-B light: Photons in this range are known for sun burning, leads to photo carcinogenesis and photo aging.
- UV-A light being comparatively longer in wave length has less detrimental effects.

TABLE 26.1 Range of Electromagnetic Spectrum (Bolton, 2010)

Wave Length (nm-Nanometer (10^{-9}), Unless Stated	Name of Wave
100–200 nm	Vacuum UV (VUV)
200–280 nm	UV C
280–315 nm	UV B
315–380 nm	UV A
380–780 nm	Visible light
780 nm–1.4 μm	IR A
1.4 μm–3.0 μm	IR B
3.0 μm–1.0 μm	IR C

During time of survey, it was observed that most of the solid materials like paper documents, textiles materials, etc., becomes brittle in nature and changes its color as it ages. It is well known fact that microorganisms (Kowalski, 2009) may be responsible for the same. Same is also observed in water and air medium which are affected by unwanted growth of microorganism. So, at first identification of the types of micro-organism are needed which are grown on subject materials. If micro-organism is identified by experimentation, then it can be tries to remove or control for safety purpose.

Some microorganisms were identified through microscope in laboratory studies at Jadavpur University (Table 26.2).

As an illumination engineer it is necessary to find out ways and means for destroying microbes like fungi, algae, bacteria, etc., with the application of light. An effort has been made by using UV radiation which is available in artificial light sources specially mercury vapor discharge type light sources.

In this chapter various types of lamps, i.e., incandescent lamp, Conventional Fluorescent Tube, CFL, and LED (both warm and cool white) and UV-C lamps are experimented to find out the UV-C ranges in it (Table 26.3).

Experimentation has been carried out to measure the amount of UV-C of the lamp which is mounted horizontally from top, facing downward. The test results at various distances are as follows (Table 26.4 and Figure 26.1).

TABLE 26.2 Few Common Microorganism and UV Needed to Combat (Philips Catalogue, n.d.)

Microorganism Under Discussion	Dose Needed to Kill with UV (J/m²)
Mold Spores	
Aspergillus niger	1,320
Rhizopus nigricans	1,110
Penicillium digitatum	440
Virus	
MS-2 Coliphase	186
Rotavirus	81
Hepatitis A	73
Polio virus	58
Influenza virus	36
Bacteria	
Salmonella typhimurium	80
Bacillus anthracis	45.2
Vibrio chlolerae	35
Staphylococcus aureus	26

TABLE 26.3 UV-C Output from Various Artificial Light Sources

Distance (ft.)	Lamp Specification with UV-C (Microwatt/cm2) Range						
	Incandescent (40 Watt)	FTL (40 watt)	CFL Cool (8 watt)	CFL Warm (8 watt)	LED Cool (8 watt)	LED Warm (8 watt)	UV-C Lamp (8 watt)
2	2.1	0.9	0.6	0.5	0.3	0.3	91
4	0.9	0.2	0.3	0.3	0.2	0.1	38
6	0.7	0.1	0.2	0.1	0	0	18
8	0.5	0	0.1	0.1	0	0	7

Mathematical modeling through MATLAB simulation can be done as follows:

- **Along the Vertical Axis from Lamp:**
  $UV_{ver} = 584.08e^{-0.454x}$, Where x is the distance of separation.
  Thus the generalized equation can be drawn as $UV_{ver} = A_v \times e^{-b_v x}$.
- **Along the Horizontal Axis from Lamp:**
  $UV_{hor} = 239.27{-}^{0.455x}$, Where x is the distance of separation.
  Thus, the generalized equation can be drawn as $UV_{hor} = A_h \times e^{-b_h x}$.

TABLE 26.4 Distance vs. UV-C Ranges of Test Lamp

Position	SL. No.	Distance (ft.)	UV-C ($\mu W/cm^2$)
Vertical (from the center of the lamp)	1	1	153
	2	2	95
	3	3	57
	4	4	40
Horizontal (from the center of the lamp)	1	1	148
	2	2	91
	3	3	55
	4	4	38
	5	5	27
	6	6	18
	7	7	11
	8	8	7
	9	9	4
	10	10	2

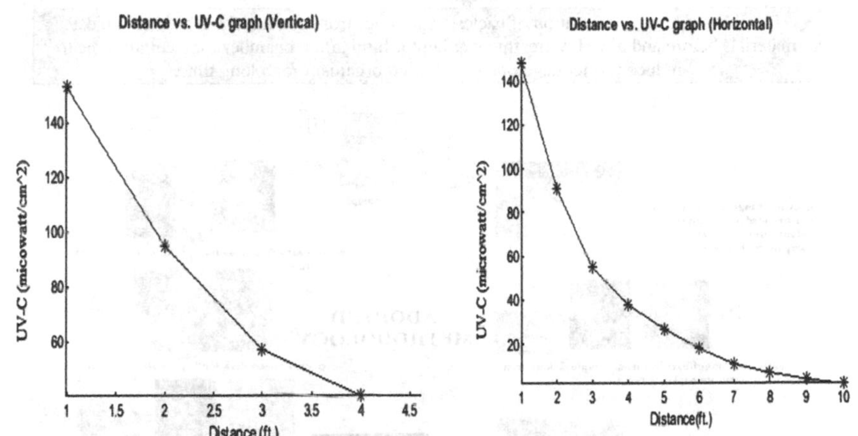

FIGURE 26.1 Nature of curve (decaying) for UV-C with distance.

The nature of decay of UV-C is same and values of "b_v" and "b_h" attained is almost same. Values of "A_v" and "A_h" are not same as the through of light is not the same in all directions due to its linear shape. The intensity of factor "A_v" and "A_h" will vary with the type, size, shape, and intensity of radiations of the lamps.

Thus UV-C lamp employed for this experimental purpose was T5 Slim Line of 4 W, 6,400 K, 306 microwatt/m^2 output.

This experiment attempts to give a brief idea about how UV radiation destroys or control microbes like fungi, algae, and bacteria with specific kind, intensity, and exposure time.

26.3 EXPERIMENTAL STEPS

The experimental steps are shown in Figure 26.2.

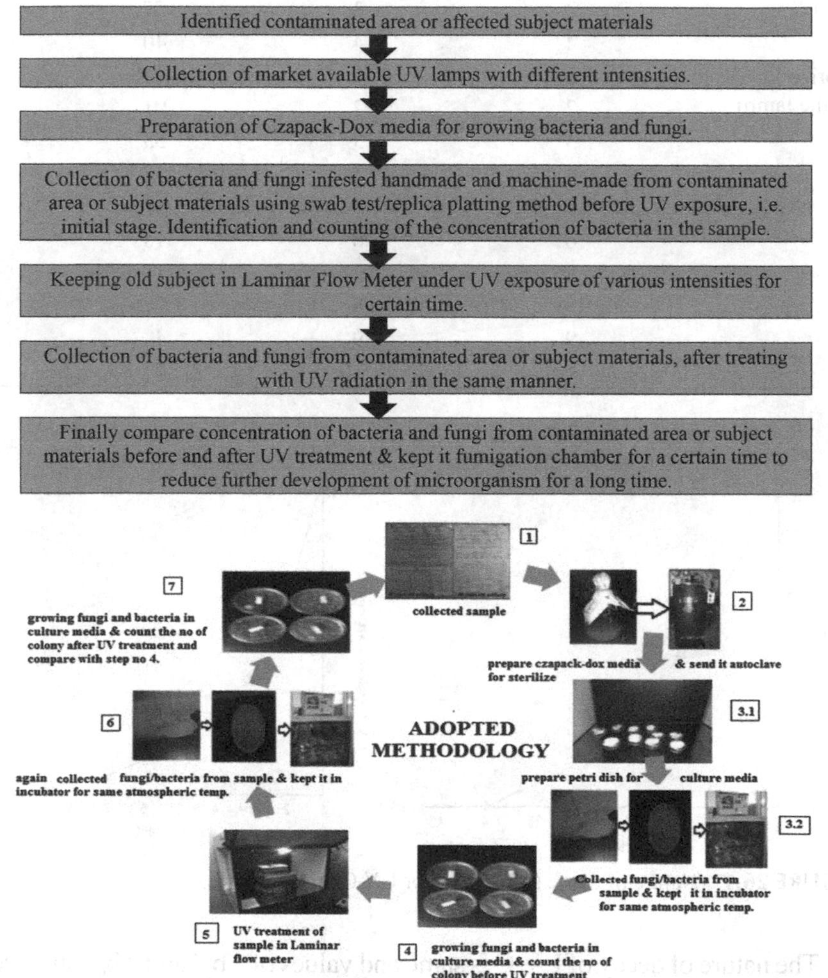

FIGURE 26.2 Procedure of collection of microorganisms and reduction by UV application.

26.4 NECESSARY INSTRUMENTS TO CARRY OUT EXPERIMENT

1. **Laminar Flow Meter:** This apparatus is laboratory-based equipment used for germicidal operation incorporating UV lamps. This

machine is not in use to keep the interior of the work zone clean and decontaminated.

2. **Autoclave:** It is a steam sterilizer machine. It is mainly use steam under pressure to kill harmful microorganism like bacteria, viruses, fungi, and spores. The items are heated to an appropriate sterilization temperature for a specific amount of time.
3. **Incubators:** Incubator is a device which is used to grow and maintain cell or microbiological cultures. This instrument maintains humidity, optimal temperature, and other constraints (Figure 26.3).

FIGURE 26.3 Incubator chamber.

26.5 EXPERIMENTAL DATA

From our earlier laboratory studies at Jadavpur University—under School of Illumination Science and Engineering (SISED) in collaboration with School of Environmental Studies of Jadavpur University upon Bacteria results are as follows:

26.6 OBSERVATIONS

For the above studies it has been observed that as the intensity and duration increase effectiveness of reduction for bacteria's colony increases for both cases:

1. UV Treatment Using Swab Test (Table 26.5):

TABLE 26.5 Result of Swab Test Method (Anubrata and Kamalika, 2019)

Sample	Initial No. of Bacterial Colonies	Intensity of UV lamp	Duration (hr.)	Dose (mJ/cm^2)	Final no. of Bacterial colonies	Reduction (%)	Observations
I.A	15	612	3	1.836	6	60.00*[a]	Time remaining same with double intensity reduction is only 15%
I.A	15	306	3	0.918	8	46.66*[a]	
I.A	60	612	6	3.672	17	71.66*[b]	By increasing the Duration of time with higher intensity further 11.66% improvement
I.A	60	306	6	1.836	31	48.33*[c]	Not much effective with lower intensity
I.B	9	612	3	1.836	3	66.66*[d]	Time remaining same with double intensity reduction is 20% in case of lower initial concentration
I.B	9	306	3	0.918	5	44.44*[d]	
I.B	15	612	6	3.672	4	73.33*[e]	By increasing the Duration of time with higher intensity further 10% improvement
I.B	15	306	6	1.836	7	53.33*[f]	Not much effective with lower intensity
II	17	612	3	1.836	6	64.70*[g]	Time remaining same with double intensity reduction is more than 20% in case of lower initial concentration
II	17	306	3	0.918	10	41.17*[g]	
II	25	612	6	3.672	6	76.00*[h]	By increasing the Duration of time with higher intensity further 11.3% improvement
II	25	306	6	1.836	13	48.00*[i]	Not much effective with lower intensity

2. UV Treatment Using Replica Platting Method (Table 26.6):

TABLE 26.6 Result of Replica Platting Method (Anubrata and Kamalika, 2019)

Sample	Initial no. of Bacterial colonies	Intensity of UV lamp	Duration (hr.)	Dose (mJ/cm²)	Final no. of Bacterial Colonies	Reduction (%)	Observations
I.A	37	612	3	1.836	14	62.16*[a]	Time remaining same with double intensity reduction is only 13.5%
I.A	37	306	3	0.918	19	48.64*[a]	
I.A	60	612	6	3.672	15	75.00*[b]	By increasing the Duration of time with higher intensity further 11.66% improvement
I.A	60	306	6	1.836	28	53.33*[c]	Not much effective with lower intensity
I.B	29	612	3	1.836	10	65.51*[d]	Time remaining same with double intensity reduction is 17% in case of lower initial concentration
I.B	29	306	3	0.918	15	48.27*[d]	
I.B	55	612	6	3.672	12	78.18*[e]	By increasing the Duration of time with higher intensity further 13% improvement
I.B	55	306	6	1.836	25	54.54*[f]	Not much effective with lower intensity
II	21	612	3	1.836	7	66.66*[g]	Time remaining same with double intensity reduction is 20% in case of lower initial concentration
II	21	306	3	0.918	11	47.61*[g]	
II	33	612	6	3.672	7	78.78*[h]	By increasing the Duration of time with higher intensity further 12.12% improvement
II	33	306	6	1.836	16	51.51*[i]	Not much effective with lower intensity

- Time remaining same with double intensity reduction is only 13 to 15%
- By increasing the Duration of time at higher intensity further 12% improvement.
- Not much effective with lower intensity.
- Time remaining same with double intensity reduction is 16 to 20% in case of lower initial concentration.
- By increasing the Duration of time with higher intensity further 10–12% improvement.
- Not much effective with lower intensity.
- Time remaining same with double intensity reduction is only 20%.
- By increasing the Duration of time at higher intensity further 11–12% improvement.
- Not much effective with lower intensity.

26.7 RESULT ANALYSIS

Conventional mercury vapor discharge lamps are rich source of UV. Even newly introduced. LED lamps also content UV-C. Thus, an array of UV lamps can be able to sanitize indoor area. The arrangement can be designed based on the type, size, and density of virus as well as size of the place, etc.

Since UV-C is carcinogenetic for the long time exposure and it can damage eyes restriction of human movement in the place must be maintained.

From our earlier studies it appeared that during ancient time people in Egypt used to treat through Phototherapy by introduction of UV-A and bio extract – Psolaran (Ammi majus). At present synthetically developed Amotosalen (S-59) along with UV may alleviate respiratory suffocation. Here TiO_2 has been reported to act as a catalyst.

From earlier studies it has also been found that UV has been successful upon H_1N_1, Ebola, etc., virus with UV irradiation of 2 mJ/cm^2 of 222 nm and 4 J/m^2 (0 to 30 s) of 254 nm, respectively.

26.8 CASE STUDIES (DAVID, N.D.)

Under the guidance of David Brenner, PhD, Director of the Center for Radiological (Columbia University) believes they may have identified a new, low-cost solution to eradicating airborne viruses in indoor public areas. Here reports has been attached in Table 26.7.

TABLE 26.7 Some Information from the International Study (David, n.d.; Misc, n.d.)

Disease	Causative Agent	Organism Persistence	Symptoms	Effective on Person to Person	Infective Dose (Aerosol)	Incubation Period	Mortality
Anthrax	*Bacillus anthracis*, infected animals are responsible for transmitting to humans.	Life of Spores may be more than 40 years	Fever, cough, discomfort in respiratory track, etc.	No	10,000– 50 000 spores	1–6 d	High if symptoms occur.
Plague (i) Bubonic, (ii) Pneumonic, (iii) Septicemic	*Yersinia pestis*	Can persist in soil for about 24 days.	Cough and cold followed by High fever, Dyspepsia, etc.	Yes	<100 organisms	1–10 bacteria, more for aerosol (100–20,000) Treatment	High, unless treated immediately.
Ebola virus disease (i) Bundibugyo virus, (ii) Ebola virus, (iii) Sudan virus, (iv) Taï Forest virus	Filoviridae family, genus Ebolavirus *caused by* plants, arthropods, birds, etc.	Stable	Intense weakness, muscle pain, headache, sore throat, vomiting, diarrhea, rash, impaired kidney, and liver functions	Yes	i) particles of <5–10 μm ii) droplets of diameters >20 μm	1–21 d, avg. 8–10d	90% fatality

26.9 CONCLUSION

From the above studies it can be inferred that UV can be utilized in controlled manner to sanitize the infected areas and large scale water purification (Adrita et al., 2020) for rural area. Further studies can also be made with Photo therapy along with chemical aids as stated above for the diseases where Antibiotic fails. In this case development of catalyst with the help of upcoming Nanotechnology will be a great solution.

KEYWORDS

- **COVID-19 virus**
- **germicidal effect**
- **photo biological effect**
- **ultraviolet**

REFERENCES

Adrita, S., Antalina, S., Soumya, G., Ratan, M., & Kamalika, G., (2020). "Development of UV aided water purification system powered by Solar energy for remote villages." *International Conference on Sustainable Water Resources Management Under Changed Climate*. Organized by School of Water Resources Engineering, Jadavpur University, Kolkata in Collaboration with Department of Geography, Women College Calcutta and NIT Durgapur.

Anubrata, M., & Kamalika, G., (2019). "Studies on germicidal benefit of ultra violet ray upon old paper documents." *Light& Engineering Journal, 24*(4), 17–24. ([ISSN 0236-2945]. Scopus Indexed & SCI, Peer Reviewed & UGC Approved. (Included UGC Care List: Group-A).

Bolton, J. R., (2010). *Ultraviolet Applications Handbook* (3rd edn.). (Updated) // Bolton Photo sciences Inc.

David, B. *Misc*. publications of PhD, Director of the Center for Radiological Research and Professor at Center for Radiological Research at Columbia University Irving Medical Center. http://www.columbia.edu/~djb3/ (accessed on 11 January 2022).

Kowalski, W., (2009). *Ultraviolet Germicidal Irradiation Handbook UVGI for Air and Surface Disinfection* (p. 501). Springer Heidelberg Dordrecht London New York.

Misc. publications of National Library of Medicine, US. https://collections.nlm.nih.gov/ (accessed on 11 January 2022).

Philips Catalogue, Philips lighting B.V. *UV Health and Wellness*. www.philips.com/uvpurification (accessed on 08 December 2021).

Zaria, G., (2020). BBC survey. "*There's Only One Type that Can reliably Inactive COVID19- and it's Extremely Dangerous.*" https://www.bbc.com/future/article/20200327-can-you-kill-coronavirus-with-uv-light (accessed on 11 January 2022).

INDEX

A

Abnormal
conditions, 198
overheating, 271
Absolute mean brightness error (AMBE), 3, 8, 9, 12, 13, 16
Accelerometer chip, 148
Accounting
information, 350–353, 357, 360
process, 349, 350, 359, 360
professionals, 349, 352, 359
software, 349–352, 355–360
Accuracy, 9, 14, 15, 26, 28, 29, 60, 62–64
handwritten text recognition tool, 20
phase-shift information, 205
value, 19, 29, 32
AC-DC-AC conversion, 260, 262
Achievability, 188
Activation function, 11, 27, 64, 74, 76
Actual
gradient value, 24
non-vessel pixels, 9
raw pixels, 60
vessel pixels, 9
Adagrad, 24, 28–32
Adam optimizer, 24, 28, 74
Adaptive
cruise control (ACC), 291
gamma correction (AGC), 4–7, 13–17
histogram equalization (AHE), 4, 6
Additive white gaussian noise (AWGN), 335
Adenine, 322
Adequate irrigation, 188
Ad-hoc
networks, 290–292, 294
vehicle network, 289
Administrative bodies, 48
Advanced technologies, 146, 165
Affordability, 185
Agricultural
applications, 220
automation, 153
backgrounds, 187
sector, 188, 220
Agro nutrition, 188, 194
alert (ANA), 1, 69, 71, 74, 78, 79, 187–190, 194
Ailment symptoms, 16
Alert
center, 158, 159
message, 158, 159, 161–164
Algorithm, 5, 21, 24, 28, 30, 31, 36–38, 40, 41, 49, 50, 59, 60, 62, 71, 74, 100, 102, 107, 109, 111–113, 116, 119–124, 126, 132–135, 139, 200, 201, 227, 228, 254, 277–281, 284–287, 294, 300, 305–310, 319, 324–330, 335, 336
Alkaline chemistry, 157
Amazon, 87, 88, 92, 93, 96, 110
Ambient temperature, 238
Ambulance, 158, 159
Amusement park, 96
Analysis
brushless DC motor, 229, 234
DC-DC converter, 229, 231
discussions, 89, 103
expanded examination revelation (noteworthy bits of knowledge), 89
improved client encounters purchaser ventures, 90
simpler progressively secure approaches (pay on the web), 91
solar pv array, 228, 230
Analytical
calculations, 291
E-commerce, 93
Anatomical artifacts, 4
Anchor
node, 69–73, 75–78
street, 291
Android

application, 148
interface, 148
Anti
aliasing technique, 25
allergy drops, 170
antibiotic drugs, 170
jamming authorization, 334
Application oriented, 5, 19
network architecture, 22
Arbitrary
combinations, 17
design, 364
Architecture, 63
Arduino, 152, 153, 159, 164–166, 175, 177, 178, 180, 181, 184, 190, 193, 194
boards, 152, 180
IDE, 152, 178, 190
micro-controller, 166
board, 152
nano, 164, 178, 180
ultrasonic sensor techniques, 175
uno microcontroller, 165
Arithmetic
coding, 330
encoding
algorithm, 327
technique, 320, 327
Arthritis, 172–175, 183
Artifacts, 4, 62
Artificial
feed-forward neural network, 70
intelligence, 36, 44
light sources, 385
neural network, 22, 79
Assessment procedure, 183
Asymmetric
algorithm, 285
cryptography, 327
key, 278, 320, 323, 325, 330
algorithm, 278, 323
cryptographic scheme, 325
cryptography, 309, 323
encryption, 320
Athletes, 148
Atrophy, 10
Attractive dashboards, 102
Audible range, 180
Audio steganography, 308
Authentication, 276, 283, 285
problem, 20
Authorized
user, 276, 306, 307, 316
vehicles, 290
Auto-encoders, 60
Autoclave, 389
Automated, 188
disease analysis system, 3
learning algorithms, 21
perception, 59
retinal
analysis system development, 4
diagnostic system, 4
image analysis system, 4
torque, 257
Average
accuracy, 69, 73, 74, 76, 78, 79
range error, 69, 71, 73, 74, 77, 79
training, 14
accuracy, 14

B

Background diabetic retinopathy, 10
Bacteria, 383–385, 387, 389
Ball bearings, 377
Barker code, 334, 336
Base
isolation, 377
technique, 378, 379
station (BS), 70
Baseline intraocular pressure, 173
Batch size, 12, 22, 23, 25, 29–32, 63
Battery backup, 158
Bayes net, 21
Bearing fault, 204
Behavioristic, 85
Better predictive analytics, 86
Bi-directional communication, 102
Big data, 81, 84, 85, 89, 90, 92, 93, 95–105, 320
analytics, 84, 85, 89, 93, 103
amalgamation, 89
companies, 84
environment, 97
solutions, 103
validation tools, 97
Bilinear interpolation, 6

Bio-medical images, 5
Biocide effects, 384
Biological
  DNA sequences, 320
  molecules, 320
Biomedical image data enhancement, 7, 73
Bit error rate (BER), 334, 337–339, 344
BLDC motor, 219–222, 227–229, 234, 235
Blindness, 171
Blinking LEDs, 38
Blood vessels, 4
Body temperature (BT), 145, 148, 149
Boltzmann constant, 241
Bond degradation, 379
Boost converter, 239, 243
Bottom-hat transformation, 8
Braced frames, 378
Broadcasting, 302
Broken rotor bar, 204
Business
  analytics, 97
  atmosphere, 351
  brand value, 84
  communications, 276
  development, 87
  endeavors, 91
  environment, 96, 97
  fundamental concern, 91
  opportunities, 97
  performance, 97, 351
  plan, 165
  transactions, 95
Buttery maculopathy, 10
Buyer demand pattern, 120
Buzzer, 179–181, 184
Bygone models, 60

C

Cake pops, 86
Canny edge detector, 41, 43
Capacitor, 223, 226, 244
  equivalent series resistor, 226
Capillary action technique, 174
Carbon-zinc, 157
Carcinogenetic, 392
Cardio-vascular diseases, 4
Carrier media, 316
Causation multiple messages, 290
Cell state, 52
Center
  average methods, 247
  gravity (COG), 247
Centroid defuzzification, 247
Ceramic antenna, 152
Change error (CE), 61, 246, 247
Channels of communication, 277
Character
  classification methods, 20
  recognition, 21, 22
Charging-discharging, 226
Chemical
  analysis, 189
  fertilizers, 187, 188
  laboratory, 187, 189, 194
Child trafficking, 162, 164
Chinese remainder theorem, 320, 321, 325, 327, 328, 330
Chip payment cards, 276
Chronic problem, 170
Cipher text, 277, 281, 283–285, 305, 306, 308, 316, 324, 328
Class investigation center point, 86
Classification
  accuracy, 19
  efficiency measurement, 10
  evaluation, 5
  quality assessment (CQA), 8
Classy fuzzy logic MPPT controller, 237
Client, 83–87, 92, 93, 109, 110, 119, 120, 131, 278, 283, 358
  dedication, 84
  execution variable, 84
  hello, 277, 278
Climatic parameters, 188
Cloud, 86, 109–112, 119–123, 127, 131, 132, 134, 140, 158, 163, 166
  architecture, 134
  computing, 109–112, 119–123, 127, 131–134, 140
    background, 111
    classification of load balancing technique, 121
    cloud system, 121, 132
    load balancing, 121, 132
  data centers, 120
  environment, 112

system, 110, 132
technology, 121
Clustering, 4, 37
Coaching classes, 162
Coal industry, 70
Coast guards, 159, 163
Code division multiple access (CDMA), 333, 334, 338–344
Coefficients of,
correlation, 199
wavelet, 200
Coif, 199, 200, 204
Coiflets, 199, 200, 204
Color histogram, 60
Colossal data, 89, 90
Column
beam joints, 378
sum, 112, 113
Combat viruses, 384
Commercial companies, 176
Communication
channel, 275, 278
experience, 98
network, 306, 307
process, 284
science, 276
Communities, 100
Community transmission, 48, 49
Compact gadgets, 84
Comparative analysis, 21, 62, 97, 119
Competent
network, 19
scheduling technique, 127
Competitive market, 95, 97, 103
Completion times, 126, 135
Complex
variants, 60
vision problems, 60
Complimentary
delivery, 86
transportation, 87
Comprehensive
eye examination, 173
wireless network, 289
Compression
algorithms, 277
permutation, 280
spring, 182
technique, 327
Computational
characteristics, 321
complexity, 22, 29, 31
science, 276
software, 204
time, 65
vision, 35–38, 40, 41, 44, 59, 60, 63, 149
algorithms, 35
methods, 41
techniques, 37
Computerization, 188
accounting, 357
Conditional random fields, 5
Confidential, 276, 277, 328, 330
data, 316
Confusion matrix, 62, 64
Congestion filter, 257
Congruent equation variable, 321
Connectionist temporal classification (CTC), 20
Consecutive nodes, 23
Constant voltage methods, 227
Consumer
loyalty, 84
oriented applications, 132
Contemporary literatures, 3
Content analysis, 97
Continuous conduction mode, 229, 231
Contrast
enhancement, 6, 8, 73
limited adaptive histogram equalization (CLAHE), 4–6, 12–17
Control
input values, 246
pitch accurator system (CSA), 257
Convenience of research, 353
Conventional
DC motors, 220
energy
crisis, 237
resources, 237
fluorescent tube (CFL), 385, 386
mercury vapor discharge lamps, 392
motors, 220
neural network, 61
Converter, 219, 221–229, 231–235, 238, 239, 243, 244, 249, 250
duty cycle, 239

Convolution
  layers, 11, 61
  neural network (CNN), 4, 5, 10, 11, 13, 15–17, 21, 60–62, 66, 67
    architecture, 10, 13
Cornea, 175
Correlation
  analysis, 202, 204
  coefficient, 198, 199, 202–212, 215
  technique, 202
Corresponding output matrix, 76
Cost reduction, 103
Country-wide prediction, 49
Covariance matrix, 202
Cover writing, 307
COVID-19, 47–50, 52, 53, 55, 56, 255, 384, 394
  virus, 48, 52, 394
Creo elements, 182
Critical bending moment, 380
Cross-cutting investigation applications, 85
Cross-entropy, 61
  loss function, 11
Cryptanalysis, 308
Cryptographic, 276–278, 292, 305, 308, 309, 319–325, 327–329
  algorithms, 282
  asymmetric key cryptography, 309
  hashing, 309
  keys, 277, 278, 282
  symmetric key cryptography, 308
Cryptosystem, 308, 320, 321, 323, 326
  using knapsack, 330
Cumbersome, 105, 349, 350
Cumulative distribution function (CDF), 6
Current
  input, 51, 52
  ripple, 225
  signals, 197, 199, 202, 239
  signature analysis, 197
Customary datasets, 85
Customer, 85, 89–91, 96, 98–103, 123, 132, 350–352, 356, 358, 360
  feedback, 102
  requirements, 99
  satisfaction, 98, 350–352, 358
  tendencies, 90
  value proposition, 98
  wellbeing, 98
Customized recommendations, 87
Cyber-crime, 306
Cytosine, 322

D

Dashboard camera, 36
Data
  accessibility, 99
  acquisition
    system, 204, 205
    task, 205
  analytics, 85, 102, 105
  augmentation, 12, 63
  bank, 84
  breaches, 105
  centers, 120
  encryption, 275, 287, 316, 330
  encryption standard (DES), 275, 280, 281, 284–287, 305, 306, 316
    algorithm, 284, 286, 305, 306
    decryption algorithm, 281
  innovation, 92
  management, 99
  preprocessing, 52
  reception, 70, 72
  sets, 50, 60
  traffic, 96, 103, 285
  transmission, 70, 72, 277, 298, 320
Database, 5, 21, 25, 26, 149, 158, 161, 162
  workstation, 111
Dataset, 9, 10, 12, 16, 17, 21, 25, 36, 50, 52, 62, 64–66, 75
  description, 9, 25
DC
  DC
    boost converter, 238, 239, 243, 244, 246, 248–250
    converters, 220, 239, 243
  link capacitance, 226
  motor, 181
  type electrical energy, 238
Dead load (DL), 366, 369
Debugging, 153
Decentralized, 302
Decision making, 38, 93, 349, 351, 360
  process, 95, 97
Decryption, 275, 277, 279, 281, 282, 284, 305–308, 310, 316, 320, 322, 323, 325, 327–329
  process, 323, 325, 328

Deep
architecture, 5, 19, 22
classification
evaluation metrics, 5
experiments, 5
metric evaluation, 13
process, 19
ConvNet, 62
learning, 5, 12, 17, 20–22, 27, 29, 35–40, 44, 59–62, 66
application, 20
architecture, 5, 22
environment, 20
methods, 21
model, 39, 40, 60, 66, 67
result verification, 17
techniques, 21
neural network (DNN), 5, 10–12, 19, 20, 22, 31, 32, 61
architecture, 11, 19
implementation requirements, 12
training, 12
structural benefits, 61
Defense sector, 165
Defuzzification, 245, 247
Degree of,
accuracy, 49
acidity, 189
agreeability, 353, 355
Delivery, 98, 99, 101, 109, 131, 173, 175, 219, 291
channel, 101
Demand forecasting, 101
Demographic detail, 101
Denoising, 203
performances, 199
Deoxyribonucleic acid (DNA), 319–323, 327–330, 384
cryptography, 320, 321, 327, 328
encoding, 320, 327, 330
Dependability, 92, 123
Detection, 21, 37, 39, 59, 60, 62, 103, 178, 189, 197, 198, 215, 301, 334, 335
Detrimental effects, 385
Device detection range, 178
Diabetes, 3
retinopathy, 10
Diffie-Hellman
algorithm, 278, 279
design, 284, 285
key, 284, 286
exchange algorithm, 275, 281, 286
exchange protocol, 278
Merkle algorithm, 278
Digit
recognition, 19–22, 31, 32
system, 21
Digital
currencies, 276
image processing, 6
India, 56
object recognition, 59
people, 101
retinal image for vessel extraction, 9
technology, 99
Digitization, 20, 99
Digitized
handwritten documents, 20
system, 99
Dilating drops, 170
Diode (D), 244, 246, 247, 253, 258, 285, 286, 289, 292, 299, 323, 363, 370, 371, 392
Direct sequence spread spectrum (DSSS), 333, 334, 344
CDMA
communication system, 334
receiver system, 334
Directional greedy routing (DGR), 291
Discrete
logarithmic
calculations, 286
problem, 284
wavelet transform (DWT), 197, 199–201, 215
Disintegration, 36
Disk shaped masks, 7, 73
Distant objects, 149
Distinctive assortment, 83
Distorted, 61, 339, 344
continuous signal analysis, 199
stator currents, 199
Distraction of mind, 184
Distributed
channel activities, 99
nature, 290
system, 120, 123, 127, 133
District laboratory, 190

Documents study, 97
Domain, 5, 19, 20, 32, 70, 103, 197, 200, 202, 203, 250, 320, 334
DoS attack, 290
Drinking, 39
Drive, 9, 10, 12–16, 219
  train model, 258
Driver attention
  monitor, 41
  tracking, 38
Drop bottle holder device, 169, 172
Dropbox, 110
Dropout, 22, 23, 25, 27, 29–32
  hyperparameter, 25
  value, 23, 27, 29, 31, 32
Drug, 170–172, 174–177, 182, 184, 185
  installation, 172, 176
  schedule, 172
Dry eye symptoms, 170
Ductility, 378, 379
Durability, 170, 183
Duration of time, 392
Duty cycle, 239, 243, 244, 249, 250
Dynamic
  algorithm, 111, 121, 122, 133
  load balancing algorithms, 111
  performances, 234
  state performance, 229

E

Earthquakeresisting structures, 364
Ease of implementation, 234
Easy eye dropper, 172, 175–177, 182, 185
E-commerce, 93
  business, 83
  organizations, 85
  sites, 83
  websites, 89
Economic
  condition, 101
  indicators, 101
  operation, 253
Edge
  detection, 37, 60
  node based greedy routing (EBGR), 290
  segmentation, 37
Effective enhancement technique quantitatively, 5
Efficiency, 4, 5, 16, 19, 20, 24, 36, 38, 41, 61, 71, 103, 110, 164, 219–222, 234, 235, 267, 271, 284, 358–360
Elastic modulus, 364
Electric
  circuits, 152
  consumption, 178
  devices, 271
  fault, 199
  grid network, 256
  power, 254
  science, 276
  torque, 260
  vehicles (EVs), 239, 250
    charging stations (EVCSs), 239
Electromagnetic
  radiations, 189
  spectrum, 384
  torque, 229, 234
  wave, 384
Electron
  charge, 241
  design, 184
Electronic
  aids, 184
  arduino, 185
  commerce, 84, 276
  commutation, 220, 221, 227, 228, 231
    method, 228
  components, 155
  devices, 84, 183, 220
  gadgets, 84
  model, 155, 170, 175, 182
  technology, 148, 170
Elliptic curve, 285
  discrete logarithm problem, 286
E-mail, 148
Emergency
  alert systems, 165
  responders, 149
Empirical
  research, 97
  studies, 85
Encoded encrypted image, 309
Encrypted, 275, 277–282, 284–286, 305–308, 310, 316, 319, 320, 322, 323, 325, 327, 328
  data, 307, 320, 327
  message, 310

purpose, 323
End-to-end latency, 290
Energy
consumption, 122
dissipating, 378
devices, 378, 379
seepage, 255
storage unit (ESU), 239, 243, 249, 250
Enhancement techniques, 3, 5, 8, 12, 15–17
Enterprises, 103
Entity coordination, 292
Environmental parameters, 70
Epicenter, 364
Epochs, 14, 21, 32, 63–65, 67
Error
backpropagation algorithm, 61
matrix, 76
percentage, 56
prone, 20
Evaluation metric, 5, 8, 9, 14, 19, 26
deep classification metrics, 9
IQA metrics, 8
Eve teasers, 159
Execution measurements, 83
Existence of scaling function, 203
Expansion
boards, 152
permutation, 280
Experimental
phase, 59
set up, 204
data acquisition system, 205
Explicit node, 123
Explorative study, 86
imagining future Wal-Mart, 88
Wal-Mart versus the opposition, 87
Exponential
dampers, 378
growth, 104
External
attacks, 292
communication, 356, 357
Extraction process, 200
Eye
ball center, 173
drop
applicator, 173, 174
bottle holder, 170, 172, 175
installation, 171
dropper, 169, 170, 173, 176, 178, 180
examinations, 170
lids, 175
piece arrangements, 176

F

Facial recognition, 60
False
negative (FN), 9, 26, 63
positive (FP), 9, 26, 63
recognized non-vessel pixels, 9
Fast fourier transform (FFT), 197–200
Fault
detection, 197, 200
diagnosis, 199
prediction, 198
tolerance, 110, 122
Feasibility study, 189
accuracy, 190
cost, 189
lifetime, 190
time consumption, 190
Feature extraction, 203
Fiber
confinement, 379
reinforced concrete, 379
Final
cipher text, 327
permutation, 281
Finance, 96
information, 352, 359, 360
portfolios, 132
statements, 349, 350
Finest scale approximation, 200
Fine-tuned layers tested, 61
Fire
breakout, 163
brigade, 158, 163
fighters, 160, 161, 163
Firmness abnormality, 367
Firmware, 153
Fitting explicit requirements, 83
Flora micro-controller, 165
Forefront advancements, 91
Foreseeing conduct, 83
Forget gate, 50, 51
Forums, 100
Fossils energy, 253

Fovea, 4, 10
Frequency
bandwidth, 198
domain signal, 203
information, 198
Friction bearings, 377
Fungi, 385, 387, 389
Fuzzification, 245, 247
Fuzzy
control, 245
logic controller (FLC), 238, 245, 250
MPPT, 245
controller, 238, 239, 245, 246, 248, 249
de-fuzzification, 247
fuzzification, 245
inference engine, 246

G

Gaming, 112
Gamma
correction, 4, 6, 12
Gate signals, 228
Gaussian-filtering, 37
Generators, 254, 255, 257, 262, 285
Genetic material, 384
Geographical
routing protocol, 291
statistics, 363
Germicidal
agent, 384
effect, 384, 394
operation, 388
Giant companies, 96
Gigantic data, 89
Glaucoma, 3, 171, 173, 182, 184, 185
sufferers, 171
Global
big data market growth, 105
business scenario, 358
casualty, 48
energy crisis, 220
environment, 254, 320
market, 88, 96, 254
pandemic, 47, 48
positioning system (GPS), 147–149, 152, 164, 166
module, 152, 164
system for mobile communications (GSM), 148, 149, 153, 154, 164, 166, 190
module, 149, 154, 164
warming, 238
Wind Energy Council (GWEC), 255
Google
drive, 110
online colab cloud environment, 12, 27
tracks customized ads, 87
trends online tools, 84
Government agencies, 44, 105
Gradient
descent, 24
exploding problem, 50
Gray scale, 40
image, 40
Greedy traffic-aware routing, 291
Green channel images, 12
Greenhouse
farming, 188
gas emission, 238
Grid
frequency, 254
network, 257, 271
operation, 255
reliability, 254
Ground movements, 364, 367
Guanine, 322

H

Hadoop, 97
Hall
effect signal position sensors, 222
sensor signal values, 228
Hand
tremors, 172
written character, 20
Handheld
devices, 84
gadgets, 84
Handpicked features, 22
Handwritten
character, 21
digits, 25, 26
recognition, 20, 21, 32
text recognition tool, 20
Harassment, 151, 162, 163

Hardware requirement, 151
9V battery source, 157
arduino UNO microcontroller, 152
breadboard, 155
jumper wires, 155
NEO 6m GPS module, 152
SIM800L GPRS-GSM module antennae, 153
switches, 156
Harmful cryptanalyst, 319
Harmonic, 257
changes, 257
Hashing, 309
Hassle-free supply chain management, 103
Heads up display (HUD), 35, 37
Health issues, 238
Healthcare, 105
Heart rate (HR), 148
variability (HRV), 148
Heat, 70, 164
Hemorrhages, 4
Heritage computing, 351
Heterogeneity, 134
Heterogeneous
nodes, 79
sensor systems, 37
Heuristic method, 245
Hidden
layer, 21–24, 27–32, 69, 70, 74, 76
state, 51, 52
High
chance of elliptic curve discrete (ECDH), 286
end architecture, 20
frequency signal components, 201
gain converters, 220, 221
quality analysis, 60
voltage applications, 243
Higher image representations, 60
Histogram oriented gradients (HOG), 38
Historical
data, 100
information, 50
Homogeneous virtual machines, 120
Hop information, 71
Horizontal displacement, 378
Horrendous incidents, 163
Horticulture, 188
Hough lane detection, 40
Household applications, 220
Human
error, 35, 36
experiences, 59
intervention, 165
observer perception, 8
to-computer interaction, 147
visual perception based metric, 16
Humidity, 389
Hungarian method, 112, 113, 115, 116, 136, 139
Hydraulic shock absorbers, 378
Hyper
parameter optimization, 19, 22, 27, 32
architecture, 19
batch size evaluation, 29
dropout evaluation, 29
graphical representation (hyper parameter evaluation), 30
learning rate evaluation, 27
number of hidden layers, 28
optimization algorithm evaluation, 27
parameter, 20–23, 31
tuning, 23
serious economy, 92
tension, 3

I

ICMP messages, 293, 296, 298
Ideal customer profile (ICP), 84
Image
acquisition, 306
classification, 39, 44, 59–61
data, 4, 5, 10, 12–14, 16, 17, 21
dataset, 4, 5, 10, 12–14, 17
enhancement, 3–5, 8–10, 12, 16
enhancement techniques, 6
adaptive gamma correction (AGC) enhancement, 6
contrast limited adaptive histogram enhancement, 6
morphological operation enhancement, 7
net data set, 39, 59
processing, 7, 9, 25, 35, 37, 40, 44, 62, 67, 73, 305, 306
systems, 25, 306

techniques, 40
quality
assessment (IQA), 3, 5, 8, 11–13, 15–17
selection, 309
recognition, 60
segmentation methods, 37
steganography, 305, 306, 308, 316
technique, 305
transformation, 4, 6
Immediate rescue operations, 162
Implementation requirements, 5, 27
Improvisation, 185
Incandescent lamp, 385
Incentives, 98
Incremental conductance, 227
Incubators, 389
Independent-installation, 185
In-depth understanding, 52
Indian market, 174, 175, 184
Individual
class probability, 23
parameter, 22
Induction
generator, 257
machine, 215
stator, 198
motor, 197–199, 202–206, 215, 220
Inductor, 223, 225, 229, 231, 244
Industrialization, 220
Industries, 21, 44, 103, 105, 198, 220
Inference engine, 245, 246
Influenzas, 50
Information
biological system, 86
dashboards, 102
development, 83
driven
advertising, 92
markets, 92
transferring module, 291
Infrared (IR), 177–181, 184, 384, 385
obstacle sensor module, 178
photodiode, 178
radiations, 178
receivers, 178
transmitter, 178
Initial permutation, 280
Innovative eye drop, 172
Input
gate, 50–52
image, 309
layer, 23, 24, 27, 69, 74
Installation procedure, 174, 176
Instillation attempts, 173
Instant customer communication, 102
Instantaneous power, 228, 246
Institutional review board (IRB), 183
Instructive assortments, 91
Insulation breakdown, 271
Integrated infrastructure, 36
Integrity, 276, 328
Intelligence
business purposes, 103
gathering, 277
transport systems, 291
video processing, 60
Intensity, 6, 7, 12, 25, 178, 363, 364, 366, 387, 389, 392
Interceptors, 285
Intermediate computations, 23
Internal
attacks, 292
communication, 359
threaded passageway, 173
Internet, 84, 86, 109, 131, 146, 147, 153, 305–307, 316, 319, 320, 351
business firms, 83
thinking (IoT), 147, 319, 320, 330
Intra business communication, 350–352
Inventory, 95, 99, 103
Inverse discrete wavelet transform (IDWT), 201
Inverter, 204, 220, 222, 262
Investment decisions, 350
Iron-oxide, 189
Irradiation, 219, 228–231, 234, 235, 238, 239, 241, 242, 245, 248, 249, 392
level, 229–231, 234, 235
Isolated workstations, 110
Itchiness-relieving, 170
I-V characteristics, 241

J

Jamming condition, 333, 334
Jumpers connecting wires, 190

K

Kelvin, 241
Kernel functions, 61
Key exchange, 276–278, 280, 286, 308, 323
  algorithm, 278
K-means clustering, 37
Knapsack
  algorithm, 325, 327, 328
  private key, 328
  public key, 327
K-nearest neighbor (KNN), 21

L

Laboratory equipment, 388
Laminar flow meter, 388
Large
  scale water purification, 394
  span structures, 364
Latency, 37, 40
Lateral drift, 370
Launching new products, 86
Leakage inductance, 221
Leaning rate, 23
Learning
  algorithm, 22–24, 31, 59
  rate, 22, 24, 27–32, 61, 63
    evaluation, 28
    value, 24
Least significant bit (LSB), 308–310, 316
  algorithm, 309
Legitimate prerequisites, 92
Light, 8, 37, 70, 73, 134, 149, 163, 174, 178, 185, 188, 192, 220, 254, 360, 383–385, 387
  detection and ranging (LiDAR), 37
Lightweight, 172, 183
Line segment extraction, 37
Linearity, 204
Linguistic
  diversity, 349, 350, 352, 359, 360
  inputs, 245
  mélange, 350
  variables, 246, 247
Liquidity analysis, 350
Literature review, 61, 96, 291
  data preprocessing augmentation, 63
  dataset description, 62
  model implementation, 63
  proposed work, 62
  training fine-tuning, 63
Lithium-ion, 157
Live load (LL), 366
Load
  balance, 107, 109, 111, 112, 116, 117, 119–123, 126, 127, 131, 133–135, 139, 140
    min-min (LBMM), 112, 116, 123, 126, 127, 134, 135, 139
    stratagem, 123
  balancers, 123
  resistor (RL), 244
  torque, 229, 234
  voltage requirement, 243
Localization, 61, 70, 71
Location
  detection, 178
  tracker, 145, 146, 158, 162–165
    device, 145, 146
Logarithmic loss, 64
Logistics, 101, 105
Log-loss curves, 30
Long short
  term memory (LSTM), 20, 21, 50
  time memory, 56
Long stretch accomplishment, 91
Loss function, 24
Low frequency signal components, 201
Low narrowband jamming, 334
Low voltage ride through (LVRT), 254, 262
Lubricating drops, 170
Lucidity, 352
Luminance pixels, 4
Luminosity, 188

M

Machine
  fault analysis, 199
  learning, 21, 22, 24, 25, 36, 37, 44, 56, 61, 100–102
    algorithms, 102
    problems, 21
    systems, 25
    tool, 105
Magic wristbands, 96
Magnitude variations, 202

Makespan, 134, 135, 139
Making strategic decisions, 102
Malformed font style, 21
Malicious node, 289, 293, 296, 299, 302
  isolation, 299
Mammogram images, 5
Mammoth, 49
Man-in
  middle attack, 278
  the-middle attack, 285
Manual intercessions, 188
Manufacturing performance, 89
Market
  basket analysis, 101
  identification, 101
  penetration, 102
  segmentation, 101, 102
Marketing
  decisions, 85
  departments, 101
MasterCard, 87
Mathematical
  algorithms, 277
  design, 220
  manipulation, 201
  model, 50, 102, 250
  tools, 200
MATLAB, 195, 199, 206, 220, 221, 228, 234, 235, 241, 254, 257, 333–335, 337, 344, 386
Matplotlib, 52
Maximum
  average sensitivity, 16
  permissible drift, 371, 373, 374, 376
  power point, 238
    tracking (MPPT), 221, 227, 231, 237–239, 243, 245–250
  row-sum, 136–138
Mean
  integrated systems, 99
  of maximum (MOM), 247
  squared error (MSE), 8, 203
Measurement syringe, 183
Mechanical
  fault, 197, 198
    detection, 215
  stresses, 257, 261
  torque, 227, 258, 260
  type faults, 215
Media sharing, 112
Medial canthi area, 175
Medical, 4, 9, 10, 44, 48, 53
  image segmentation, 10
  professionals, 4
Message
  authentication, 278
  communication, 285
Method, 113, 131, 136, 200, 305
  current signal reconstruction (decomposition of wavelet), 201
  evaluation, 9
  selection of optimal mother wavelet, 203
  statistical analysis, 202
  wavelet analysis, 200
Methodology, 50, 157, 175, 293, 352, 384
  long short term memory (LSTM), 50
Metric
  of authentication, 292
  values, 5, 15, 16, 29
Metropolitan city, 149, 150
Micro aneurysms, 4
Microbes, 385, 387
Microbiological cultures, 389
Microcontroller, 146, 152, 153, 190
Microgrids, 254
Microprocessor, 146
Microsoft office, 110
Migration time, 122
Military communications, 276
Min time, 135
Mines environment, 70
Mini-batch gradient descent, 24
Minimum
  completion time (MCT), 112, 117, 123, 126, 140
  permissible values, 257
Mining, 44, 70
Min-min (MM), 112, 116, 123, 126, 127, 134, 135, 139
  algorithm, 134
  schedule, 134
    algorithm (MM), 112, 123
    technique, 112, 123
Mirror arrangement, 183
Misaligned rotor, 204
Mobile, 36, 84, 157–59, 162, 164, 193, 290, 307
  ad-hoc network (MANET), 289

Model
accuracy rate, 62
design, 63
learning phase, 24
Moderate damage risk zone, 366
Modern
cryptology, 276
retail business, 103
Modified National Institute of Standards Technology (MNIST), 21, 25, 26
database, 25
Modular transformations, 327
Moisture sensor, 190
Molecular transformation, 384
Momentresisting frame, 378, 379
Momentum, 24
Monolithic prediction, 49
Morphological
enhanced, 5, 16
abnormal images, 15
image, 16
techniques, 5
test images, 16
operation, 3, 4, 7, 16, 73
operators, 4, 12
Morphology, 13, 17
Mother wavelet, 197, 198–200, 202–204, 206–212, 215
selection, 215
Motor, 177, 181, 182, 185, 198, 199, 204, 206, 215, 219, 220, 226–229, 232–235
torque, 234
Multi lingual hand writing recognition, 20
Multidimensional linear data, 61
Multi-language, 360
facility, 351
software, 352
Multilayer
nonlinear transformation, 61
perceptron, 21
Multi-level component analysis, 202
Multilingual
computing, 351
facility, 351
information management, 351
Multi-media division, 123
Multiple
heterogeneous sensors, 37
path routing technique, 293
Multiprocessing techniques, 38
Multistoried
building, 160
structure, 378

N

Naïve-Bayes, 21
Nanotechnology, 394
Narrowband jamming, 334, 337
environment, 334
Natural damping, 379
NavShoe, 149
device, 149
Negative
big (NB), 246, 247
small (NS), 246, 247
Neighborhood
climate deviations, 86
nodes, 293
Network
architecture, 20, 27, 31, 63
communication, 299
execution resources, 12, 27
failures, 164
hyper parameter, 23
optimization, 31, 32
learning, 22
performance, 20, 23, 29, 32
process, 23
security, 306, 319
throughput, 299
Neural
network, 20, 23, 25, 52, 60, 62, 69, 74
layers, 23
parameters, 20
Neurons, 22, 23, 25, 31, 62
New revenue streams, 96
Nickel
cadmium, 157
metal hydride, 157
Nitrogen, 189, 191, 194
Node, 23, 27, 70, 71, 73–76, 78, 111, 112, 120, 122, 123, 132, 133, 139, 140, 289–294, 296, 298, 302, 320
localization, 69–71, 77–79
transmission range, 290
Noise, 4, 37, 40, 220, 334, 335, 337

Nominal management endeavor, 121
Nonconventional energy, 238
Non-digital platforms, 100
Non-electronic, 169, 170, 172, 175, 182–185
  design, 183
  device, 169, 170, 184
  eye dropper, 183, 184
  model, 170, 175, 182
Non-horizontal line, 21
Non-human understandable characters, 21
Nonlinear controllers, 245
Non-orthogonal, 199
Non-popularity, 172
Non-readable text, 308
Non-repudiation, 276
  characteristics, 323
Non-standard hieroglyphs, 276
Non-stationary signals, 197, 199
Non-traditional methods, 134
Normal
  encryption algorithms, 287
  photo diodes, 178
  recurrence of vibration, 364
Normalization
  algorithm, 25
  graph cut segmentation scheme, 38
Noteworthy speed, 83
Novel corona virus, 48, 383
Nucleic acids, 322, 323
Numerical
  analysis, 353
  data samples, 204
Nutrient composition, 189

O

Object
  classification, 60
  detection, 44
Objective function, 22, 31
Ocular
  condition, 170, 171
  disturbance, 170
  drug, 169, 172, 184, 185
  surface, 171, 173, 177, 184
Onboard
  computer, 36
  processing environment, 38
On-demand
  network, 132
  routing protocol, 291
One-way communication, 102
Online
  business frameworks, 91
  office software, 123
  payment, 102
  software, 112
  storage, 112, 123
  strategic policies, 89
  transaction, 102
Operating
  model, 98
  radio, 39
  temperature, 241
  units, 105
  voltage range, 178
Opportunistic load balancing algorithm (OLB), 112, 123, 135
Oppressive loaded system, 112
OptiAid, 175
Optic disc, 4, 10
Opticare eye drop dispenser, 174
Optimal
  energy, 254
  mother wavelet, 198, 199, 202–204, 215
  parameter settings, 102
  resource exploitation, 121
  temperature, 389
Optimization, 11, 19, 20, 22, 24, 27–32, 60, 254, 334
  algorithm, 11, 24, 27–30, 32
  business model, 99
  hyper parameter, 20
    values, 31
Optimizer, 24, 28
Optimum
  mother wavelet, 204, 205
  productivity, 188
Organic
  framework, 90
  matter, 189
  molecules, 384
Organization, 20, 86, 88, 89, 93, 95–97, 291
  challenges, 97
Orthogonal wavelet, 199, 200
  decomposition, 199

Oscillating signal, 180
Oscillations, 227
Outdoor augmenting, 149
Output
gate, 50, 52
layer, 22, 23, 27, 31, 69, 74
level, 178
Overlapping image patches, 13

P

Packet loss, 299, 301
Pairwise co-prime positive integer, 321
Pandemic, 48, 49, 52, 53, 56, 383
Paper documents, 385
Parameter
optimization, 19, 22
search space, 22
specific learning rates, 24
Parasitic capacitance, 227
Passive underlying vibration control, 377
Path establishment, 302
Pathological
artifacts, 4
samples, 10, 15, 16
symptoms, 15
Pay per use, 111
Payment
flexibility, 98
patterns, 102
P-box permutation, 281
Peak signal noise ratio (PSNR), 3, 8, 12, 13, 16
Perfect forward secrecy, 285
Performance, 27, 122, 302, 344
analysis, 21, 334
evaluation, 29, 292
indices, 237
management, 97
measurement, 15, 292
index, 16
system (dynamic state), 229
Permutations, 285
Personalization, 84, 92, 93
guidance, 132
Petabytes, 86
Phosphorous, 189, 191, 194
Photo
aging, 385
biological effect, 383, 394
carcinogenesis, 385
diodes, 178
therapy, 392, 394
transistors, 178
Physical
abused, 151
machine, 110
games, 132
systems, 110
workstations, 120
Pigment epithelium, 10
Pigmentary epithelial atrophy, 10
Pitch
control, 254, 259–261, 266–271
speed, 257
Pixel, 9, 10, 25–27, 61, 62, 310
intensity levels, 6
Plain text, 277, 283, 324
Plant monitoring, 188
system, 188
Plastic hinge formation, 365
Plateau region, 24
Police, 158, 159, 162, 163
Pollution, 220, 238
Pooling of sources, 110
Popular transposition techniques, 324
Population density, 48, 49, 53
Portfolio management, 349, 350
Position routing mechanism, 291
Positive
big (PB), 246, 247
small (PS), 246, 247
Post-operative
cases, 171, 184, 185
drugs, 170
Potassium, 188, 189, 191, 194
Potential customers, 100–102
Pothole, 36
contours, 42, 43
data function, 37
detection, 37, 38, 41, 44
detector, 38, 41, 43
images, 38
tracking system, 37
Power
consumption, 185, 219
grid network, 255
output, 234
quality, 256, 257, 265, 267, 270, 271

regulators, 180, 185
spectral density (PSD), 333–335, 337–344
system, 253, 254, 257
Precautionary measures, 52
Predicting, 47, 49, 50, 52, 53, 55, 56, 61, 69, 70, 72, 76, 77
trends, 100
Preliminaries, 321
asymmetric key cryptosystem, 323
chinese remainder theorem, 321
DNA encoding, 321
image enhancement evaluation, 16
rail fence technique, 324
stage, 3
XOR algebraic operation for DNA sequence, 322
Pre-operative, 170
Pre-process, 37, 38
image datasets, 10
input function, 63
Pressure, 70, 171, 176, 389
lowering drugs, 170
Pre-trained ResNet18, 62
Price
comparisons, 99
fluctuation, 101
optimization, 101
services, 103
strategies, 101
Prime number, 279, 281, 284, 285
Privacy enhancement technologies, 292
Private key cryptography, 308
Probabilities
density function, 6
distribution, 4
of occurrence, 6
Process analysis, 97
Procurement methods, 103
Product performance, 102
Professional accounting terms, 358, 360
Proper
energy management, 250
secured supervision, 157, 165
technologies, 20
Proposed
method, 112, 135
algorithm, 136
model, 74
ANN model development, 74
model training and testing, 75
system model, 38
driver attention monitor, 39
pothole detector, 40
technique, 293
working steps (proposed work), 293
Protocols, 110, 183, 277, 278, 290, 291
Prototype wavelet function, 203
Psychographic, 85
Public
communication channels, 277
key, 281, 284, 286
cryptography, 278, 309, 323, 325, 327
encryption, 307
utility accounting, 351
Pulsating output, 220
Punching resistance, 379
Purine
adenine, 322
guanine, 322
Push-button switches, 156
PV array output voltage, 243
P-V characteristics, 241, 242
PV module, 238–243, 249
Pyrimidine
cytosine, 322
thymine, 322
Python programming, 52

Q

Quadrature mirror filters, 201
Quake vibration, 377
Qualitative
analysis, 60
evaluation, 8
Quantitative assessment, 173
Quantum computing, 105

R

Rail fence technique, 320, 324, 327, 330
Random
dynamic mapping, 310
forest, 21
overlapping patches, 12
variables, 6
Range
algorithms, 71

localization algorithms, 71
positioning methods, 71
Rapid flexibility, 110
Raspberry pi, 62
platform, 62
Rate of transmission, 47, 53, 56
Reaching behind, 39
Reactive
power, 257, 262, 264, 267, 269, 270
routing protocol, 291
Real
life dataset, 4
time
location, 145, 147, 148, 151, 157
monitoring, 35
world
applications, 61
situations, 184
Reasonable price, 98
Received signal strength (RSS), 70
Receiver
initiated, 121, 133
category algorithms, 133
operator characteristic graph, 14
start up category methods, 121
Reconstructed
image, 10, 62
program, 198, 199
signal constituents, 202
synthesis, 201
waveform, 207–212
Rectified linear unit (ReLU), 11, 27, 63, 64, 74, 76
Rectifier unit, 262
Recurrent neural network (RNN), 20, 50, 60
Redness-relieving drops, 170
Reduced power losses, 234
Region, 6, 40, 43, 100, 101, 384
of interest (ROI), 40–43
Regional
language, 194, 349–353, 355–360
facility, 349–352, 355–360
transliteration, 360
Regressive motion sensors, 60
Regularity, 203
Regularization method, 25, 30, 31
Reinforced concrete frame, 363, 379, 380
Related work, 85, 146, 334
GPS-SMS child tracking system (smart phone), 147
IoT wearable smart health monitoring system, 148
pedestrian tracking (shoe-mounted inertial sensors), 149
smart IoT device (child safety-tracking), 146
smart wearable bluetooth fitness tracker, 148
temperature regulating timepieces, 149
Renewable
energy
power stations, 238
resources, 238
sources, 219, 220, 238, 254
power, 255
Rescue operation, 159, 160
Residual eye drops, 173
ResNet, 63–67
ResNet-110, 63
ResNet-18, 62–65, 67
ResNet-50, 62–67
model, 62
Resolution variable, 124
Resource
allocations, 120
parity, 292
pooling, 110
utilization, 122
Respiratory suffocation, 392
Respondents personal information, 353
Response time, 122
Results discussion, 40, 52, 183, 294, 337
simulation design, 294
Retail
business, 98, 99, 103
industry, 96, 100–103, 105
locations, 98
market, 350, 360
sector, 96, 98, 103
Retailer website, 102
Retinal
artifacts, 7, 73
fundus, 3, 5, 16
images, 3–5, 16
image, 4, 5, 8–10, 12, 73
enhancement tool, 12
surface, 4

vessel segmentation dataset, 9
structures, 12
vessel segmentation, 3–5, 16
Revenue model, 97, 98
Revolutionary contribution, 98
RGB color sensor, 190
Ribonucleic acid (RNA), 384
Road, 35, 36, 44, 88, 295
accidents, 35, 36
camera, 36
conditions, 35, 36, 40
side unit (RSU), 290, 293
surface, 36
sweeping vehicle automation, 291
Roadside
infrastructure, 293
units (rsus), 290, 293, 294, 296
Robotics, 44
Robust security system, 319
Role of,
big data in retail, 95, 98
retail business model implementing big data, 99
traditional retail business model, 98
Root mean squared error (RMSE), 69, 71, 72, 74, 76, 78, 79, 198, 199, 203–212, 215
Rotor
position, 228
speed, 226, 229, 234, 258–260
unbalance, 210
Route
discovery process, 291
overhead, 299, 302
Routers, 290
Row-sum, 112
Rubber bearings, 377

S

Safe driving, 39
Sample data, 22, 204
Sanitization, 383
Saturation current, 241
S-box, 281
substitution, 280
Scalability, 110, 122, 134
Scaling, 61, 91, 200
School of Illumination Science Engineering (SISED), 389
Scikit learn machine-learning analysis, 52
Scratched training, 61
Screw system, 169, 176
Season, 100, 101
wise trends, 100
Secondary data, 98
Security, 92, 105, 123, 145, 146, 151, 162, 163, 165, 166, 273, 276, 286, 287, 291, 292, 305–309, 316, 319–321, 323, 330, 334
devices, 330
measures, 105
protection, 292
sector improvement communication, 333
stricken laws, 92
Segmentation, 3–5, 9, 16, 21, 37, 38, 60, 102
accuracy, 16
barriers, 102
Seismic
dampers, 378, 379
ground vibrations, 364
load, 364–366, 376, 377, 379, 380
performance, 378
tremors, 364
vibration, 377, 379
waves, 364
zone, 365, 379
factor, 365, 376, 377
Seismicity, 364, 380
Self-configuration, 289, 302
feature, 290
Sender
beginner category, 121
initiated, 121, 133
Senior citizens, 101
Sensitivity, 9, 14–16, 102
analysis, 102
Sensor node, 69–78, 295
SEPIC, 219, 221–223, 229, 231, 234, 235
buck boost converter, 219, 234
Sequence assemblage, 124
Series-parallel resistances, 239
Service provides on-demand, 110
Shear walls, 378
Shopper information, 85
Short term fourier transform (STFT), 198–200
Shunt resistance, 241

Sigmoid
function, 51, 52
output, 52
Signal
compression, 197
decomposition, 197, 200, 201
noising, 197
processing, 199, 200
Simple trackers, 158
Single
blinded technique, 183
phase voltage unbalance, 204
Sinusoidal
excitations, 364
waveform, 262, 265, 270
Skewness, 21
Slab-column connection, 379
Small
business enterprises, 103
key length, 286
pocket sized device, 172
Smart phone, 105, 147, 148, 188
camera, 38
Smooth power generation, 260
Social
distancing, 48
media, 100
websites, 101
Socio-economic, 85, 92
benefits, 105
problems, 238
Sodium hypo-chloride, 383
Soil surface, 36
Solar, 219, 220, 222, 228, 229, 231, 234, 238, 239, 254
energy, 238
irradiation
availability, 219
levels, 234
panel, 220
power generation technology, 220
Sophisticated algorithms, 105
SOS button, 146, 158, 159, 161–166
Sound
component, 185
retail business system, 98
Space studies, 44
Sparse interpretation, 61
Special
database 1, 25
database 3, 25
Speed
control, 254
convergence to zero, 203
Spores, 389
Sportspersons, 148
Spring frameworks, 377
SPV array output power, 234
Square root of the variance, 203
Stacksparse coding, 61
Stakeholders, 349, 350, 359
Standard
deviation (SD), 72, 203
test conditions (STC), 238, 241–243
State
farm dataset, 39
of-the-art, 20
Static
algorithms, 111, 122
compensator (STATCOM), 253, 254, 257, 260, 262, 264–271
infrastructure, 290
load balancing algorithms, 111
system algorithm, 121
Stationary signal analysis, 198
Statistical
analysis, 56, 97, 202
error
IQA metric, 8
metrics, 8, 16
parameters, 202
Stator winding fault, 204
Steganography, 305–308, 316
Stego image, 307, 316
Sterilization
indoor areas, 384
temperature, 389
Stochastic
approximation, 24
gradient descent (SGD), 11, 24, 30, 31
momentum gradient descent (SGDM), 63
Storage, 20, 109, 110, 131, 132, 220, 320
Store extension, 88
Storey
displacement, 367, 369–375, 379
drift, 367, 370–372, 374, 375, 379

Story drift, 364, 367, 370, 371, 373, 374, 376–380
Strategic
decision-making, 105
marketing decision, 102
planning, 100
Strong visual models, 60
Structural
elements, 8, 73, 378
integrity, 10
Substantial growth, 254
Support vector machine (SVM), 21
Sustainable economic development, 103
Switches, 156
pulse frequency, 225
Sybil, 290, 292, 293, 296, 300, 302
attack, 290, 292, 293, 296, 300, 302
vanets, 292
Symlet5 mother wavelet, 215
Symlets (Sym), 199, 200, 204, 207
Symmetric, 121, 133, 307, 308, 320
algorithms, 285
encryption, 307
key
algorithm, 285, 323
encryption algorithms, 287
encryption, 320
Symmetry, 203
System
communications, 132
congruence, 321
design, 190, 191, 222, 335
block diagram, 191
calculation of voltage current ripples (convertor), 225
circuit diagram, 191
control of proposed system, 227
DC link capacitance of VSI, 226
DC-DC convertor, 222
hardware required, 190
irradiation levels, 229
P and O MPPT algorithm, 227
PV panel, 222
software required, 190
water pump, 227
model, 39, 40, 71
morphological operation enhancement, 73
network model, 71
useful definitions, 71
stability, 255
topology, 290
Systematic
analysis, 61
cost model, 98
Systemic illness, 176

T

Tablet, 84
Tabular validation, 53
Tanh function, 51, 52
Target
marketing, 103
revenue, 98
Tear duct, 171
Technique comparison, 299, 300
Technological
advancement, 103
availability, 188
penetration, 104
Telecommunications, 96
company, 96
Temporary fertility, 187
Tensor flow, 12, 27, 37, 39, 62
Tenterhooks, 161
Terabytes, 86
Test
accuracy, 14, 16, 22, 26, 27
input matrix, 75, 76
sensor node, 76
Text steganography, 308
Textiles materials, 385
Texting, 39
Theoretical background, 22
classifier optimization, 22
deep neural network (DNN), 22
network hyper parameters, 23
batch size, 25
dropout, 25
hidden layers, 24
learning rate, 24
optimizer, 24
network layers, 23
hidden layer, 23
input layer, 23
output layer, 23
Thermal voltage, 241

Three-layered architecture, 97
Thresholding schemes, 199
Throat cavity, 171
Throughput, 121–123, 293, 297–300, 302
Thymine, 322
Time
  complexity, 22, 32
  consuming, 5, 20, 24
    process, 187
  difference of arrival (TDOA), 69–72, 78, 79
  duration, 198, 229, 234
  frame, 98
Tophat bottom-hat transformations, 7, 73
Topical
  drugs, 170, 171
  hypotensive medication, 173
  therapy, 170
Tracking
  devices, 149
  problem, 151
Traditional
  buck-boost converters, 220
  business data storage, 110
  cryptography system, 320
  cryptosystems, 321
  instillation, 173
  retail business, 99
    model, 102
  voltage control techniques, 262
  working, 257
Traffic
  counting, 60
  rules, 36
  scenario, 60
Trafficked, 151
Training, 10, 12, 31, 64–66
  details, 5
Transducer, 180
Transfer learning, 39, 40, 61, 62
Transient voltages-currents, 257
Transmission
  lines, 267, 271
  process, 319, 320
  system handling operators (TSO), 255
Transposition technique, 324
Trapdoor function, 285
Trend
  forecasting algorithms, 101
  identification, 100
True
  negative (TN), 9, 26, 63
  positive (TP), 9, 26, 63
Tubular eyedropper, 174
Tunable isolator, 377
Turbine generator, 267
  set, 265
Turbulence, 257
Twitter notification, 148
Two-dimensional signals, 306
Two-phase scheduling, 135

U

Ultraviolet, 384, 394
  ray, 384
Unassigned jobs, 136, 138
Unauthorized
  access, 276, 319
  user, 276, 307, 309
Unbalanced
  cost matrix, 112
  loads, 120
  matrix, 112, 117, 135, 139
  resource usage, 120
  rotor, 204
Uncommitted component analysis, 334
Unencrypted data, 277
U-net
  model, 10
  type connections, 10
Uneven illumination, 4
Union territories, 49
Unpredictable workloads, 120
Unreadable format, 307
Upsampling, 201
Urban business area, 160
Used cases, 160
  child safety, 162
  fire fighter safety, 160
  fishermen safety, 161
  women safety, 162
UV radiation, 385, 387

V

V2I communications, 292
Validation, 12, 14, 22, 26, 27, 30, 31, 53, 55, 62, 64–66, 97
  accuracy, 26
  phase, 26

set, 26
Value information, 86
Vanishing moments number, 203
Variable
pitch technology, 254
rotation speed, 254
speed, 255, 257, 260, 264, 268, 271
Vascular abnormalities, 10
Vehicle
ad-hoc networks, 290
automation, 291
communication, 291
to vehicle (V2V), 290, 292, 295
traffic density, 291
Vernacular language, 352
Versatile
base separation framework, 377
infrastructure, 289
Vertical acceleration, 364
Vessel
classification, 4, 5, 9, 14, 16
identification, 10
probability map, 5
segmentation, 3–5, 16, 17
Vibratory style, 364
Video
frame, 306
processing, 38
spilling applications, 92
steganography, 308
Virtual
machine, 119–121, 132
monitor, 121, 132
Virtualization
concept, 121
techniques, 132
Viruses, 384, 389, 392
Visual
clues, 38
impairment, 169, 183, 184
information
IQA metric, 8
metrics, 8
perception IQA metrics, 16
Vital body parameters, 146, 149, 165
Voltage, 180, 205, 206, 212, 219–222, 226, 228–230, 239–244, 246, 248–250, 254, 257, 262, 268–271
gain, 221
ripple, 226
sag, 257
swell, 257, 269, 271
switching inverters (VSI), 219–222, 226–228, 234
Volt-second balance equation, 223, 225

W

Walmart, 86–88, 93
labs examinations, 86
stores, 86
Wastage of,
drug, 171–173, 183, 184
quantity, 170, 172
medicine, 172
Water pumping
applications, 219, 220
system, 219, 221, 234
Wavelet, 197–200, 203, 204, 206
analysis, 199, 203
coefficients, 197–200, 206
decomposition, 200, 201
family, 200
functions, 199, 200, 206–212
transform, 197, 199, 200, 204, 205
Wearable
device, 151, 157
tracker, 146
Weather, 36, 161, 164
Web
browsing history, 101
business, 85, 91
field, 85
life information, 86
showcasing systems, 83
Websites replacing branch agencies, 110
Weighting distribution (WD), 6, 7
Wind
energy
assimilation, 253
technology, 254
generation, 253, 255
installation capacity, 253
generator, 254, 255
outturns, 257
intermittency, 254
power, 254, 255, 260
power in India, 255

integration of wind power (existing grid network), 256
proportional control pitch actuator system (CSA), 257
shear, 257
turbine, 254–257, 260, 264, 268
Wireless
connection, 148
links, 70
sensor network (WSN), 69–71, 78, 79
Workstation, 110, 111, 120, 121, 132, 133
consolidation, 120
utilization, 120
World
carbon footprint, 254
electricity consumption, 254
Health Organization (WHO), 48, 50
Wormhole attack, 290

X

XOR operation, 322, 323

Z

Zero padding, 201
Zigzag cipher, 324

9781774639979